M & SONS,
& LONDON,
FACTURERS.

BITION, London, 1885.

INTERCHANGEABLE MATERIAL FOR REPAIRS.

atalogue of Interchangeable Material
" Watches
UPON APPLICATION.

PRIZE MEDALS.

ALL GRADES
OF
ENGLISH
LEVER WATCHES,
KEYLESS
AND
NON-KEYLESS.

iii

Colour Plate 26. **Animated Pocket Watch – Cycling.** See page 364

Colour Plate 24 Opposite. **Novelty Butterfly Form Watch, Swiss.** Spring-loaded hinged wings open to reveal watch. See page 355

POCKET WATCHES
19th & 20th Century

Alan Shenton

ANTIQUE COLLECTORS' CLUB

©1995 Alan Shenton
Reprinted 2008
World copyright reserved

ISBN 978-1-85149-211-8

The right of Alan Shenton to be identified as author of this work has been asserted by him in accordance with the Copyright, Designs and Patents Act 1988

All rights reserved. No part of this publication may be reproduced, stored in a retrieval system, or transmitted in any form or by any means electronic, mechanical, photocopying, recording or otherwise, without the prior permission of the publisher

British Library Cataloguing-in-Publication Data
A catalogue record for this book is available from the British Library

Printed in China
for the Antique Collectors' Club Ltd., Woodbridge, Suffolk

Contents

Colour Plates 6

Acknowledgements 13

Introduction 15

Chapter 1	Verge Watches	18
Chapter 2	Cylinder Watches	38
Chapter 3	Lever Pocket Watches	
	Part 1 – English	70
	Part 2 – Swiss	152
Chapter 4	American Lever Watches	192
Chapter 5	Duplex Watches	215
Chapter 6	Pin Pallet Lever Watches	229
Chapter 7	Roskopf Watches	259
Chapter 8	Chronographs	277
Chapter 9	Novelty Watches	341
Chapter 10	Repeating Watches	379
Chapter 11	Pocket Alarm Watches	382
Chapter 12	Eight Day Watches	404

Index 427

Colour Plate 1. **Thil à Genève**. Gold 14 carat open face case. Early bar movement engraved Thil à Genève. See page 42

Colour Plates 2 and 3. **Anonymous.** Lady's silver open face fob watch hallmarked Birmingham **1886** (?1896). See page 64

Colour Plates 4 and 5. **James McCabe, Royal Exchange, London.** 18 carat gold full hunter engine-turned case hallmarked London **1864.** See page 89

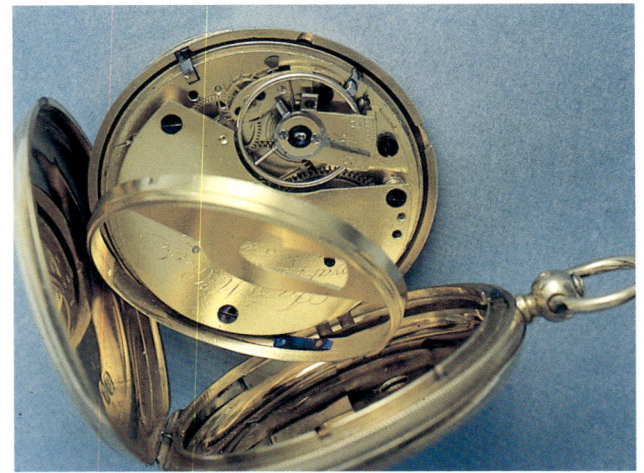

Colour Plates 6 and 7. **Thomas Prest, Chigwell, Essex.** Silver open face single bottom case hallmarked London **1845.** Prest's keyless winding. See page 81

Colour Plate 8. **George Wilson, 29 Middlegate, Penrith.** Silver open face double bottomed engine-turned back, snap-on bezel, hallmarked London **1885.** See page 111

Colour Plate 9. **VANGUARD, American Waltham Watch Co, Waltham, Mass, USA.** 14 ct gold filled open face Dennison case. See page 204

Colour Plate 10. **SOUVENIR EUROPEAN WAR, Robert Ingersoll Bros.** Nickel chrome on brass open face. See page 252

Colour Plate 11. **G Rosskopf & Co Patent (Vittori & Cie), La Chaux de Fonds, Switzerland.** Nickel open face case. See page 265

Colour Plate 12. **Roskopf Système Patent** Nickel open face case. See page 271

Colour Plate 13. **Messaggero, Echappement Roskopf.** Nickel open face case. See page 272

Colour Plates 14 and 15. **Anon. Switzerland.** Double sided centre seconds chronograph stop-watch, oxydised steel case. See page 331

Colour Plate 16. **Waltham Watch Co., Waltham, USA**. 6 seconds stop-watch timer, chrome on base metal. See pages 338-339

Colour Plate 17. **Charles Frodsham, 27 South Moulton Street, London W.** Micrometer chronograph for timing flight of projectiles, nickel open face case. See pages 339-340

Colour Plate 18. **Skeletonised Reverse Movement.** Oxydised steel open face, Borgel case. See page 343

Colour Plate 19. **Twenty-four Hour Dial.** Nickel open face case. See page 343

Colour Plate 20. **Instrument Panel Watch.** See page 344

Colour Plate 21. Travelling watches. See pages 348-349

Colour Plates 22 and 23. **Digital Pocket Watch, Swiss.** Oxydised open face steel case. See page 353

Colour Plate 24. **Calendar Pocket Watch, Swiss.** See page 361

Colour Plate 25. **Animated Pocket Watch – Scouts Jamboree, Smith's Industries England.** Chrome open face case. See pages 366-367

Colour Plate 26. **Time Teacher Pocket Watch (Smith's Industries)**. Orange plastic open face case. See page 368

Colour Plate 27. **Cigarrillos Excelsior, Courvoisier Frères, Chaux de Fonds, Switzerland.** Gilt open face case. See pages 370-371

Colour Plate 28. **Commemorative – Joffre/Kitchener, Swiss.** Silver open face case. See page 375

Colour Plate 29. **Cigarette Lighter, Swiss.** Brass engine-turned case. Flint petrol lighter fitted with watch movement. See page 377

Acknowledgements

The author wishes to acknowledge the very considerable help of the many watch dealers and collectors in providing the opportunity for examining and photographing their watches. Without this co-operation the compilation of this book would not have been possible.

A very special mention to the London auction houses Christie's South Kensington, Phillips and Sotheby's, Pieces of Time and William L. Scolnik, and at all times the always helpful interest of my friend Tommy White FBHI, still much 'with it' at ninety years of age!

Above all to my wife Rita, for many hours of patient typing and computerising for publication – my love and thanks.

Alan Shenton

Introduction

In order to take an intelligent interest in any subject it is essential to acquire a sound background knowledge. The horologist who wishes to take his pleasure from the study of watches will need to have some insight into the techniques, materials and the social factors that influenced the decoration of the cases, dials etc., an overall picture of the evolution and developments of the mechanism, together with an awareness of the skilled workmanship and knowledge of the past craftsmen responsible for the finest watches.

As an inexperienced collector, albeit having acquired the requisite background knowledge, it is often very difficult when confronted with the task of appraising a watch to assess general merit while at the same time being alert to the many details that require careful scrutiny prior to embarking upon a purchase.

A carefully evolved system is required – a system that eventually becomes second nature and requires no conscious thought; a system that avoids arriving home only to discover that in the excitement of making a purchase the illmatching hands or hair line cracks on the dial were not observed! In other words, a mental cataloguing to avoid falling into the trap pointed out by Sherlock Holmes of seeing without observing!

It would appear to be logical to commence with the outside and to work inwards. The following comments are intended only as guidelines and to alert a newcomer to some of the pitfalls, not to answer all the queries. The compiling and format of such a checklist will eventually depend on individual approach and specific area of interest.

The Case
This is helpful in assessing overall quality. Is it of gold, silver, pinchbeck, enamel or what? Is it engraved, engine-turned, pierced or repoussé? If it is made of gold or silver the hallmark should be carefully noted as this may be of considerable help where date, quality, and assay office are concerned. One of the traditional pocket size reference works is *Bradbury's Book of Hallmarks* which gives marks of British and Irish as well as foreign imported silver and gold plate. *Pocket Edition Jackson's Hallmarks,* edited by Ian Pickford, is based on the new, revised edition of *Jackson's Silver and Gold Marks of England, Scotland and Ireland* and includes a selection of over 1,000 makers whose work is of particular interest. On occasion the casemaker's initials and case number also appear and further documentation can then be found in the standard reference works, although it must be said that as only initials appear it is often quite speculative to attribute these to specific casemakers. Usually only educated guesses can be made and it is not often that there is any other corroborative evidence.

Shape of the case – is it an unusual shape? Is it abnormally thick or thin? Are there any unusual features regarding the pendant or bow? Is it a replacement or of an inferior metal? Does the length of the pendant indicate that there should be an outer case that is now missing? Is the bow freely pivoting on a special safety Dennison bow? In other words, are there any unusual features and, if so, what is their significance? (It is assumed that the intensive preliminary homework will have given the requisite knowledge as to what is usual and what is unusual!) The possibility that case and movement do not match should be considered – is the assay office and date of hallmark consistent with the date of the movement or known dates of maker? Is it a later case? Are the case and movement number compatible, or are there any noticeable adaptions to the case to take that particular movement? If the watch is an English lever should there be a dustcap? Is there evidence of a cap stud remaining in the top plate or the hole where it might have been? Do bear in mind that some Swiss watches were manufactured in imitation of English styles and also had dust covers. For example, Samuel's Swiss levers and a number of cheap Swiss watches imported by Sir John Bennett that looked superficially like an English lever.

General condition should be taken into consideration. Are the hinges of the case sound and not strained? Does the bezel fit well? Is the case unacceptably bruised? Does it have its original glass, for example original bull's-eye for a verge? Has the centre cannon pinion been filed to allow a lower glass to be fitted?

The Dial
It is most important that an enamel dial is not damaged in any way. Chips or hairline cracks are virtually impossible to have repaired satisfactorily or at a price commensurate to a modestly valued watch. Cracks can be temporarily cleaned but as new dirt accumulates they become only too visible. Are there any unusual features of the general layout that give some indication of interesting technical variations in the movement? Is it an alarm. calendar, special purpose watch? Inscriptions as to maker or retailer can usually be pursued in the standard reference works giving working dates etc.

The hands require scrutiny to ascertain whether they are of brass, gold or steel – whether they are commensurate with the date of the watch in style – whether they are modern replacements and of the correct length. Are they all present (ie minute, hour and in some instances seconds)? It would not be the first time that a purchase was made and in the excitement their lack overlooked! Lack of hands can sometimes indicate a problem. Has the seconds hand just fallen off or is it missing because the pivot to the seconds arbor been broken off by careless handling?

Whether in working order or not is the first consideration. Following that attention should be turned to evaluating whether it is a simple example of the type of watch under consideration or whether it has any additional interesting features. Obviously complications such as repeat, alarm, or calendar work are highly desirable. Less obvious, but of importance when taking stock of the watch generally, is whether the balance is of steel, brass, gold or compensated in any way. Is it free sprung or has it a regulator? Is the method of regulation of particular interest? Is the movement adjusted and to how many positions? Has it a patent pinion or other safety feature? Is the fusee chain intact?

The Escapement
The type of escapement can be of vital interest and it is essential to have a good knowledge of the various types. Is it a pin pallet, ratched tooth or club tooth lever? Is it rack lever, a duplex or a cylinder? Is it of English, Continental or American origin? Having found the category into which a particular escapement falls there are of course many variations. Vigilance and close observation is of great importance. Numerous bargains have been had because the vendor has not paid close enough attention to assessing the escapement. Tourbillons have been purchased for virtually nothing when the vendor threw it to one side as it was the odd man out in his box of verge watches!

General
If the movement is exceptionally rare or unique, this should override the presence of faults or damage.

One final piece of advice – do not run before you can walk. In order not to be overwhelmed with facts, figures, dates etc., it is better to concentrate on one particular type of watch and, when that is fully mastered with all it variants, then it is time to move on and widen the circle to include a further type of watch. Museums are excellent for viewing watches but generally speaking they do display the more exotic and rare examples. Sensible and courteous use of salerooms can be of more benefit. Careful handling of other people's watches is essential and often recourse to the owner or more expert companion when opening cases is appreciated by all concerned. Generally speaking it is a hazardous procedure to pass valuable watches from hand to hand and the adoption of the habit of placing it on a neutral surface for the other party to pick up saves many an argument.

Lastly, try to form an opinion as to why the watch is being offered for sale and at that particular price (especially if a 'bargain' price) and do not forget that a watch that is not in working order can have more than one problem that does not become apparent until the obvious fault is rectified!

As well as learning by observation and experience, a few well selected reference books are essential tools for any collector. The selected bibliographies appearing at the end of chapters provide suitable titles for further reading. (9th ed. Britten refers to the ninth edition of *Old Clocks and Watches and their Makers* by F.J. Britten.) For anyone not conversant with technical terminology, nomenclature of watch parts, case, hand or buttons styles etc., the purchase of a copy of the current *Watch and Clock Encylopedia* by Donald De Carle should not be delayed. It will prove to be a constant companion.

CHAPTER ONE

Verge Watches

Much has been written on the early history of the watch and in any event most of the examples employing this escapement fall outside the period under discussion. It was felt necessary, however, to make passing reference in order to place the late examples in their true context.

It used to be accepted that watchmaking commenced in Nurnberg in 1510 when Peter Henlein (1480-1542), a locksmith, succeeded in making a watch that could be carried in purse or pocket. This achievement was recorded by Johann Cocleus (1479-1522) in a book published in 1512 entitled *Pomponius Mela's Cosmography*. However, Professor Enrico Morpurgo has produced documentary evidence from the State Archives of Modena in the Ducal Chancellery of a letter from Jacopo Trotti, ambassador at Milan. Dated 19 July 1488, the letter refers to the making of three watches, two of which were to strike on a bell and were intended to be part of three costly garments richly decorated with pearls. Certainly an important area of watchmaking developed in Germany at the beginning of the sixteenth century to be followed fifty years later by a considerable industry at Blois in France. These areas were to retain their dominant position until Christian Huygens (1629-1695) invented the spiral balance (January 1675) and it was successfully applied by Dr Robert Hooke (1635-1702) as recorded in his diary entry for February 1675. The application and exploitation of this discovery by Thomas Tompion (1638-1713) and Daniel Quare (1632-1724) thereby placed English watchmaking in the dominant position which was to be maintained until the end of the nineteenth century when overtaken by the mass-production methods emanating from the USA and Switzerland.

There had been over this period a gradual evolution of a portable timepiece that was more than adequate for the everyday requirements of the ordinary man. A good watch was capable of an accuracy of one to two minutes per day with some fluctuation depending upon prevailing conditions such as temperature changes, thickening of oil, general cleanliness or whether the owner was of a sedentary disposition or an active horseman! Gradually there had emerged a standard form of robust verge escapement watch which was to survive until the end of the nineteenth century in spite of its acknowledged shortcomings - being a frictional rest recoil escapement and competition from the duplex and lever escapements. The average Englishman continued to prefer his key wound verge watch that was heavy in pocket and solid in hand!

There were, however, some concessions necessitated by changing fashion and need to maintain a competitive purchase price. The earlier English verge watch had had a great deal of decoration to both movement and case, but gradually it became more functional, the only continuing concession being the ornamental piercing and complicated designs of the balance cocks. The ornate pierced pillars of the first two centuries gave way to simple cylindrical pillars. Enamel dials, more easily read, prevailed over the earlier champlevé dial. Eventually the single case largely superseded the pair-case and plain silver, gold or pinchbeck cases with less fussy adornment prevailed. On the Continent, however, there developed a preference for a more elegant watch. This was made possible by exploiting the potential of the cylinder escapement (a horizontal escapement), thus allowing the development of a slimmer watch. Abraham Breguet in particular was one of the eminent makers able to display this elegance to the full. Thinner watches became the norm on the Continent.

The supreme standard of craftsmanship achieved by the eminent makers satisfied the prevailing market in the seventeenth century and indeed a steady demand for quality workmanship continued. There was, however, an increasing demand for more reasonably priced watches. In order to meet this demand it was necessary to introduce mass-production methods. Frédéric Japy (1749-1812) introduced into his manufactory machines that were capable of producing verge *ébauches* in quantity. In fact by 1801 he was producing 100,000 *ébauches* annually. Eventually each country developed its own centres for the making of *ébauches* ready for finishing by the individual watchmakers. In this country the main centres were to be found at Prescot, Liverpool and Coventry.

It is hoped to demonstrate by means of the watches appearing in the following illustrations that there is still much to appreciate in these comparatively late examples of verge watches - there is certainly sufficient variation of detail to provide interesting study. There is also the considerable advantage that the asking prices are often still within the means of the average collector.

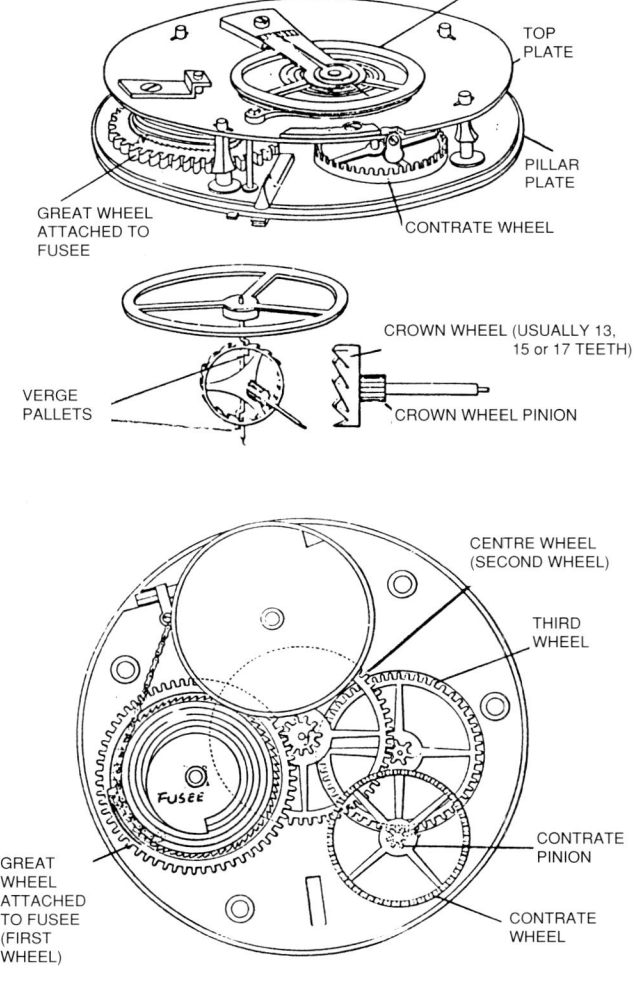

Verge watch movement

Verge Watches

PLATE 1A

PLATE 1
John Thomas, 55 St James Street, Covent Garden, London

Silver pair-case hallmarked London **1800.** Casemaker WL (Pigot's Directory 1823 lists casemaker William Linsley, 69 Banner Street, St Lukes, London). White enamel dial, black arabic numerals, fine minute markings and gold hands. Cylindrical pillars. Fusee chain. Single barrel flange. Flat steel undersprung balance. Tompion regulator with silver dial engraved 1-4. (Note engraved hand on top plate as pointer.) Movement engraved John Thomas London 1535.
Key size 4. Diameter 55mm

PLATE 2

PLATE 1B

John Thomas is listed in 9th ed Britten as goldsmith & jeweller, 1787-1815. Watch by Jn Thomas London numbered 5071 in the City Museum, St Albans.

Thomas Tompion, often referred to as the Father of English clockmaking, was baptised in 1639 and died in 1713. His style of regulator comprised a segmented rack following the outer coil of the balance spring. This was geared to a small wheel carrying a turning square. The watch was regulated by altering the position of the curb pins by the use of a key. The index dial was calibrated in order to assess how much to shorten or lengthen the effective length of the spring.
This watch has been included as a typical early 19th century example for purposes of comparison with the later watches appearing on the following pages.

PLATE 2
James Graham, 85 Piccadilly, London

Open face double bottom pinchbeck case, engine turned body. Casemaker's mark CM. White enamel dial, black roman numerals, gold arrow hands. Cylindrical pillars. Fusee chain with decorative pierced chain guard escutcheon. Single barrel flange. Flat steel three arm balance visible under small engraved cock. Regulator index mounted around the jewelled endstone of balance staff with scale engraved on the balance cock.
Key size 4. Diameter 49.3mm

James Graham is listed in 9th ed Britten as clock and watch-maker active 1800-1806 – bankrupt 1804.

Verge Watches

PLATE 3A

PLATE 3B

PLATE 3
Vale & Kenyon

Silver pair-case hallmarked **1803.** SA (possibly Stephen Adams). White enamel dial, broad black roman numerals, gold spade hands. Silver dustcap pierced to frame *(it is unusual to have a silver dustcap)*. Finely decorated and pierced cock of unusual design spanning the back of the movement. The dustcap retained by a small turn button on centre of the cock. Especially lavishly engraved, ie dustcap, back plate of movement with makers' names (Vale & Kenyon London No 1617) engraved on detachable plate over the fusee. Cylindrical pillars. Fusee chain. Steel three arm flat rim undersprung balance with Tompion style regulator.
Key size 8. Diameter 59.5mm

The Coventry firm of Vale & Rotherham was founded by Samuel Vale in 1747 - by this date Rotherham was a partner in the firm. There is no record of Vale and Kenyon. Vale & Co recorded in Holdens Triennial Directory 1807 Vol II as watch manufacturers, Spon Street, Coventry.

PLATE 4
James Hargraves, Sivan Lane, Bawtry, Notts.

Silver pair-case hallmarked Birmingham **1803.** Case-maker's mark WR. White enamel dial with black roman numerals, gold spade hands. Cylindrical pillars. Fusee chain. Single flange barrel. Flat steel undersprung balance with a transitional style of regulator – the small blued steel dial operating index on the barrel side of the finely pierced geometrical design of the cock. Top plate engraved over the barrel position Jas Hargraves Bawtry.
Diameter 55.5mm

According to Yorkshire Clockmakers *by Brian Loomes and* Clock and Watchmakers of Nottinghamshire *by Harold H Mather, James Hargraves was born in 1768 and died in 1835 aged sixty-seven years. He was married in 1797 to Sara Hibbard and established as a watch and clockmaker by 1807.*

PLATE 4

Verge Watches

PLATE 5A

PLATE 5B

PLATE 5
Boutevile & Norton, 158 Aldersgate Street, London

Silver half hunter single bottomed case hallmarked London **1805**. Casemaker's mark NS. White enamel half hunter dial, black arabic numerals, double spade hands. Winding through the dial. Cylindrical barrel. Fusee chain. Barrel bridge engraved Boutevile & Norton London No 540. Flat steel undersprung balance with Bosley type regulator. Engraved cock of grotesque design.
Key size 5. Diameter 51mm

Boutevile & Norton, 159 Aldersgate Street, London listed 9th ed Britten 1802-4 at 158 Aldersgate Street and 1810-19 at 175 Aldersgate Street. It is worth remembering that there is always the possibility that premises have been renumbered and that the person in question has not actually moved a few doors up!

Joseph Bosley was apprenticed 7 July 1718 to William Cartright for a period of seven years. Free 17 January 1725. According to Some Account of the Worshipful Company of Clockmakers by William A Overall 1881 on 6 March 1755 the 'Court was called to consider of a Patent lately applied for by Joseph Bosley, Watchmaker of Leadenhall (described in 1761 as opposite East India House) and said to be for an improvement in the making of Watches.' They resolved 'That the Master and Wardens be desired to call to them such of the Assistants as they think fit, and enquire into the proceedings relating to it, and take proper measures to prevent it being carried into operation, and the Renter Warden to defray the charges of such an application'. The Clockmaker's Company at this time took the view that the granting of patents (monopolies) to individuals could in certain circumstances be constrictive to the trade in general and therefore tended to be conservative. The Renter Warden 'was desired to buy one of the Watches made according to the new Patent' in April.

Joseph Bosley applied for a patent (Patent No 698) on 1 March 1755. This was presented on 21 June 1755.
1st For increasing the number of teeth in small pinions throughout the whole movement of repeating and other watches. The pinions consequently become larger and the wheel that leads them goes farther from the centre. A wheel and pinion more than commonly used, is necessary to prevent the Watch going down before the usual time, but each wheel leading its pinion so much farther from the centre lessens the friction. The balance wheel goes the contrary way.
2nd A new invented slide, which slide has no wheel attached to it. The Index turns upon a brass socket, and points to an arch of a circle, divided with the word faster on one end, and the word slower on the other, and the Index may be made with a cock to keep it down, or with screws, or with springs.

The Bosley regulator is therefore a pivoted lever underneath the balance cock and able to move concentrically with the balance arbor. The short arm with two index pins embraces the outer end of the balance spring whilst the long arm moves across the scape which is engraved on the top plate. It is the precursor of the modern index.

> Bosley, Joseph, Watch-maker, *opposite the East India-house, Leadenhall-street*, has obtained a Patent for making Repeating and other Watches on a new principle, which occasions a lighter friction throughout the whole work, and carries a greater power to the ballance: and for a new Slider, which regulates the motions of Watches with more certainty, and is more intelligible than that in common use.   HOLDEN'S LONDON DIRECTORY 1799

PLATE 5C

Verge Watches

PLATE 6A

PLATE 6B

index to engraved calibrations on top plate. Engraved Recordon Charing Cross No 1476. Finely engraved and pierced cock of floral design.
Key size 4. Diameter 45mm

Compare with Francis Berguer c.1820 (Plate 13). Baillie Vol 1 lists Louis Recordon as succeeding Emery in 38 Cockspur Street (Charing Cross), London. It is worth noting that he retired in 1796 but the firm continued in name after this date.

PLATE 6
Robert Westmore, 179 Friargate, Preston

Silver pair-case hallmarked Chester **1806.** Casemaker's mark HA (possibly Hugh Adamson known to be working 27 Highfield Street, Liverpool 1800-25). Cream enamel dial, black roman numerals, gold skeletonised spade hands. Cylindrical pillars. Fusee chain. Flat steel undersprung balance with finely pierced and engraved cock and Bosley type regulator. Engraved Rt Westmore Preston No 1065.
Key size 6. Diameter 57.3mm

Robert Westmore listed Pigot's & Co Commercial Directory Lancashire 1822. Also listed in Lancashire Clockmakers *by Brian Loomes as working as silversmith and watchmaker at 178-181 Friargate between 1822-28. Freeman Preston in 1817.*

PLATE 7
Louis Recordon, 38 Cockspur Street, London

Gilt gunmetal case with bull's eye glass. Casemaker's mark TG (possibly Thomas Gibberd 1802-11). White enamel dial, black roman numerals, blued steel spade hands. Winding through the dial. Fusee chain. Polished steel flat rimmed undersprung balance with Bosley type regulator

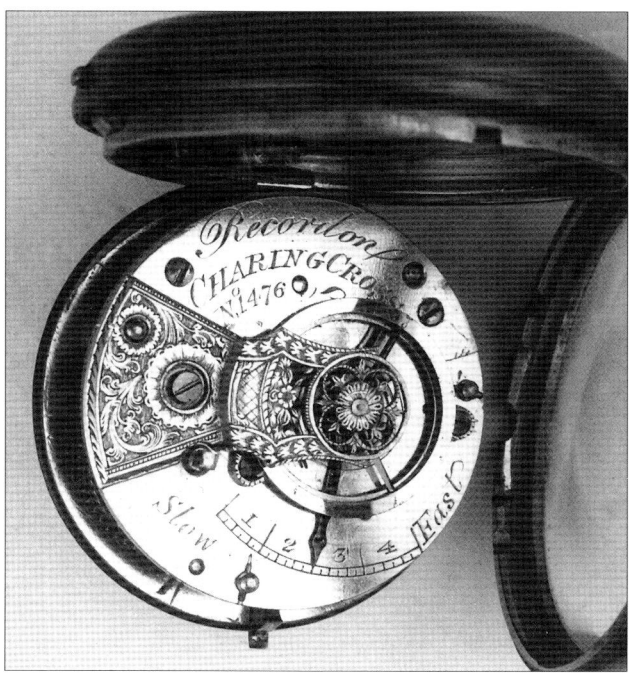

PLATE 7

23

Verge Watches

PLATE 8A

PLATE 8B

PLATE 8
William Broad, 53 Leadenhall Street, London

Silver pair-case hallmarked London **1812**. Casemaker's mark IH. White enamel dial, black arabic numerals, beetle and poker hands. Cylindrical pillars. Fusee chain. Single flange to barrel. Polished steel three arm flat rimmed undersprung balance with Bosley type regulator with engraved calibration on top plate. Finely pierced cock of foliate design. Top plate engraved Wm Broad London No 213.
Key size 6. Diameter 57mm

The dated (1823) embroidered watch 'paper' is an attractive additional feature The name appears to be William Jant and is probably that of a previous owner. 9th ed Britten lists William Broad, 53 Leadenhall Street, London as being active 1792-1836 as a clock and watchmaker having been apprenticed to Samuel Norton, musician and gaining his freedom of the Clockmakers Company 6 February 1792.

PLATE 9
Alexander Copeland, 113 Leadenhall Street, London

Silver pair-case hallmarked London **1812**. Casemaker's mark RG and pendant maker TE (possibly Thomas Emmett, 31 Great Sutton Street, Clerkenwell). White enamel dial and inset seconds dial inscribed Copeland London 317, black roman numerals, gold spade hands. Cylindrical pillars. Fusee chain. Single flange to barrel. Polished steel flat three arm undersprung balance with finely pierced cock incorporating foliate and bird design. Bosley type regulator with engraved scale on top plate. Engraved Copeland Leadenhall London No 317 on top plate. A lever at the 23 minute position operates a stop acting on the edge of the crown wheel.
Key size 6. Diameter 6mm.

PLATE 8C

Verge Watches

PLATE 9A

PLATE 9B

9th ed Britten lists Alexander Copeland as son of Alexander Copeland, St Botolph, Aldgate, a mariner and apprenticed to Jno Wontner Watchmaker Minories on 14 August 1801, turned over to Daniel Ward, Merchant Taylor Company and free 3 April 1809, liveryman 1810, working between 1800 and 1815 at 113 Leadenhall Street. Listed in directories as watch and clockmaker.

PLATE 10
John Doughty, London

Silver pair-case hallmarked London **1815.** Casemaker's mark WF. White enamel dial, black roman numerals, gold spade hands. Fusee chain. Polished flat steel undersprung balance with silvered Tompion type regulator and engraved hand as pointer. Barrel bridge engraved John Doughty 820 with London on the foot of the finely pierced cock of floral and dragon design.
Diameter 56.7mm

No record of John Doughty working around these dates in London has been found — only of a Thomas. It is always an additional point of interest to have some biographical notes relating to the maker, but this is not essential when assessing a purchase The engraving on this watch was sufficient reason for acquisition. There is always the possibility of learning more of John Doughty at a later date.

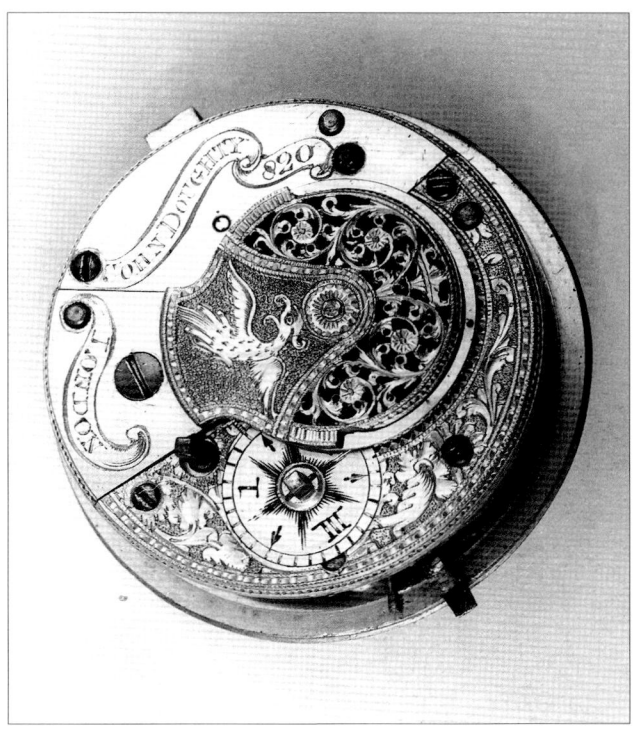

PLATE 10

Verge Watches

Plate 11a

Plate 11b

Plate 11c

Plate 11
John Prince, 9 Dorvill Row
Hammersmith, London

Silver pair-case hallmarked London **1818.** Casemaker's mark MC. White enamel dial, black roman numerals, decorative pierced spade hands. Cylindrical pillars. Fusee chain. Single barrel flange. Polished steel three arm balance with Bosley style regulator. Gilt movement engraved Prince Hammersmith No 392, **Key size 7. Diameter 51.8mm**

9th ed Britten records Prince at King Street in 1826 and 20 Dorvill Row 1832-8, but the watchpaper states 'Watch and Clockmaker to the Duke of Devonshire' with an address of 9 Dorvill Row.

PLATE 12
William Gravell & Son, 49 St John Street, West Smithfield, London

Silver double bottomed case with milled edge to body. Hallmarked **1819**. Casemaker's mark IR (a casemaker Isaac Russell, 5 Richmond, St Lukes, London listed in Pigot's Directory 1823). White enamel dial, black roman numerals, gold spade hands. Dust cover with blued steel slide engraved Gravell & Son London No 12573 repeated on the barrel bridge. Cylindrical pillars. Fusee chain. Flat steel undersprung balance of rounded edge with Bosley type regulator. Finely pierced cock of floral design with diamond endstone to balance staff.

9th ed Britten lists William Gravell Senior 1810-42, William Gravell Junior 1810-51.

PLATE 12

PLATE 13A

PLATE 13B

PLATE 13
Francis Berguer, 201 High Holborn, London

Gilt gunmetal case. Casemaker's mark P. White enamel dial, black roman numerals, gold spade hands. Cylindrical pillars. Fusee chain. Flat rimmed steel undersprung balance with Bosley type regulator to engraved index on top plate. Engraved F Berguer Holborn No 859 on barrel bridge. **Key size 5. Diameter 47.5mm**

9th ed Britten lists Francis Berguer as working 1815-20, High Holborn, London.

Verge Watches

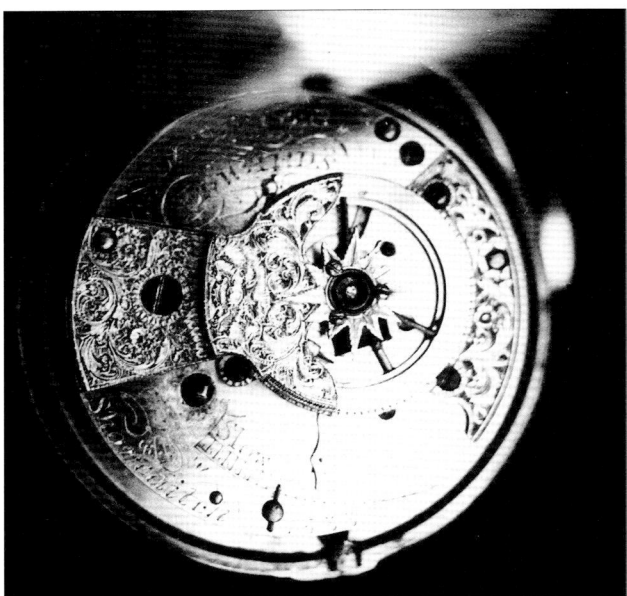

PLATE 14

PLATE 14
Lormier & Edwards, 17 Shoreditch, London

Silver pair-case hallmarked London **1820.** Casemaker's mark WL (possibly William Linsley, 69 Banner Street, St Lukes London). White enamel dial, black roman numerals, broad gold spade hands. Yellow tinted bull's eye glass. Cylindrical pillars. Fusee chain. Flat rimmed three armed steel undersprung balance with Bosley type regulator with engraved scale on top plate. Fine engraved balance cock of floral design (with surrounding cock shaped to simulate a star) with diamond endstone to balance arbor. Top plate engraved Lormier Edwards Shoreditch.
Key size 7. Diameter 56.5mm

9th ed Britten lists David Lorimer (Lormier) at this address 1805-1821 becoming, according to one source, Lormier & Edwards after this date, while a directory gives the partnership dates as 1806-1826. Benjamin Edwards is listed alone 1826-1870.

PLATE 15
William Fothergall, London

Movement. Cylindrical pillars. Fusee chain. Backplate engraved Wm Fothergall London 1244. Finely pierced cock of grotesque design with flat steel three arm balance undersprung balance with silver indicating dial for Tompion type regulator.
Diameter 46.5mm

Again this watch movement has a particularly fine cock and engraving. Careful study will demonstrate that these cocks are never identical in every feature.

PLATE 16
William Hanson, London

Movement. White enamel dial, small black arabic numerals, gold spade hands. Cylindrical pillars. Fusee

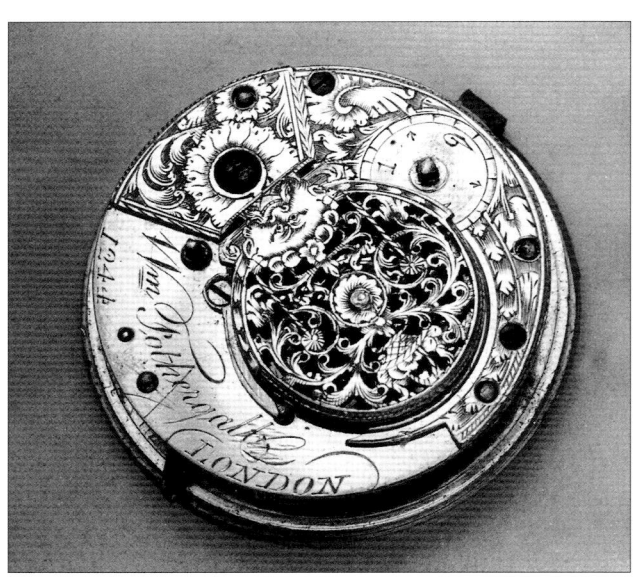

PLATE 15

PLATE 16

chain. Flat steel three arm undersprung balance. Engraved and pierced balance cock with two birds in a foliate design. Back plate engraved Wm Hanson London 4086.
Diameter 40.7mm

9th ed Britten's lists William Hanson Windsor 1820.

PLATE 17
William Wood, 60 Middleton Street, Clerkenwell, London

Movement. White enamel dial, small black roman numerals, gold spade hands. Cylindrical pillars. Fusee chain. Flat steel three arm undersprung balance with silver regulator dial engraved I to IIII. Back plate engraved Wm Wood London No 2951.
Diameter 46.3mm

9th ed Britten lists William Wood as being active at this address 1826-1842.

PLATE 18
George White, Trongate, Glasgow

Silver pair-case hallmarked Birmingham **1821**. Case-maker's mark V & Co (Vale & Co later Vale and Rotherham Coventry). White enamel dial, black roman numerals, gold spade hands. Cylindrical pillars. Fusee chain. Flat steel three arm undersprung balance. Finely pierced cock of grotesque design. Blued steel Tompion style regulator dial with unusual transitional index pointer

PLATE 17

operated by disc type regulator. Balance plate engraved Geo White Glasgow 24950.
Key size 6. Diameter 57.5mm

9th ed Britten lists George White at this address 1824-49 and states that there is an example (watch) of his work in New York University.

PLATE 18A

PLATE 18B

Verge Watches

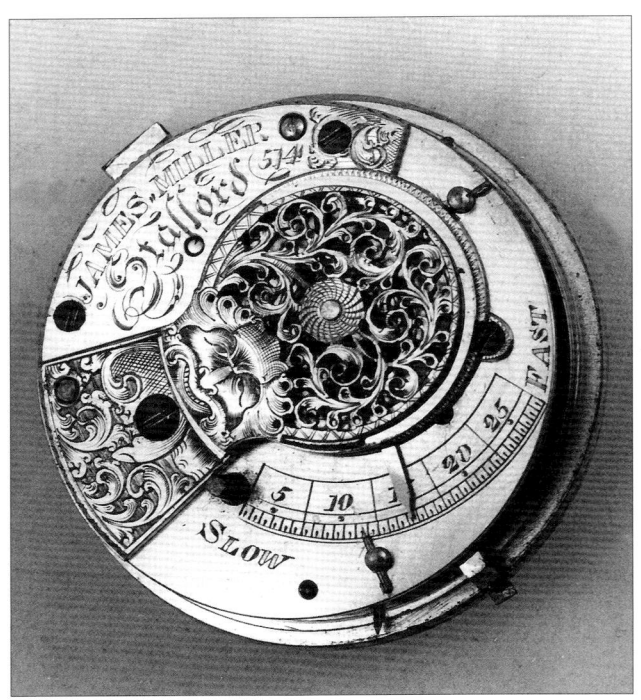

PLATE 19

PLATE 20

PLATE 19
James Miller, Stafford

Silver pair-case hallmarked Birmingham **1823**. Casemaker's mark WB (possibly William Brown, Well Street, Coventry listed Pigot's Directory 1822). White enamel dial, black roman numerals, gold spade hands. Cylindrical pillars. Fusee chain. Single barrel flange. Polished steel balance under pierced balance cock of foliate and grotesque design (note mask at base of cock) with Bosley type regulator with calibrated scale on top plate. Barrel bridge engraved James Miller Stafford 514.
Key size 7. Diameter 57.8mm

PLATE 20
Martin Roper, Little Dockray, Penrith, Cumberland

Silver pair-case hallmarked London **1825**. Casemaker's mark CM (possibly Charles Maston, 18 Red Lion Street, Clerkenwell who entered his mark 9 August 1824. White enamel dial, black roman numerals, gold spade hands. Cylindrical pillars. Fusee chain. Flat steel balance under pierced cock of floral and grotesque design with Bosley type regulator. Engraved Martin Roper Penrith No 8968.
Key size 7. Diameter 57.5mm

John Penfold in his book Watchmakers of Cumberland *(1977) gives the following interesting details concerning this maker. Martin Roper was born in 1774, the son of Joseph Roper, tailor of Penrith and Margaret his wife. He was baptised on 4 September 1774 George Martin, but that seems to be the only time the first Christian name was used. It appears that he set up on his own about 1812. His place of work was Little Dockray between Blue Bell Yard and Camalts Yard except for a brief period in 1820 when he was described as being in Middlegate. This could be a descriptive error as his workshop and cottage combined stood at the head of Little Dockray where it joins Middlegate on a site more recently occupied by Messrs Pears and Boulas. He was wont to display a few watches of his make in his kitchen window. He lived alone and never married. He retired about 1845. In the 1851 Census he is listed as residing at lodgings as an outdoor pauper at the house of Launderess at Netherend. He died December 1851.*

Verge Watches

PLATE 21A

PLATE 21
William Smith, 8 Rawsthorne Street, Goswell Road, London

Silver double bottomed hunter case hallmarked London **1835**. Casemaker's mark AT (possibly A J Thickbroom, 10 Galway Street, St Lukes London). White enamel dial, black narrow roman numerals, gold spade hands. Fusee chain. Broad flat rimmed polished steel undersprung balance with Bosley type regulator. Plain triangular balance cock with small diamond endstone to barrel arbor.

PLATE 21B

Barrel bridge engraved William Smith AD 1836 with No 1830 on top plate.
Diameter 48.4mm

The Post Office Directory for 1836 records this maker as watch, clock and chronometer and springmaker at this address but the same Directory for 1844 lists him at No 27 Lombard Street.

PLATE 22
Mary Metcalfe, Cockermouth

Movement. Cylindrical pillars. Fusee chain. Round fusee stop stud. Finely pierced and engraved cock with flat three arm undersprung balance with Bosley regulator to engraved index marks on top plate. Cock of foliate design with grotesque mask. Barrel plate engraved Mary Metcalfe Cockermouth 28872. Dial side of pillar plate stamped 19/15.
Diameter 44.4mm

Loomes in Watch and Clockmakers of the World *lists a Mary Metcalfe Maryport 1823 - this is a town near Cockermouth. Penfold in* The Clockmakers of Cumberland *groups the Barwise, Mitchell and Metcalfe family. Mary was related to the noted John Barwise. She commenced clockmaking in Maryport around 1825 but returned to Cockermouth in the 1830s. On her father's death in 1844 she took over the business but died in 1849.*

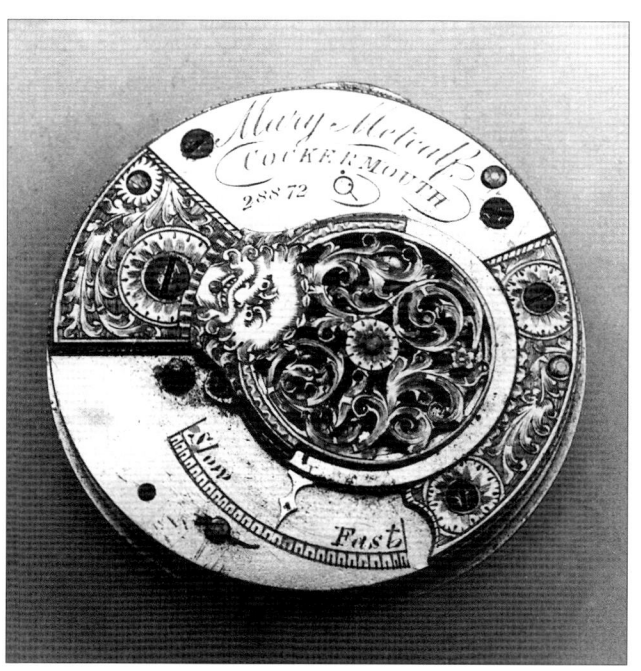

PLATE 22

31

Verge Watches

PLATE 23A

PLATE 23C

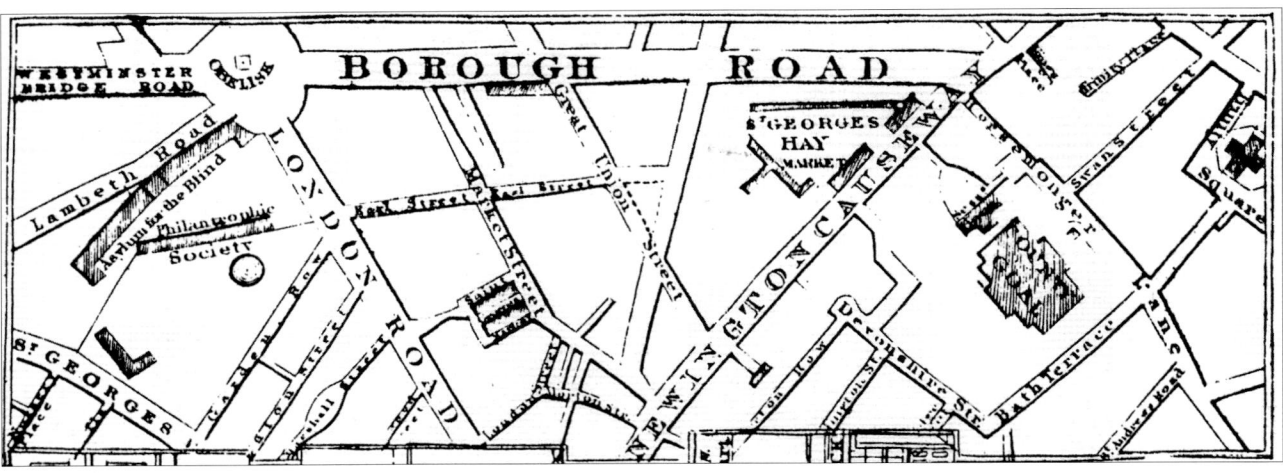

PLATE 23B

PLATE 23
Francis Pearce, 9 Newington Causeway, London

Silver open face double bottomed case hallmarked London **1841.** Casemaker's mark WC (possible William Carter, 22 Galway Street, Bath Street, St Lukes, London). Pendant hallmarked **1840.** Maker IW. Silver engine turned and engraved dial, applied gold roman numerals, gold spade hands. Fusee. No barrel flange. Mainspring setting up under dial. Plain triangular cock with flat steel undersprung balance with Bosley type regulator Engraved on top plate Frs Pearce Newington Causeway No 6908.

See map for location of Newington Causeway - between Borough High Street and Elephant and Castle. Francis Pearce is listed as silversmith in Pigot's directory 1844.

Plate 24
Sophia Harris, London

Silver single bottomed case hallmarked London **1841**. Casemaker's mark CH (directories list two possible case makers – Charles Harte, 21 Wynyatt Street, Goswell Road, Clerkenwell; Charles Hubbard, 9 Poole Terrace, City Road. Further details of these two makers can be found in *Gold and Silversmiths* by John Culme). White enamel dial, black roman numerals, gold moon style hands. Cylindrical pillars. Fusee chain. Flat steel under-sprung balance with plain cock and Bosley type regulator. Engraved on barrel bridges Sophia Harris London 1842.
Key size 5. Diameter 43mm

Is 1842 the date and Sophia Harris the owner? Or is Sophia Harris a widow carrying on her husband's business and 1842 the movement number? This is often a controversial discussion.

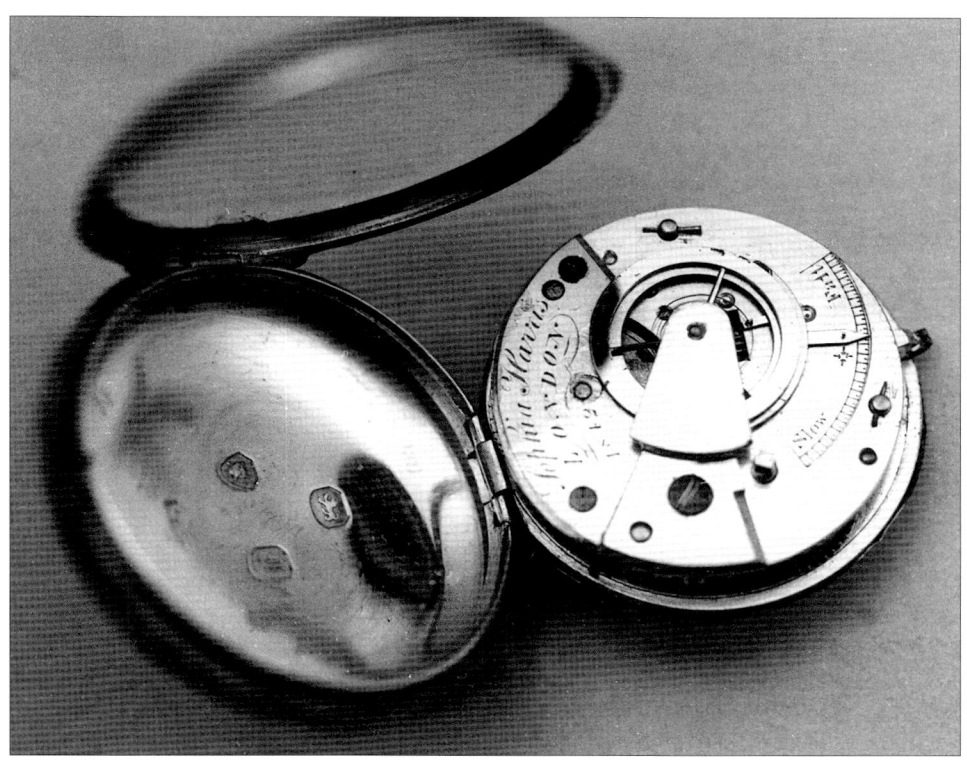

Plate 24

Verge Watches

PLATE 25A

PLATE 25B

PLATE 25
William Feltham, Tavern Street, Stowmarket

Silver pair-case hallmarked Birmingham **1843.** Casemaker's mark R & S, J & SS. White enamel annular dial, heavy black roman numerals, gold spade hands. Engraved brass centre floral design and the brass edge to the dial engraved 'Keep Me Clean and Use Me Well and I to You the Truth Will Tell' (an adage that appears on many watches but always adds a little interest). Cylindrical pillars. Fusee chain. Flat rimmed steel undersprung balance with Bosley type regulator. Finely pierced balance of foliage and grotesque design. The variations of these grotesque masks are of interest in themselves. Engraved on barrel bridge Barnard Girling. W Feltham. Stowmarket and No 7634 on top plate.
Diameter 58.5mm

William Feltham was in business as a watchmaker and acted as the Registrar of Marriages from 1844. He worked in Tavern Street, Stowmarket from 1838 to at least 1855. Barnard Girling appears to have been the owner but White's Suffolk Directory for the period only lists a Samuel Girling Gardener and Nurseryman.

PLATE 25C

Plate 26
Samuel Carmen, Broad Street, Harleston, Norfolk

Open face gilt case. Casemaker's mark WC (possibly William Carter, 22 Galway Street, City Road, London). Decorated and engine turned gold dial, raised gold roman numerals, blued steel hands. Cylindrical pillars. Fusee chain. Flat broad rimmed steel three armed undersprung balance with Bosley type regulator. Plain triangular balance cock with endstone to balance arbor. Barrel cock engraved Samuel Carmen Harleston.
Diameter 42.3mm

Brian Loomes in Watch and Clockmakers of the World *lists Samuel Carmen as working at Broad Street, Harleston in Norfolk between 1830 and 1875.*

Plate 27
Septimus Miles, Ludgate Street, London

Open faced double bottomed case hallmarked London **1851.** Casemaker's mark RO/JE (possibly Richard James Oliver and John Evan Edwards, 19 Galway Street. Pigot's Directory gives them as working at this address in 1852 and Oliver alone in 1860). White enamel dial, black roman numerals, gold spade hands. Dustcap engraved Septs Miles Ludgate Street 3495 with blued steel spring slide pierced to frame a narrow decorated balance cock incorporating both floral and grotesque design with diamond endstone to balance arbor. Flat rimmed polished steel three armed balance (arms triangular in cross section) and Bosley type index to engraved calibration on top

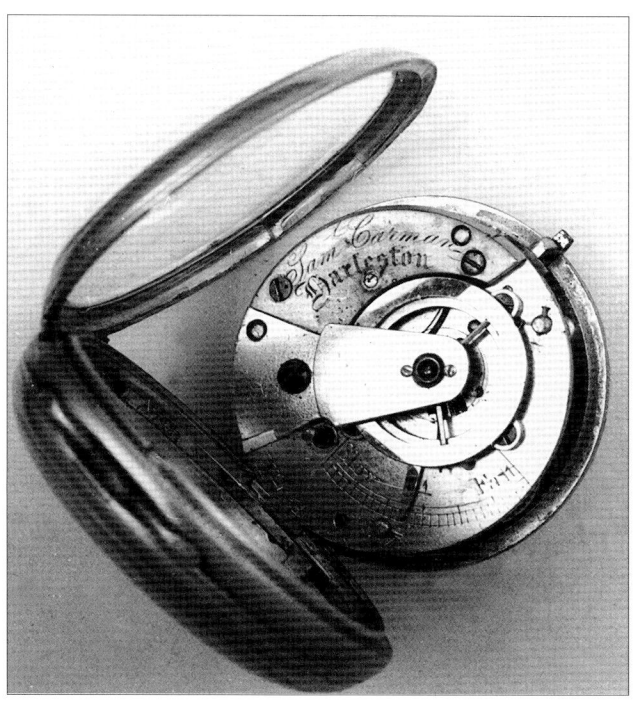

PLATE 26

plate. Cylindrical pillars. Fusee chain. Barrel bridge signed Septs Miles London 3495.
Key size 5. Diameter 52.6mm

9th ed Britten lists Septimus Miles as active 1794-1847, liveryman of the Clockmakers Company 1810. There are no details of Septimus' apprenticeship in the Register of Apprentices for the Clockmakers Company but an Arthur Fellenberg Miles was apprenticed to his father Septimus of Carter Lane, Doctors' Commons, watchmaker for seven years on 4 September 1843.

PLATE 27

Verge Watches

PLATE 28

PLATE 29

PLATE 28
F Parriott, Henley

Silver pair-case hallmarked London **1857**. Casemaker's mark RO/JE (possibly Richard James Oliver and John Evan Edwards listed in Pigot's Directory as working at 19 Galway Street, St Lukes as Oliver and Edwards 1852 and Richard Oliver 1860. John Culme in his *Directory of Gold and Silversmiths* states that Richard James Oliver was first at this address in 1845 where he was joined in 1846 by Edwards. They dissolved the partnership on 16 February 1859). White enamel dial, black roman numerals and gold spade hands. Dustcap with blued steel slide pierced to fit an unpierced cock of floral design with diamond endstone. Dustcap maker's mark I x R. Cylindrical pillars. Fusee chain and pierced chain escutcheon. Flat polished steel three armed undersprung balance with Bosley type regulator to calibrated scale engraved on top plate. Balance bridge engraved F Parriott, Henly.
Diameter 52mm

PLATE 29
James Mitchell, Mid Street, Keith, Banff, Scotland

Silver pair-case hallmarked London **1860**. Casemaker's mark ED (possibly Mrs E Dyer, 11 Gloucester Street, St John Street Road, London EC. Elizabeth Dyer's mark was registered 26 November 1851 after her husband died). White enamel dial, broad black roman numerals, broad gold spade hands. Cylindrical pillars. Fusee and chain flat rimmed three arm steel balance Tompion style regulator with silver dial with engraved hand acting as pointer. Engraved and pierced cock of foliage design. Barrel bridge engraved Mitchell Keith 11720.
Key size 7. Diameter 55.5mm

James Mitchell, Mid Street, Keith, Banff, Scotland is listed in both Kelly's Directory for 1860 as well as Brian Loomes' Watch and Clockmakers of the World.

PLATE 30
Louis Frary, Rue du Rhone, Geneva, Switzerland

Open face gunmetal case with enamelled back and marcasite decoration around bezel. Casemaker's mark stamped Leyton Bonijol. White enamel dial inscribed Louis Frary, arabic numerals. Winding through the dial at 2 o'clock position. Cylindrical French style fusee and chain with single flange to barrel. Adjustable potence. Three arm brass balance. The balance cock pierced to form letters FRARY. Steel cocqueret. Silver regulator dial. Top plate engraved L Frary A Geneve.
Key size 3. Diameter 47.7mm

Louis Frary c.1825 listed as fabricant et marchant d'horlogerie. See Watches of Fantasy 1790-1850 *by Oswald Patrizzi.*

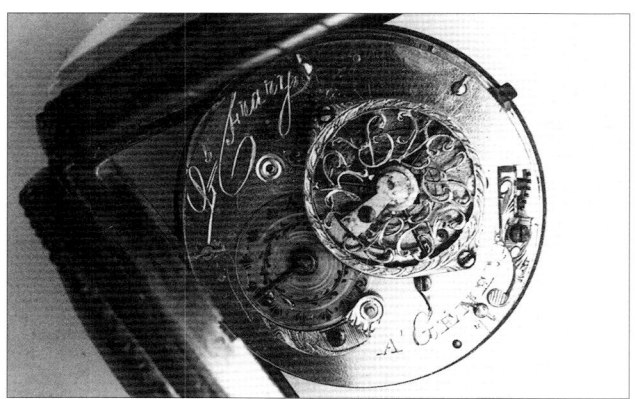

PLATE 30

Verge Watches

PLATE 31A

PLATE 31B

PLATE 31
Anonymous – French

Movement. White enamel dial, black roman numerals, Breguet style blued steel hands. Winding through dial. Hexagonal pillars. Fusee chain. Engraved design to edge of backplate. Three armed brass undersprung balance. Finely pierced balance cock of French style with steel cocqueret. Silver disc regulator dial engraved Advance Retard. Movement fitted with sprung dust cover.
Diameter 47mm

PLATE 32

PLATE 32
Anonymous – French

Movement. Winding through dial. Cylindrical pillars. French style fusee and chain. Mercurial gilding to plates. Gold balance and scapewheel. French style of decorated cock. Blued steel index dial regulator inlaid with brass. Steel work under dial polished and finished to high standard.
Diameter 44.5mm

PLATE 1. Cylinder escapement

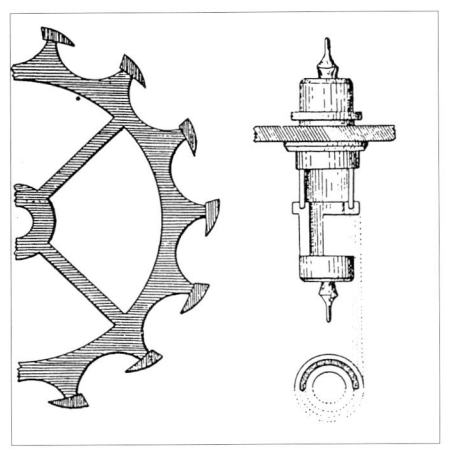

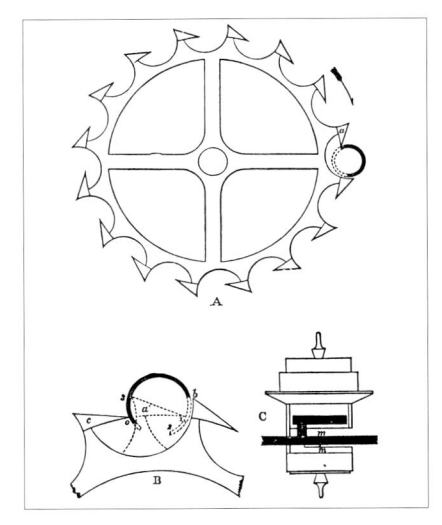

CHAPTER TWO

Cylinder Watches

Having worked on various ideas for improving the verge escapement, Edward Booth (Edward Barlow) 1636-1717, Thomas Tompion and William Houghton took out the following patent. The recorded details are somewhat vague but not without interest.

AD 1695 September 23 - No 344

A new sort of watch or clock with the ballance wheele or swing wheele either flatt or hollow, to worke within and crosse the center of the verge or axis of the ballance or pendulum, with a new sort of teeth made like tinterhook, to move the ballance or pendulum withall, and the pallett of the axis or verge of the ballance or pendulum are to be circular, concave, and convex, or other teeth or pallett that will not goe but by the helpe of the spring to the ballance, which will make such watch or clock goe more true and exact, and be of greater vse to our subiect both at sea and land, than any other heretofore made or now used

No Specification enrolled.

George Graham (1673-1751), apprentice and subsequent partner of Tompion, made further improvements. By 1725 he had devised the cylinder escapement in its present form and by 1727 he was using it in his own watches. The following year Julian Le Roy ordered a 'watch utilising the new horizontal escapement'. Watches with similar escapements were made by Mudge and Mudge and Dutton up to the 1770s. There were undoubtedly advantages and disadvantages. The verge watch had proved to be a robust watch and easier to make. The horizontal cylinder escapement allowed thinner watches to be made as the escape wheel was now in the same plane as the rest of the movement. Furthermore, the cylinder escapement was not a dead beat frictional type, the variations in strength of the mainspring having less effect than in the case of the verge which made the inclusion of the fusee much less important. It should be noted, however, that the fact that a cylinder escapement is of the frictional rest type is sometimes overlooked by the repairer who replaces the mainspring with a stronger mainspring in the mistaken belief that the going of the watch will be improved, whereas in fact the reverse is the case.

The early cylinders were made of steel with a brass escape wheel but the brass was soon found to have a serious abrasive effect on the cylinder and the escape wheel was changed to steel. Urban Jurgensen (1776-1830) was the first to use steel for the escape wheel. The cylinder escapement was further refined c.1764 with the use by John Arnold, John Elliott and Abraham Breguet of a ruby cylinder and a steel wheel. Later James Ashley cut the escape wheel teeth on three different levels in order to reduce the wear on the cylinder by three times. At first the cylinder escapement was fitted to the same movement used in the verge fusee – full plate and undersprung balance. The English cylinder had the plugs made in two parts, the cylinder being bushed with brass and the steel plugs fitted

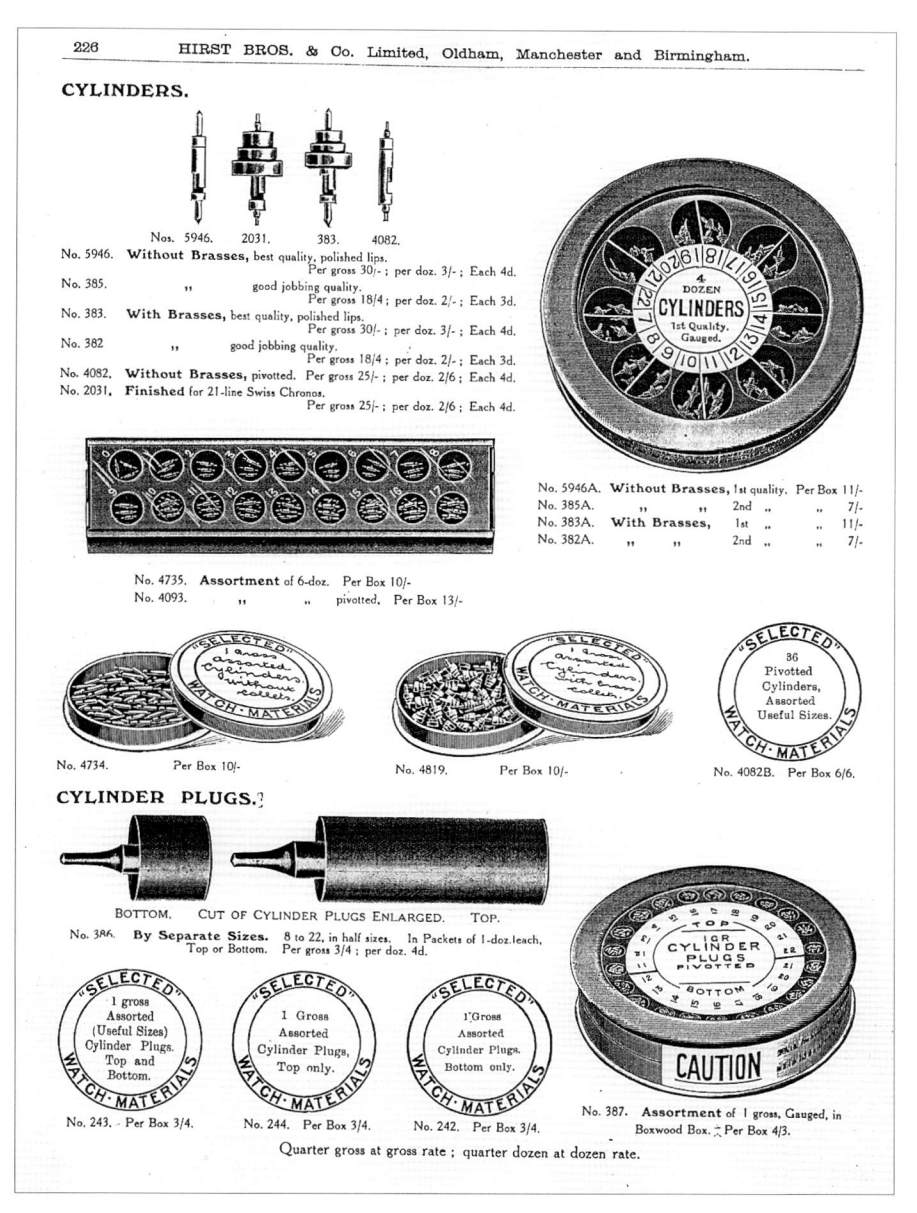

Plate 2. *Fourth Wide Awake Catalogue*, c.1916

into brass. The acceptance and development of the cylinder escapement was far more widespread on the Continent. At an early stage they had recognised the advantages of using a steel cylinder. The advent of the Lepine bar movement in which the top plate and the train in separate cocks (bars) screwed to the top plate was certainly advantageous. The inclusion in the Geneva cylinder movement of the Maltese cross form of stop work, the abolishing of the fusee and the use of the going barrel and the smaller faster balance wheel requiring a smaller balance cock all contributed to a smaller and thinner watch.

Readers are advised to consult for further study an excellent major reference on cylinder escapements in *Clock and Watch Escapement* by WJ Gazeley and also the excellent *L'Echappement à Cylindre (1720-1950)* by Henry L Belmont. This book provides the background to the manufacture, work-force and their methods of producing the cylinder in the Le Haut-Doubs area during the nineteenth century. It includes details of the invention in 1869 of the machines for cutting the scape wheels etc. An excellent word picture of this region is portrayed by Charles Allix in his book *Carriage Clocks;* Chapter VI covers the factories of the Doubs and the development of Japy at Beaucourt.

39

Cylinder Watches

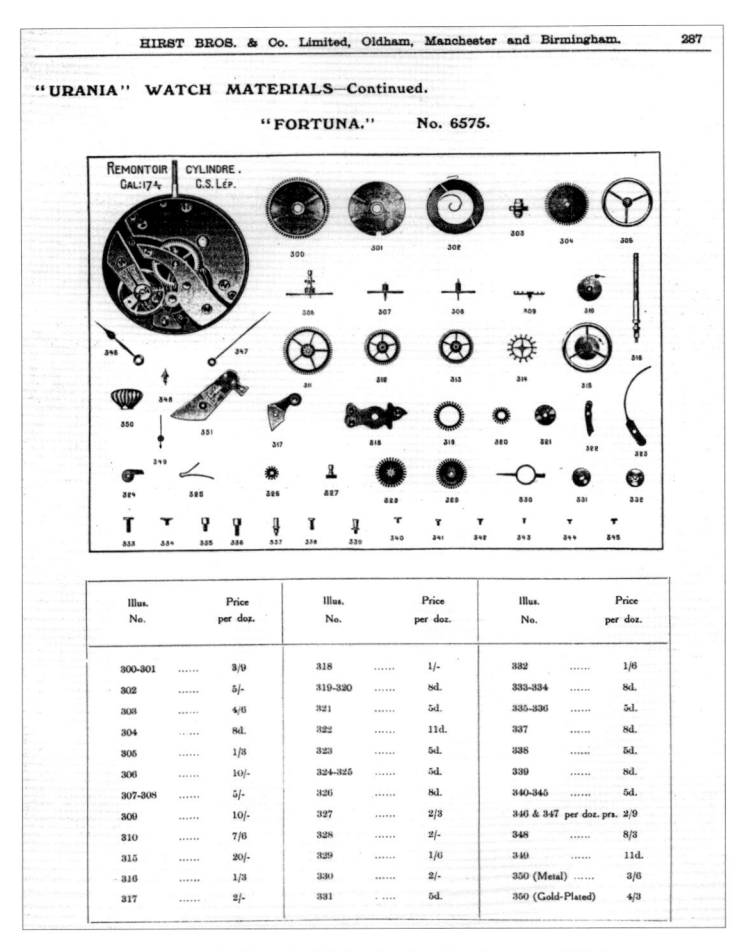

Plate 3. *Fourth Wide Awake Catalogue*, c.1916

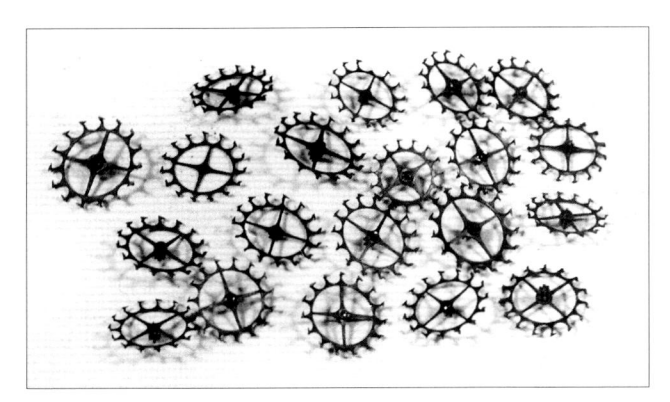

Plate 4. Selection of cylinder escape wheels

Plate 5. French manufacturer of cylinder escapement wheels

Plate 6
John Grant, 75 Fleet Street, London

Silver engine-turned hunter double bottomed case hallmarked London **1822**. Casemaker HG (Horatio Gough, 23 Coppice Row, Clerkenwell – PO Directory 1823/4). Dustcap with visible bell-shaped cock through cover. White enamel dial with three copper feet to secure dial inscribed 5053 upper centre, black roman numerals, gilt spade hands, inset dial. Key winding. Top plate engraved Grant London No 5053. Full plate fusee movement with cylindrical pillars. 8 jewels. Cylinder escapement.
Key size 6. Diameter 55mm

Cylinder Watches

PLATE 6A

PLATE 6B

PLATE 7
E Raffin, à Genève, Switzerland

Lady's silver open face case, hinged back bezel and gilt cuvette. Silvered engine-turned dial inscribed E Raffin à Genève, black roman numerals, blued steel spade hands. Key wound and set. Early bar movement engraved E Raffin à Genève on bottom plate. 6 jewels. Cylinder escapement. Steel cylinder. Gold three arm balance and blued oversprung balance spring. Blued index on balance cock engraved RA.

Key size 3. Diameter 40.4mm

According to ébauche details Tardy Dictionnaire *this appears to date from 1800-1820.*

PLATE 7A

PLATE 7B

41

Cylinder Watches

PLATE 8A

PLATE 8B

COLOUR PLATE 1 (page 6) and PLATE 8
Thil, à Genève

Gold 14 carat open face case, straight ribbed pattern to body, hinged back bezel and cuvette. Small monogram (5.6mm) on back of case with motto CRESE CRESE on either side of triple turreted castle – the centre turret flying a pendant. Gold engine-turned and florally decorated dial, blued steel hands, inset seconds dial. Key wound and set. Early bar movement engraved Thil à Genève. 6 jewels. Cylinder escapement. Brass three arm balance and blued steel oversprung balance spring. Polished steel index on balance cock engraved S F.
Key size 3. Diameter 45mm

Thil à Genève, Fabricants d'Horlogerie Cite et Corraterie 1830-1835, listed in Watches of Fantasy *by Patrizzi and Sturm.*

PLATE 9
Hy Matthey

Gilt silver full hunter case, cuvette engraved Echappement à Cylindre Huit Trous à Rubis. Inset back cover stamped HMD Argent. White enamel dial secured with two copper feet with dog screws, narrow black roman numerals, gilt moon hands. Key wound and set. Early bar cylinder movement. Brass three arm brass balance and blued oversprung balance spring. Steel cylinder. Polished index on balance cock.
Key size 4. Diameter 47.7mm

PLATE 9

Cylinder Watches

PLATE 10A

PLATE 10
Henriot, à Genève

Silver engine-turned open face case, hinged back bezel and cuvette. Inside back cover stamped 1185 and 19478 and I in diamond shape. White enamel dial secured with two copper feet with dog screws inscribed Henriot à Genève, gilt thin Breguet style hands. Key wound and set. 6 jewels. Bar cylinder movement. Engraved Henriot à Genève. Steel scape wheel. Brass three arm balance and blued oversprung balance spring. Polished index on cock engraved SF. Complete with silver twisted rope lady's Albert in original black leather presentation case.

Key size 4. Diameter 42.4mm

Henriot Fabricant horlogie environs 1820 listed Watches of Fantasy *by Patrizzi and Sturm. 9th ed Britten lists c.1785. Example with tinted gold case by this maker in Victoria and Albert Museum.*

PLATE 10B

Cylinder Watches

PLATE 11A

PLATE 11B

PLATE 12A

PLATE 12B

44

Cylinder Watches

PLATE 13A

PLATE 13B

PLATE 11
Klaftenburger, 457 Regent Street, London

Lady's silver full hunter case with gold hinges to back and front covers, hinged cuvette engraved Klaftenburger 457 Regent St No 8569. White enamel dial, black roman numerals, gold spade hands. Inset seconds dial between VII and VIII position. Pendant winding – flat topped button engraved with arrow to indicate direction of winding. Patek Philippe early keyless winding with wolf tooth winding wheel, pull to set hands. Provision for a male key square retained on barrel arbor. 10 jewels. Bar cylinder movement. Gold three arm balance. Blued steel oversprung spring. Pointer index on balance cock engraved SLOW FAST.
Diameter 40mm

Adrien Philippe was particularly interested in stem winding systems and took out several patents to protect his designs. His association with Patek came about through his successful exhibiting at the Paris Exhibition of 1844 of his slim stem winding watches. The variations to be found would readily form the basis of a specialist collection.
Keyless Mechanisms *by Vaudrey Mercer Horological Journal January 1987.* Patek Philippe Genève *by Martin Huber Alan Banbery.*

PLATE 12
Anonymous

Silver open face 0.935 case, Swiss hallmark, snap-on bezel, hinged engine-turned back and cuvette. White enamel dial, two copper feet and hip screws, black roman numerals, gold spade hands, inset sunk seconds dial. Keyless wound with push piece to set hands at 57 minute position. 10 jewels. Bar movement with steel cylinder and scape wheel. Flat rimmed gilt three arm balance with blued steel oversprung balance spring. Polished index on balance cock engraved SF. *Unusual winding and click work.*
Diameter 47.8mm

PLATE 13
J Myers, Chaux de Fonds, Switzerland

Silver open face case, hinged back, bezel and cuvette. Cuvette engraved J Myers, Westminster Road. White enamel dial inscribed J Myers Chaux de Fonds, two copper feet and dog screws, black roman numerals, narrow blued steel spade hands, inset sunk seconds dial. Key wound and set. 6 jewels. Cylinder bar movement. Steel cylinder and scape wheel. Three arm gold balance. Blued steel oversprung balance spring. Index on balance cock engraved FAST SLOW. *Unusual winding and click work.*
Key size 4. Diameter 46mm

John Myers listed at 135 Westminster Bridge Road as watch and clockmaker in Kelly's Directories for 1865, 1866 and 1868 and at 131 & 135 in 1870. By 1880 John Myers & Co. Still appeared in the 1890 Directory at this address.

Cylinder Watches

PLATE 14A

PLATE 14
Muller, Genève, Switzerland

Silver open face case with gold hinges to back, bezel and cuvette. Cream enamel dial inscribed Muller Genève, two copper feet and hip screws, narrow black roman numerals, blued steel Breguet style moon hands, inset sunk seconds. Key wound and set. 10 jewels. Early bar movement engraved Muller Genève. Steel cylinder and scape wheel. Three arm brass balance. Blued steel oversprung balance spring.
Key size 6. Diameter 45.1mm

J Muller, Fabricant et marchand d'horlogerie, Genève, 1830-1840.

PLATE 14B

Cylinder Watches

PLATE 15A

PLATE 15B

PLATE 15C

PLATE 15
Baume, Genève

Lady's silver open face highly decorated case, hinged back bezel and cuvette. Inside back cover stamped CB with leaf motif either side. White enamel dial inscribed Baume Genève, two copper feet with dog screws, black roman numerals, blued steel narrow spade hands. Key wound and set. 10 jewel. Cylinder bar movement engraved on bottom plate BAUME GENEVE motif of CJB and 927. Steel cylinder and scape wheel. Brass three arm balance. Blued steel oversprung balance spring. Polished index on to balance cock engraved FAST SLOW.
Key size 4. Diameter 42.2mm

An excellent chronology of Baume is given in J. Culme's Directory of Gold and Silversmiths: 'This, the English branch of a firm founded in Geneva by Ernest Françillon in 1834, was established in London by Celestin Baume in 1844/48. Joined by Joseph Lezard and by 1852 listed as Baume & Lezard Manufacturers and importers of Geneva watches at 75 Hatton Garden. Dissolution of partnership 25 March 1872. C Baume continued at the same address where he is listed as a manufacturer, importer and exporter of watches until 1876. Thereafter continued at the same address by his nephew Arthur Baume'. Baume exhibited at the Berne 1857, International Exhibition 1862, London 1882 and 1885 and Chicago 1893. This Directory covers the period 1838 to 1914 and is an invaluable reference source for information on the watch 'manufacturers' and retailers of this period.

Cylinder Watches

PLATE 16A

PLATE 16B

```
Schweizerische Fabrik- und Handelsmarken
publizirt im Schweizerischen Handelsamtsblatt Nr. 15, vom 21. Februar 1884.
Marques suisses de fabrique et de commerce
publiées dans la Feuille officielle suisse du commerce N° 15, du 21 février 1884.

Vom eidg. Markenamt vollzogene Eintragungen:
Enregistrements effectués par le Bureau fédéral des marques:

Le 16 février 1884, à neuf heures avant-midi.

No 1101.

André Mathey, fabricant,
La Ferrière.

ANDRÉ
MATHEY

Mouvements de montres or et argent.
```

PLATE 16C

PLATE 16
A Mathey, La Ferrière, Switzerland

Lady's silver open face watch, snap-on bezel, hinged cuvette and engine-turned back. Inside back cover stamped AM within an encircling buckled belt stamped FINE SILVER. This is probably the registered mark of André Mathey, La Ferrière, Switzerland. Snap-on engine-turned silver dial, raised gilt roman numerals, gilt hands tipped blue. Key wound and set. 10 jewels. Cylinder bar movement stamped A Mathey. Steel cylinder and scape wheel. Brass three arm balance. Blued steel oversprung balance. Polished index on balance cock engraved FAST SLOW.

Key size 3. Diameter 41mm

Cylinder Watches

PLATE 17

PLATE 18A

PLATE 17
Bumsel, Genève

Silver full hunter, engine-turned case, hinged back, bezel and cuvette (engraved GAMEL SOUTH KENSINGTON). White enamel dial inscribed BUMSEL GENEVE secured with two copper feet with dog screws, black roman numerals, blued steel narrow spade hands, inset sunk seconds dial. Key wound and set. 10 jewels. Cylinder bar movement engraved Bumsel Genève. Steel scape wheel and cylinder. Brass three arm balance. Blued steel oversprung balance spring. Polished regulator index on balance cock engraved FAST SLOW.
Key size 5. Diameter 49mm

Michael Bumsel & Co is listed in the 1867 Directory as watch importer at 39 Houndsditch (London).

PLATE 18
Anonymous

Nickel open face case, deep hinged back, snap-on bezel. White enamel dial secured by three screws, black roman numerals, blued steel spade hands. Key wound and set with space to carry key within the deep back cover. Full plate lacquered movement. Solid brass cylinder scape wheel with steel cylinder. Three arm flat rimmed balance. Blued steel oversprung balance spring. Index on balance cock engraved Slow Fast. *Early mass-production cylinder watch, almost mint state.*
Key size 6. Diameter 57mm, 26mm thick

PLATE 18B

Cylinder Watches

PLATE 19A

PLATE 19B

BEAUCOURT WATCH MATERIALS. No. 6181.

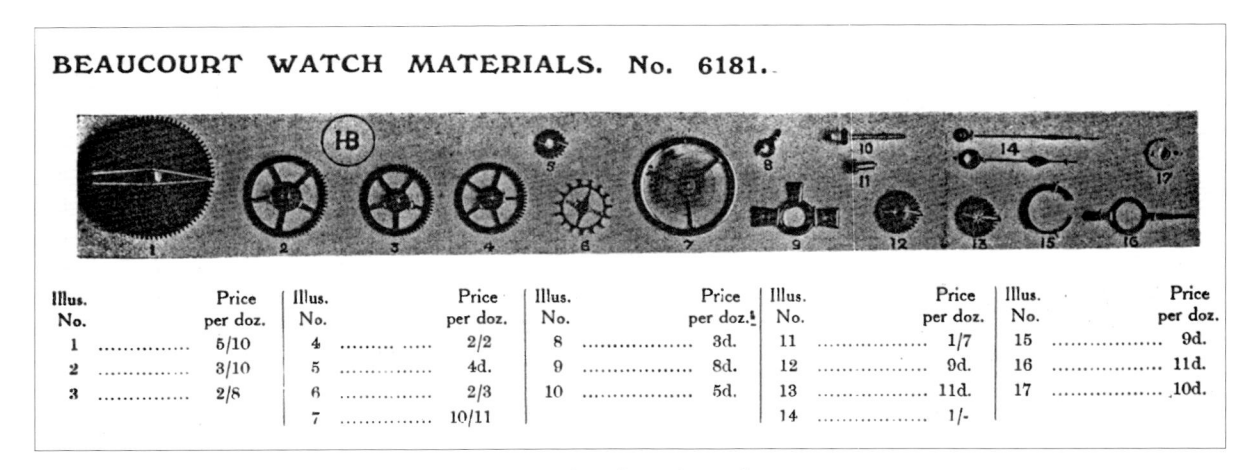

Illus. No.	Price per doz.	Illus. No.	Price per doz.	Illus. No.	Price per doz.	Illus. No.	Price per doz.	Illus. No.	Price per doz.
1	5/10	4	2/2	8	3d.	11	1/7	15	9d.
2	3/10	5	4d.	9	8d.	12	9d.	16	11d.
3	2/8	6	2/3	10	5d.	13	11d.	17	10d.
		7	10/11			14	1/-		

PLATE 19C. *Fourth Wide Awake Catalogue, c.1916*

PLATE 19
Japy, Beaucourt, France

Nickel open face case, snap-on bezel, hinged back and cuvette. White enamel dial, black roman numerals, blued steel spade hands, inset seconds dial. Key wound and set. 4 jewels. Cylinder movement. Brass three arm balance. Blued steel oversprung balance spring. Index on balance cock stamped RA SF. Bottom plate stamped with Japy trade mark.

Key size 5. Diameter 49mm

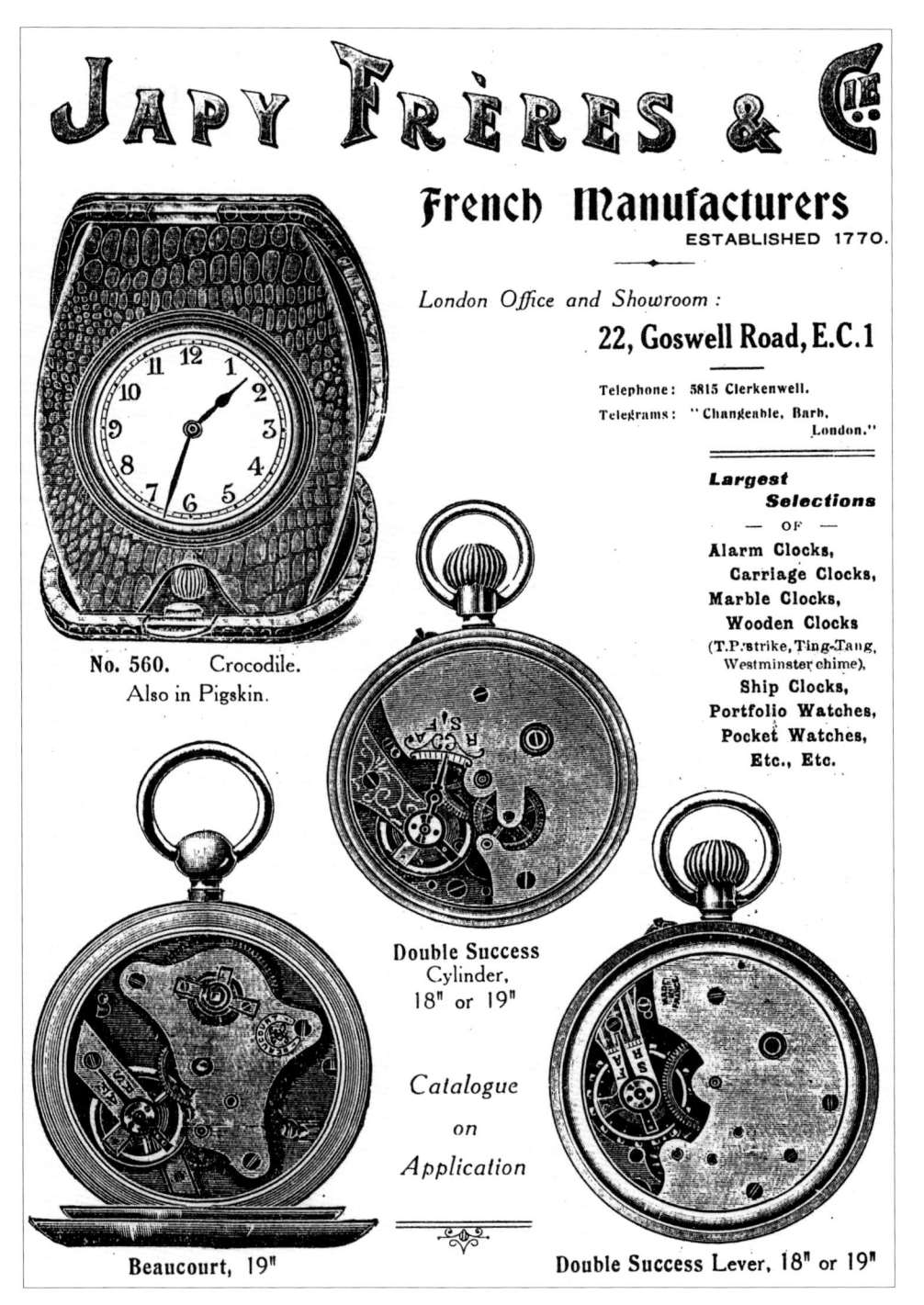

PLATE 19D. Advertisement from *Horological Journal*, 1923

Born in 1749, apprenticed first to Abraham-Louis Perrelet, Frédéric Japy founded a factory for the making of ébauches (movements) in 1771. He was one of the first to introduce machinery into his workshops and took patents out in 1799 for such machines. Rapidly expanding – 300,000 ébauches in 1813 to 500.000 in 1851 – he quickly became the main supplier of these unfinished movements to both French and Swiss finishers. Further details can be found in La Montre Française du XVIe siècle jusqu'à 1900 *by Adolphe Chapiro.*

Cylinder Watches

PLATE 20A

PLATE 20B

PLATE 21A

PLATE 21B

52

Cylinder Watches

PLATE 22A

PLATE 22B

PLATE 20
Robert's Improved Barrel

Silver open face engine-turned case, hinged back, bezel and cuvette. White enamel dial inscribed 18122, sunk centre to dial, inset sunk seconds dial, narrow black roman numerals with arabic five minute markings outside chapter ring, gold spade hands. Key wound and set. Three-quarter plate. 10 jewels. Cylinder movement. Inscribed on barrel bridge Robert's Improved Barrel. Steel cylinder and scape wheel. Brass three arm balance. Blued steel oversprung balance spring. Engraved foot to balance cock. Regulator scale engraved on top plate near centre arbor.
Key size 4. Diameter 50.5mm

Registered designs 1881 lists Louis and Edward Robert, successor to Robert Brandt & Cie, Chaux de Fonds. Henri Robert et fils listed as fabricants, Chaux de Fonds, 1882.

PLATE 21
Anonymous

Fine silver open face case, hinged back, bezel and cuvette. White enamel dial, black roman numerals, blued steel spade hands, inset seconds dial. Key wound and set. Three-quarter plate. 10 jewels. Cylinder movement engraved No 6607 on barrel bridge and FAST SLOW adjacent to centre arbor. Steel cylinder and scape wheel. Brass three arm balance. Blued steel oversprung balance spring. Index pointing across the balance rim to centre of watch. There is an engraved barrel to the watch showing Maltese cross work.
Key size 5. Diameter 50mm

PLATE 22
J Gustave, Genève

Silver open face case, hinged back, bezel and cuvette. Cream enamel dial inscribed J Gustave Genève with unusual style of black roman numerals, blued steel spade hands, inset seconds dial. Key wound and set. Top plate engraved J Gustave Genève. 6 jewels. Three-quarter cylinder movement with steel cylinder. Brass three arm balance. Blued steel overcoil balance spring. Pointer index on balance cock engraved SF.
Key size 6. Diameter 52.2mm

Cylinder Watches

PLATE 23A

PLATE 23
J Godat, Genève

Fine silver open face case. White enamel dial, black roman numerals, blued steel spade hands. Inset sunk seconds dial. Key wound and set. Silvered three-quarter plate movement inscribed J Godat Genève surrounding trade mark of a stag within a circle. 4 jewels. Cylinder escapement. Brass three arm balance. Blued steel oversprung balance spring. Index on balance cock engraved FAST SLOW. *In original red leather presentation case lined dark green velvet with cream silk lining to lid. (It is always a bonus for a collector to find a watch with matching case.)*
Key size 5. Diameter 47.5mm

The trade mark appears in the Official Trade Mark Journal applications (Class 10 Horology), application No 4563 in 1883 It appears to have been registered by Castelberg, Petitpierre and Co, 58 Holborn Viaduct, London.

PLATE 23B

PLATE 24A

PLATE 24
Nordman, Genève

Fine silver open face case, hinged back, bezel and cuvette. White enamel dial inscribed Nordman Genève, broad black roman numerals, blued steel spade hands, inset sunk seconds dial. Key wound and set. 6 jewels. Three-quarter plate cylinder movement engraved Nordman Genève No 20412. Brass three arm balance. Blued steel oversprung balance spring. Polished steel index on balance cock. Watch in round red leather carrying case.
Key size 4. Diameter 48.3mm

Compare movement with that of J Godat c.1883 (Plate 23).

PLATE 24B

Cylinder Watches

PLATE 25A

PLATE 25B

PLATE 25C

PLATE 25
Anonymous

Nickel and niello hunter case decorated with peacock on front cover and fox on back cover, gold hinges to front and back covers with a snap-on glazed cover to back of movement. White enamel dial secured with two copper feet with dog screws, thin black roman numerals and small arabic five minute interval marks outside chapter ring. Keyless wound, setting by lever at edge of bezel at 23 minutes position. 6 jewels. Three-quarter plate cylinder movement. Steel cylinder. Brass round rimmed balance.

PLATE 25D

Blued steel oversprung balance spring. Index on balance cock engraved FS AR.
Diameter 55.2mm

Cylinder Watches

PLATE 26A

PLATE 26B

PLATE 26
Platnauer

Lady's silver open face floral decorated fob watch, hallmark Birmingham **1869.** Casemaker RBB. Silver engine-turned dial with floral engraved centre, black roman numerals, blued moon steel hands. Key wound and set. Three-quarter plate inscribed Platnauer. 10 jewels. Cylinder movement. Gold three armed balance. Blued steel oversprung balance spring. Polished index on cock engraved FAST SLOW.
Key size 2. Diameter 41.2mm

Platnauer was an importer of Geneva watches. Listed as manufacturers and general merchants, Ludgate Hill and 15 St Pauls Square, Birmingham and 11 Bath Road, Bristol. Manufactory Rue Leopold Robert, 47 La Chaux de Fonds, Switzerland.

PLATE 26C

57

Cylinder Watches

PLATE 27A

PLATE 27B

PLATE 27C

681. Sydenham, C. S., [*Richard, E. A.*]. Jan. 15.

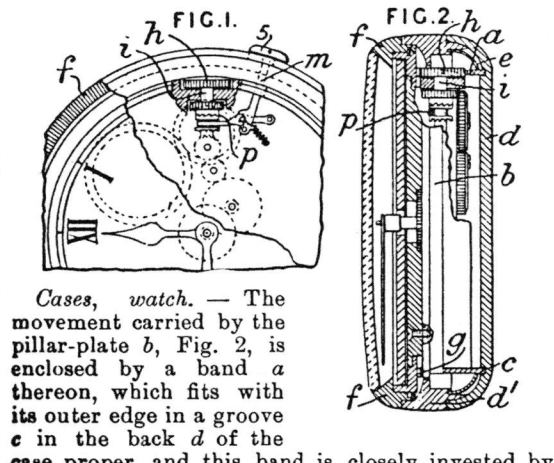

Cases, watch. — The movement carried by the pillar-plate *b*, Fig. 2, is enclosed by a band *a* thereon, which fits with its outer edge in a groove *c* in the back *d* of the case proper, and this band is closely invested by a ring *e* integral with or removable from the band *d*¹ of the case.

Keyless mechanism.—The watch is wound and the hands are set by turning the bezel *f*, Figs. 1 and 2, and therewith a ring with crown teeth *g*, which gear with a pinion *h* fixed on a radial stem *i*. The lever *m*, ending externally in a finger-piece 5, operates the clutch *p* for permitting the mainspring to be wound or the hands to be set.

PLATE 27D

Cylinder Watches

PLATE 28A

PLATE 28B

PLATE 27
Anonymous

Nickel open face case, snap-on back moulded and pierced for strap. White enamel dial, black roman numerals, blued steel spade hands, inset seconds. The bezel winding the watch in clockwise direction, hand setting by pressing pendant button before rotating bezel. Three-quarter top plate stamped S + P. 6 jewels. Cylinder movement. Steel scape wheel. Brass three arm balance. Blued steel over-sprung balance spring. Index on balance cock engraved FS AR. *An extremely rare and unusual movement.*
Diameter 45mm

UK Patent No 5851 AD 23 Sept 1829, Isaac Brown, 'Mechanism by which a watch is wound up, the winder is a circular rim, with an internal ratchet corresponding to the teeth of the barrel ratchet The winder is let into the bezel which slides round in a groove in the case. To wind the watch the winder must be turned from left to right.' See also UK Patent No 681 AD 15 Jan 1915 CS Sydenham (Richard, EA).

PLATE 28
Anonymous

Nickel open face case, snap-on bezel, hinged back and snap-on glazed cover to movement, 2/FD/EG on case and E9/40H under white enamel dial. Dial secured by two copper feet and hip screw, black roman hour numerals and arabic five minute numerals outside chapter ring, blued steel spade hands. Keyless wound, push piece to set hands at four minute position. 6 jewels. Half plate cylinder movement. Brass three armed balance. Blued steel balance spring. Index on balance cock stamped FS.
Diameter 55mm

Cylinder Watches

PLATE 29A

PLATE 29B

PLATE 29
Anonymous

Nickel open face case, snap-on bezel, hinged back cover and glazed snap-on cover to back of movement. White enamel dial secured by two copper feet and dog screws, black roman numerals, blued steel spade hands, inset seconds dial. Keyless wound, push piece at one o clock position. 6 jewels. Cylinder bar movement. Steel cylinder and scape wheel. Brass three arm balance. Blued steel oversprung balance spring. Index on balance cock engraved FAST SLOW.
Diameter 49mm

PLATE 30
Anonymous

Silver (0.800) open face case stamped Deutscher Reichs Silberstempel (mark for German silver). Gilt cuvette engraved Cylinder 10 Rubies, hinged back, bezel and cuvette. White enamel dial secured by two copper feet and hip screws, inset seconds dial, black stylised arabic numerals, Louis copper hands. Keyless wound, push piece to set hands at four minutes past position. 10 jewel. Unusual cylinder bar movement. Steel cylinder and scape wheel. Brass three arm balance. Blued steel oversprung balance spring. Index on balance cock engraved SF RA.
Diameter 47.7mm

This is an unusual movement design. Very similar design layouts are registered Liste des Dessins et Modules 1900-1914:
No 17649   19 Jan 1910   Ulysse Girard Rondey
No 18889   29 Dec 1910   Emile Juillard Parrentray
No 23069   18 Sept 1913   S Froidevaux & Block Parrentray
No 25665   9 June 1915   Emile Juillard Parrentray

PLATE 31
Le Roi

Silver (0.800) open face case, Swiss hallmark (woodcock), snap-on bezel, hinged back and cuvette. Inside back cover REGISTERED LE ROI TRADE MARK encircling King's crowned head. Milled body, engine-turned back. White enamel dial inscribed Le Roi secured by two copper feet and hip screw, black roman numerals, blued steel narrow spade hands, inset seconds. Keyless wound, push piece at four minutes past position. 6 jewels. Skeletonised three-quarter plate cylinder movement. Steel cylinder and scape wheel. Brass three arm balance. Blued steel oversprung balance spring. Pointer on balance cock engraved SF RA.
Diameter 46mm

Cylinder Watches

PLATE 30A

PLATE 30B

PLATE 31A

PLATE 31B

61

Cylinder Watches

PLATE 32A

PLATE 32B

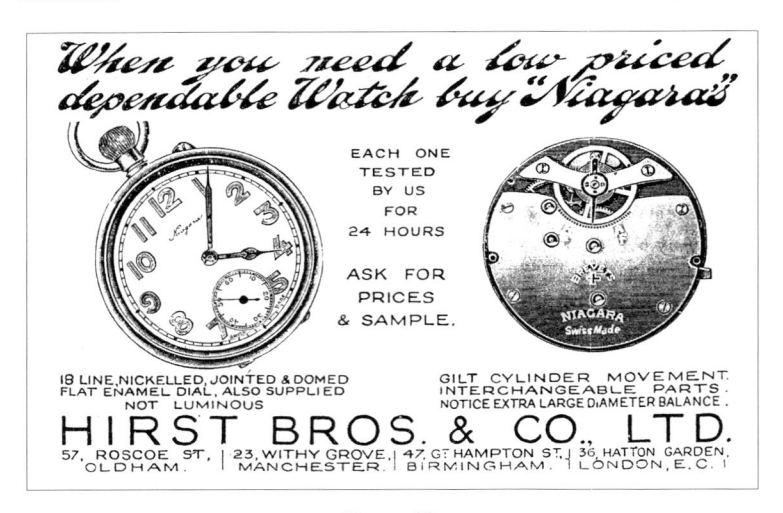

PLATE 32C

PLATE 32
Niagara

Nickel chrome plated open face case, snap-on bezel, hinged back and cuvette. White enamel dial inscribed Niagara, luminous arabic numerals and luminous fenestrated hands, inset sunk seconds dial. Keyless wound and set, push piece at four minute past position. Three-quarter plate. Top plate stamped BREVETS + NIAGARA Swiss Made. Dial plate stamped bP. 4 jewels. Flat four arm brass balance. Blued steel oversprung balance with the balance cock also acting as bridge. Index across balance rim on to main plate engraved RA SF.

Diameter 45.2mm

See Horological Journal Feb 1919 for advertisement of this watch marketed by Hirst Bros & CO Ltd, 'Gilt cylinder movement, interchangeable parts - notice extra large diameter balance'.

62

Cylinder Watches

"SPHINX"
THE WATCHMAKER'S WATCH.
Supply Restricted to Legitimate Trade.

The BEST and MOST RELIABLE

LOW PRICED WATCH PRODUCED.

A Serious Watch and not a Miniature Clock.

Machine Made.	To compete with the low priced advertised line sold at Bookstalls and by Ironmongers, Drapers and others.
Perfectly Interchangeable.	
Every Watch Guaranteed.	
Accurate Timekeeper.	
Elegant Appearance.	
Strong, Durable, Reliable.	

HANDSOME SHOW CASE, OUTSIDE ENAMELLED IRON SIGN, AND OTHER ADVERTISING MATTER FREE OF CHARGE.

H. WILLIAMSON Ltd.,
81, FARRINGDON ROAD,
LONDON, E.C.
Telephone: HOLBORN 5440.
Telegrams: "HENRY WILLIAMSON, LONDON."

PLATE 33A

PLATE 33B

PLATE 33
P Schindler, High Street, Bexhill on Sea, Sussex

Nickel open face case, snap-on back and bezel. White enamel dial secured by two copper feet and hip screws inscribed P Schindler. High Street. Bexhill on Sea, Swiss Made, black roman numerals, blued steel spade hands, inset seconds dial. Keyless wound and set, push piece at one o'clock position to set hands. 4 jewels. Three-quarter plate cylinder movement. Top plate stamped with letters SPHINX surmounted by figure of sphinx looking towards left. Steel cylinder and scape wheel. Brass three arm balance. Blued steel oversprung balance spring.
Diameter 50mm

Paul Schindler is recorded in Kelly Directory for 1897 and 1913 as watchmaker and subsequently watch and clockmaker at 3 High Street, Bexhill on Sea. H Williamson Ltd advertised the SPHINX watch in the Horological Journal *Vol VII January 1910.*

PLATE 33C

Cylinder Watches

PLATE 34A

PLATE 34B

PLATE 34C

PLATE 34
Anonymous

Nickel open face case, snap-on bezel, hinged back, snap-on inner dome recessed and pierced for central hand set knob and slotted for adjustment of index. White enamel dial, black roman numerals, blued steel spade hands, inset seconds dial. Keyless wound, pendant winding with hand setting by centre knob at back. 4 jewels. Half plate cylinder movement. Steel cylinder and scape wheel. Brass three arm balance. Blued steel oversprung balance spring. Index on balance cock, RA stamped on balance cock.
Diameter 49.4mm

A rare unusual movement – it is to be regretted that the manufacturer wished to remain anonymous and not to claim credit for the unusual design of the movement.

COLOUR PLATES 2 and 3 (page 6)
Anonymous

Lady's silver open face fob watch hallmarked Birmingham **1886** (?1896). Case stamped JW 100811 (James Walker & Son, Earlsdon Street, Earlsdon, Coventry), cuvette engraved with floral design, round screwed bow. White enamel dial with floral wreath around centre, black roman numerals, thin blue spade hands. Key wound and set. Dial plate

Cylinder Watches

PLATE 35A

PLATE 35B

stamped WF & Co within diamond. 10 jewel bar movement with cylinder scape. Blued steel oversprung three arm brass balance.
Key size 5. Diameter 40mm

William Flinn & Co was listed in the Post Office Directory for 1892 as working as a watchmaker at 15 Broadgate, Coventry.

PLATE 35
Lion's Watch

Nickel on brass open face case, snap-on bezel, hinged back. White enamel dial secured by two copper feet and hip screws inscribed Lion's Watch, black roman numerals, blued steel spade hands (minute replaced), inset seconds dial. Winding by pendant but key square and access retained from back of movement, hand setting by central button knob at back. 4 jewels. Three-quarter plate cylinder movement covered with a protective plate and pierced for index adjustment, winding and setting. Stamped with trade mark of a lion. Steel cylinder and scape wheel. Brass three arm balance. Blued steel oversprung balance spring. Index on plain balance cock.
Diameter 49mm

PLATE 35C

65

Cylinder Watches

PLATE 36A

PLATE 36B

PLATE 37A

PLATE 37B

Cylinder Watches

PLATE 38A

PLATE 38B

PLATE 36
David Perret Fils, Fabricant, Neuchatel, Switzerland

Nickel open face case, hinged back, snap-on bezel, engine-turned back with shield. White enamel dial, narrow black roman numerals, narrow blued steel hands, inset seconds dial indicating 15 second periods. Keyless wound and set by pair of turn buttons at back of movement. Back of movement covered by protective plate with round glass window to observe balance stamped Brevets SGDG and stylised fleur-de-lis trade mark enclosing letters DPF. 6 jewels. Cylinder movement. Steel cylinder and scape wheel. Blued steel oversprung balance spring. Brass three armed balance. Index on balance cock adjustable through small aperture outside balance window. *A most unusual early keyless pocket watch*
Diameter 48.5mm

Registered trade mark No 103 Marque de Fabrique de Commerce Suisse 1880, also registered No 7274 31 January 1895, D. Perret Fils & Cie, Fabricants, Neuchatel – 'Mouvement boites, cadrans et embattage de montres Transmission of marque No 103 en registered pour mouvements'.

David Perret was a member of the jury at the Paris Universal Exhibition 1878 and received an honourable mention at the National Exhibition of Horology held at Chaux de Fonds in 1881. It is interesting to note that he was granted a patent for a stem wound watch, USA Patent 378974, 6 March 1888, used in the Waltham 1888 Model with a positive set (see Bulletin National Association of Watch & Clock Collectors October 1979, p17).

PLATE 37
Belga, Numa SA Les fils de Numa Gagbenin, Tramelanlessus, Switzerland

Pinchbeck open face case, hinged back stamped NUMA inside, snap-on bezel. Square engine-turned brass dial secured by two brass dial feet and hip screws inscribed BELGA, stylised arabic numerals, blued steel spade hands, inset sunk seconds inscribed SWISS. Keyless wound. Half plate cylinder movement stamped on top plate NUMA SWISS MADE with ship's anchor ONE JEWEL ONE ADJUSTMENT. Steel cylinder and scape wheel. Brass three arm brass balance. Blued steel oversprung balance spring. Index on balance cock stamped SF RA.
Diameter 48mm

PLATE 38
Anonymous Dress Watch

Chrome on brass open face case, snap back and bezel. Oval silvered dial with pearl surround secured by two copper feet with hip screws, black arabic numerals, blued steel spade hands, inset seconds dial. Keyless wound and pull set. 4 jewels. Three-quarter plate movement stamped top plate SWISS MADE and bottom plate 4 below balance rim. Steel cylinder and scape wheel. Three arm balance. Blued steel oversprung balance spring. Index on balance cock stamped SF RA.
Diameter 49.2mm

67

Cylinder Watches

PLATE 39A

PLATE 39B

PLATE 39
Medana, Meyer & Studeli SA (Est 1885), Soleure, Switzerland

Octagonal open face dress watch, snap-on gilt brass bezel and back of mother-of-pearl. Inside back stamped M & ST SWISS MADE. Silver dial inscribed MEDANA secured by two copper feet and hip screws, black stylised arabic numerals, blued steel cathedral hands. Keyless wound and set, push piece at five minutes past position. 10 jewels. Cylinder bar movement. Steel cylinder and scape wheel. Brass three arm balance. Blued steel over-sprung balance spring. Index on balance cock engraved FS AR.

Diameter 41.1mm

Importer of Medana watches Roamer Watches England Ltd, 10/13 Newgate Street, London EC1.

PLATE 39C

PLATE 40A

PLATE 40
Medana, Meyer & Studeli SA (Est 1885), Soleure, Switzerland

Pure nickel open face dress watch, snap-on bezel, hinged back and cuvette. Silvered dial secured by two copper feet and hip screws, lilac arabic numerals, inset sunk seconds dial, blued steel arrow head hands. Keyless wound and pull set. Bottom plate stamped TG either side of balance. Top plate stamped SWISS MADE. 4 jewels. Three-quarter plate cylinder movement. Steel cylinder and scape wheel. Blued steel oversprung balance spring. Brass three arm balance. Index on balance cock inscribed SF RA.
Diameter 48.7mm

PLATE 40B

CHAPTER THREE

Lever Pocket Watches
Part 1 – English

The lever escapement for watches owes its origins to the invention of the dead-beat escapement for clocks by George Graham and the subsequent successful application of the same principles to watches in 1759 in this country by Graham's former apprentice Thomas Mudge and by Pierre Le Roy in France.

Although the lever escapement was destined subsequently to become the most widely used watch escapement for millions of watches over the next two hundred years, it did not at the time of its invention have any great impact on or significance to the watchmaking trade in general. They had always found their verge watches to be robust and serviceable, the principles were well understood and the timekeeping capabilities more than adequate for the requirements of life in the eighteenth century. It is therefore not surprising that the development of the lever lay in the hands of a small group of highly skilled technical craftsmen, men who in advance of the requirements of their contemporaries were always seeking improvements wherever they felt them possible. Thomas Mudge was a superb craftsman with the necessary skill to put his theories successfully into practice. However, the Mudge escapement was liable not to lock securely when jolted. Josiah Emery and John Leroux were able to solve this problem first by undercutting the radial face of the Mudge scape wheel and then by introducing draw to the locking face of the pallets. Some of the important names associated with this period of development are William Dutton, John Grant, George Margetts, Richard Pendleton, Robert Pennington, Francis Perigal and Louis Recordon. Readers seeking a further more detailed account are recommended to consult the very fine monograph by the late Dr Vaudrey Mercer entitled 'The Early Development of the Lever Escapement' which appeared in the *Horological Journal* in December 1963 (pp372-376) for an analysis that would be difficult to surpass.

During the same period Peter Litherland (1756-1805) was introducing and popularising the concept of the rack lever previously propounded by the Abbé Jean de Hautefeuille in 1722. Peter Litherland was granted his first patent (No 1830) in 1791, followed by No 1889 in 1792. This was for a thirty tooth scape wheel with inclined teeth to improve locking. The advantages were immediately apparent – namely secure locking and without recoil as in the Mudge escapement. No draw was required but there was the serious disadvantage of the introduction of a considerable amount of friction. The rack lever was, however, to enjoy a period of measured success. Further details can be found and studied in several articles that appeared in *Antiquarian Horology* – 'Peter Litherland & Co' by Dr Vaudrey Mercer (Volume III June 1962 p316 following); DMW Evans' letter 'Liverpool Watchmakers' (Volume IX March 1976 p705) and 'The Rack Lever' by Dr Robert Kemp (Volume XV June 1985 p375 following).

The next major step was the invention by Edward Massey of the crank roller lever with the granting of a patent (No 3854) on 17 November 1814. Subsequent stages of progressive development have become known as Massey Type levers I, II, III, IV and V. For research findings and analysis of this particular variation consult *The Massey Family* by Alan Treherne and the six patents taken out over the period 1812 to 1838. Copies of patents intended for personal study purposes can be readily obtained from the Patent Office. They do have a Search and Advisory Service at Hazlett House, 45 Southampton Buildings, Chancery Lane, London WC2A 1DR but it would be advisable to make enquiries initially as to the current fee for this service. Alternatively recourse to the Patents for Inventions – Abridgments of Specifications Class 139, Watches, Clocks and Other Timekeepers Period 1855-1930 provides the patent number and date granted. It is then but a simple matter to write to the Sales Department of the Patent Office (current address obtainable from the Chancery Lane address) and for a small fee receive a copy of the full patent. Copies of the Abridgments are held by some reference libraries but the 1979 facsimile reprint is still generally available.

Mention must be made of the work of John Savage. He invented, but apparently did not seek to patent, the two pin lever escapement that now bears his name and was

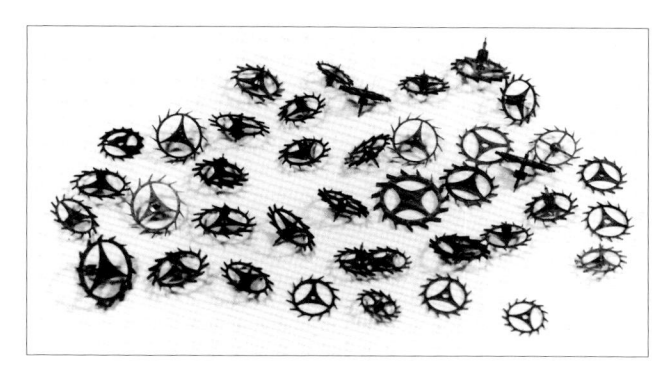

Plate 1. Examples of English pointed tooth scape wheels

Plate 2. Lever (tangential) pallets

successfully used by the better makers. The aim was to unlock the escapement on the line of centres and also to give impulse close to this point. He employed two gold pins instead of the usual single impulse pin to unlock and the use of a pin on the lever (which also acted as a guard pin) to give impulses. Perhaps one of the greatest disadvantages was that the successful operation of the Savage two pin lever required a very high standard of workmanship by the maker.

Although the table roller had established itself in the 1820s, Thomas Yates, a watchmaker of Preston, Lancaster, sought to make further improvements. These culminated in his application for a patent (No 11443) in 1846. Yates sought at first to introduce a slower train of 7,200 ½ seconds beat with the aim of less power to maintain the lower rate of vibration, less friction by adopting a heavier balance, longer hairspring, smaller mainspring and fewer scape teeth. He appointed as his main agent the manufacturer Samuel Quillam of Liverpool. It looks as if these ideas were not too successful as by 1862 the slow train had gradually been increased to 10,000 beats with an experimental trial in 1870 of 11,200 beats before shortly afterwards reverting to the original 10,000 beats. A full study of the available data has been made and recorded by Roger Carrington in *Antiquarian Horology* Volume IX 1975 p317 following.

Following the evolutionary period at the beginning of the nineteenth century there emerged a virtually standard design – the so-called English lever, namely a ratchet tooth escape wheel with tangential lever and single table roller. A further refinement was the introduction in higher grade watches of the double roller. Here the impulse and safety finger actions are separated. An interesting contemporary comment appeared in the *English Mechanic and World of Science*.

> 67045 Watch Escapement. The illustration shows the action of the double roller escapement. It will be seen that the safety-finger enters the crescent some time before the impulse pin gets to the notch; during this interval, should the hands be set back, the pallets could not trip, for the horn of the lever would be caught on the impulse pin. Double roller escapements sometimes fail to give satisfaction owing to the lever not having sufficient horn. Advantages of the double roller escapement: (1) the impulse is given nearly on the line of centres, consequently, with less friction; (2) the safety roller being less diameter, the safety-finger when in contact with it offers less resistance to the motion of the balance.
>
> B.L., Watch Repairer to the Trade

Although the pointed tooth escape wheel with tangential lever was the accepted norm in England, the Swiss in contrast adopted the club tooth form of escape wheel with a straight line lay out for pallet lever and the club tooth. This was not adopted by the English until factory methods of mass-production were instituted.

Watchmaking in this country had traditionally been carried out by a number of specialists, each responsible for the making of one particular part and working in their own small premises and homes, the 'watch' being carried from man to man in order for each stage to be completed. Clerkenwell in London and the parishes around Prescot in Lancashire were two of the main areas where such work was carried out. Movement making eventually centred around Prescot with these rough movements being sent in quantity to London, Birmingham and Coventry where the finisher (or self professed 'maker') completed and

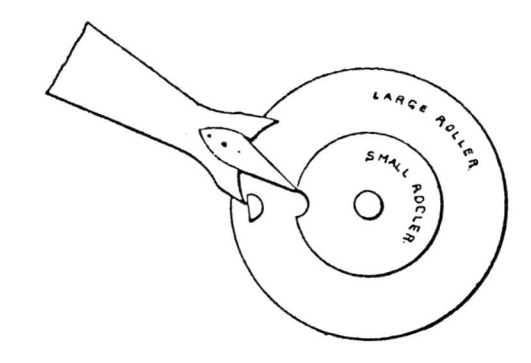

Lever Pocket Watches – English

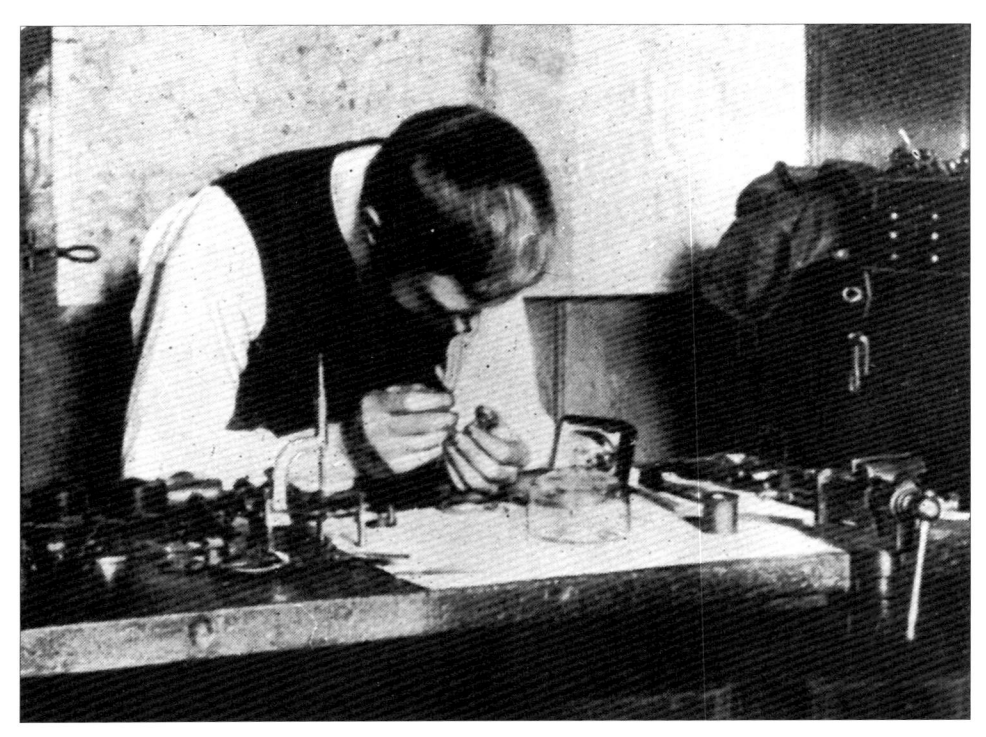

PLATE 3. The watch escapement maker at his bench at the workshops of Messrs George Oram & Son, 19 Wilmington Square, Clerkenwell, London W.C. (listed in Post Office Directory for 1890). He would be earning around £3 per week

PLATE 4. The watch jeweller at work (at Messrs George Oram & Son)

cased them. *Watchmaking in England 1760-1820* by Leonard Weiss brings together much widely scattered information on the subject which greatly assists in the understanding of the progression to the mass-production methods used later, and too late for the survival of the watch making industry in this country. However, the same cottage industry system of one man for one task still prevailed. Obviously this method could not allow for parts to be interchangeable which was essential for speedy and efficient assembly. Towards the end of the 1800s efforts were made by an enlightened few who realised that if the English watch trade was to compete with the cheaper mass-produced factory-made examples from Switzerland and America they must become more automated – an emotive period. The chapter on English Watchmaking in M. Cutmore's book *Watches 1850-1950* lists some of the early pioneers who attempted to change from the old methods in order for the trade to survive – William Ehrhardt in Birmingham (1831-1924), Rotherhams of Coventry (1880-1930 would be realistic dates in the present context), Williamsons of Coventry (1897-1931), and the Lancashire Watch Company (1888-1910), to mention a few. The latter was by far the most prolific of manufacturers. *The Lancashire Watch Company* contains an essay by Alan Smith; a facsimile copy of a trade catalogue and a description of their manufacturing methods by Henry G. Abbott. Further reading can be found in *Clocks* magazine Vol 6 June 1984 and Vol 7 September 1984 ('The Lancashire Watch Company – a Dream Destroyed' by William Ansdell). A visit to the museum in Prescot with its specialised displays would most assuredly be worth while. Articles, letters etc appearing in the horological journals for this period give further insight into the turmoil prevailing at the time. On the one hand there remained the older watchmakers resisting change and on the other there were those who realised that automation was the only way to compete with foreign imports. Petitions for trade sanctions to protect the trade and court cases against Williamsons under the Merchandise Marks Act when they were accused of using parts from their Swiss factory in 'English' watches all make interesting reading.

As well as a natural desire to disguise the fact that the retailer was not as often implied the actual maker, and also bearing in mind that the concept of marketing a brand name had not become common practice, most of the names on the watches seen dating from this period are retailers. If the case is not of gold or silver this can prove helpful as recourse to trade directories can assist with dating the watch in question. The Trade Description Act was not in force

PLATE 5. The watch finisher at work (at Messrs George Oram & Son)

and there was a great deal of literary licence and false claims. Smith's *Study of Purchase to the Guide to A Watch* (c.1900) – in reality the catalogue of the watches sold by S Smith & Sons – demonstrates the point. All the watches in this catalogue carry the name of Smith as if of their manufacture but in reality many were of Swiss origin. JW Benson's steam powered factory (1892) was in all probability only highly organised in finishing and as time went by they also bought in from other makers in this country as well as overseas. Other retailers such as Graves and Samuels would have bought in finished watches.

How do you establish who actually manufactured your watch? It is relatively simple to determine whether it is of English, Swiss or American origin, as will be shown by the examples shown in the following chapters. However, it is not always apparent which English or Swiss factory produced a given watch. To date there are no publications available that tabulate the required detail, making it readily available. Articles do appear in horological periodicals with helpful information. Study of the materials catalogues provided for the subsequent repair of these watches supplies much basic information. Inspection and comparison of actual movements demonstrate manufacturers' characteristics while some of the trade marks found on the movements are revealing. This is nevertheless a rewarding area of collecting – you acquire both watch and potential for further research!

PLATE 6A

PLATE 6
Litherland Davies & Co, Church Street, Liverpool

Silver pair-case hallmarked Chester **1811.** Casemaker NL (Nathaniel or Nicholas Lee of Liverpool). Dustcap with blued steel slide. White enamel dial, black roman numerals, gold spade hands. Key wound. Full plate rack lever movement engraved on top plate Litherland Davies & Co Liverpool 6927. Cylindrical pillars. Rack lever adjustable slide on top plate. Flat steel three arm undersprung balance with Bosley regulator with Liverpool type arrowheads to index engraving. Engraved bell-shaped cock PATENT on foot. Diamond endstone.
Diameter 55.4mm

Peter Litherland UK Patents
  1791 Oct 14 No 1830
  1792 June 12 No 1889
Antiquarian Horology
  June 1962 Volume III
  March 1976 Volume IX p705
  June 1985 Volume XV p375-386.

The numbering system started by Peter Litherland (1791) continued through all the variations of styles of the firm up until the 1860s – the highest number noted by DMW Evans (March 1976) being 42,350.c.1860.

Details of conclusion of characteristics, Liverpool Finishing Arrowhead marks, watchcock design, PATENT MOTIF, Jewelling, etc. can be found in Fusee Lever Watch *by Dr R Kemp.*

PLATE 6B

Lever Pocket Watches – English

PLATE 7B

PLATE 7
**James Moore French, 14-15 East Side,
Sweeting Alley, Royal Exchange**

Silver double bottomed case hallmarked London **1817**. Casemaker IB 433 (possible makers Jno (Junior) Baxter, 9 Waterloo Street, Radnor Street; Jonah Barnett, 43 Galway Street, St Lukes; James Bourne, 26 Red Lion Street, Clerkenwell (entered 24.3.1812). Dustcap stamped 433. Engraved bow. Silver dial with engine-turned centre (hallmarked and numbered 433), gold roman numerals and heavy gold spade hands. Key wound and set. Top plate engraved French Royal Exchange LONDON. Full plate fusee movement with Savage two pin lever escapement. Gold balance with compensation screws. Engraved bell-shaped balance (PATENT).
Key size 5. Diameter 56.5mm

French is listed as working at 15 Sweeting Alley 1808-1838. Savage introduced his escapement c.1814 but never took out patent protection. He emigrated to Canada in 1818. 'The Early Development of the Lever Escapement' by Dr Vaudrey Mercer, Horological Journal December 1963.

PLATE 7A

PLATE 8A

PLATE 8B

PLATE 8C

PLATE 8
James Perrin, Marden, Kent

Silver open faced double bottomed case with ribbed edge to body hallmarked London **1819**. Casemaker SB (?Stephen Bryan, 4 Castigna Place, Radnor Street, London, ref Pigot's Directory 1823). Dustcap stamped inside IxR. Pendant hallmarked **1818** (possibly JF, Joseph Field, 6 Red Lion Street, Clerkenwell). White enamel dial, black roman numerals, gold spade hands, inset seconds dial. Key wound and set. Full plate. 13 jewel English lever movement with fusee and chain and maintaining power. Massey Type I detached lever escapement. Pointed tooth scape wheel. Tangential pallets. Oversprung polished steel three arm balance with blued steel balance spring. Regulator index on triangular balance cock engraved SLOW FAST with graduations 1.2.3. The cover of the dustcap engraved Tho Venafs (? owner), PERRIN (? retailer).
Key size 5. Diameter 54.5mm

Edward Massey was born in Newcastle under Lyme in 1768, the eldest of eight children. His father, also named Edward, was a skilled mechanic and watch and clockmaker who had settled in Newcastle in 1763. By 1795 Edward Massey was known to be working in nearby Burslem of Arnold Bennett fame. He moved to Hanley about 1802 where his address is recorded as Water Row. When his father died in 1813 Edward Massey decided to leave the Potteries in order to find greater opportunities in fulfilling his many ambitions. He chose Coventry and is recorded as being at Ironmongers Row and Cross Cheaping, historically the market area of Coventry. Still restless to market his chronometers and marine log, Massey now turned his attention to Prescot (1818), near the busy port of Liverpool. The attraction of London, however, proved irresistible and in 1830 he joined other members of his family. In 1833 he is recorded at 28 King Street and 17 Chadwell Street, both in Clerkenwell, with a shop at 89 Strand. He died in 1852 and was buried at St James, Islington. For further biographical and technical detail see The Massey Family *by Alan Treherne, catalogue of the exhibition held at the Museum, Newcastle under Lyme in 1977.*

PLATE 9A

PLATE 9B

PLATE 9
Robert Roskell, 21 Church Street, Liverpool

Silver open face double bottomed case hallmarked Chester **1823**. Engine-turned back and milled edge to body, gold hinges. Casemaker IW (? John Widdowson, 17 Edmund Street or John Windus, 6 Limekiln Lane, Liverpool). Dustcap. White enamel dial ROBT ROSKELL LIVERPOOL, black roman numerals, gold spade hands. Key wound and set. Full plate. Movement engraved top plate LIVERPOOL 36429 ROBt ROSKELL. Fusee and chain. 7 jewelled English lever movement. Pointed tooth scape wheel. Polished tangential pallets and polished steel three arm round rimmed undersprung balance. Blued steel balance spring with Bosley type regulator to engraved Liverpool arrowhead scale on top plate. Engraved triangular balance cock of floral design with PATENT on foot.
Key size 7. Diameter 53mm

DMW Evans in his letter in Antiquarian Horology *Volume I March 1976 p705 concerning numbering of Roskell watches states that Roskell appears to have produced about 1,000 watches per year from 1800 to 1900 – the first recorded serial number 2316 hallmarked 1803 and the last recorded at the time of writing (March 1976) being 101,099 hallmarked 1900.*

PLATE 9C

Lever Pocket Watches – English

PLATE 10A

PLATE 10
John Murray (retailer), Aberdeen

Silver open face double bottomed case with plain back and engine-turned pattern to body, hallmarked London **1826**. Casemaker JJ (James Jackson, 10 Norman Street, St Lukes, London). Dustcap maker JS (Joseph Shephard, 9 Ironmongers Street, London). Off white enamel dial, black roman numerals, blued steel spade hands, inset seconds dial. Key wound and set. Full plate engraved John Murray, Aberdeen No 83. English lever movement with fusee and chain, maintaining power pointed scape wheel and tangential pallets. The escapement of Massey type III. 7 jewel movement with diamond endstones. Oversprung three arm steel round rimmed balance with blued steel balance spring. The regulator index on triangular cock engraved No 83 on foot and graduations incorporating fleur-de-lis in mid line.
Key size 4. Diameter 54.5mm

John Murray listed in Old Scottish Clockmakers *2nd ed by John Smith as active 1843.*

PLATE 10B

PLATE 11
Thomas & John Olivant, 2 Exchange Street and 1 St Mary's Gate, Manchester

Silver open faced double bottomed case hallmarked London **1827**. Casemaker TH & Co (?Thomas Hardy, 14 Rosamen Street, Clerkenwell, recorded Pigot's Directory 1826). Dustcap hinged at 10 o'clock with spring release. White enamel dial, black narrow roman numerals, blued steel spade hands, inset seconds dial. Key wound and set from back key. Classical half plate 7 jewelled English lever movement engraved on top plate. Detached lever by Thos & Jn Olivant, Manchester, No 6955. Fusee and chain maintaining power. Pointed tooth scape wheel with tangential pallets. Oversprung three arm gold balance with diamond endstone. The regulator index on balance cock engraved FAST SLOW.
Key size 5. Diameter 54.3mm

Listed in Watchmakers and Clockmakers of the World *Vol 2 2nd ed by Brian Loomes as working in Manchester 1828-1851.*

PLATE 12
Graham, London

Silver open face double bottom case hallmarked Birmingham **1829**. Casemaker VR. White enamel dial, black roman numerals, gold broad spade hands, inset seconds dial. Key wound and set. Dustcap to full plate fusee. 13 jewel movement number 4122. Pointed tooth

Lever Pocket Watches – English

PLATE 11A

scape wheel. Tangential pallet with Massey Type III roller. Steel three arm undersprung balance with three balance screws, the rim being square in cross section. Liverpool style jewelling to scape, pallet and fourth wheel arbor with diamond endstone to balance staff. Plain triangular cock. Bosley style regulator. *Altogether an unusual movement.*
Key size 5. Diameter 49.5mm

PLATE 11B

PLATE 12A

PLATE 12B

79

Lever Pocket Watches – English

PLATE 13A

PLATE 13B

PLATE 14A

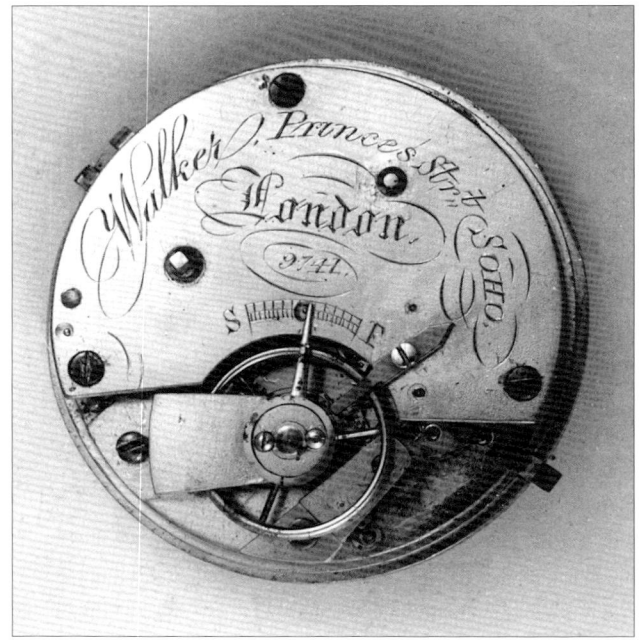

PLATE 14B

PLATE 13
Robert Summersgill, Fishergate, Preston, Lancashire

Silver open face plain double bottomed case with milled edge to body hallmarked Chester **1830**. Casemaker EK (Edward Kirkman (Kirkham) 1827-1855, Liverpool). Dustcap stamped W. Cream enamel dial with black roman numerals, gold spade hands, inset seconds dial. Key wound and set. Inscribed on top plate R Summersgill. Preston 573. Full plate. 9 jewels fusee. English lever movement with Liverpool jewelling to 4th wheel pointed tooth scape wheel with tangential pallet. Massey Type III roller. Round steel undersprung balance. The balance cock foot engraved PATENT and index engraved with characteristic Liverpool style arrows.
Key size 5. Diameter 54.5mm

Summersgill was apprenticed to Stephen Simpson of Preston 1813. Listed as working 1813-81 in Watchmakers and Clockmakers of the World *Vol 2 2nd ed by Brian Loomes.*

PLATE 14
John Walker, 40 Princes Street, Soho, London

Silver open face double bottomed engraved case hallmarked London **1835**. Casemaker EW (?Edward Walker, 46 Whiskin Street, Clerkenwell, Pigot's Directory 1832/6). Off white enamel dial, black roman numerals, Breguet style gold hands, inset seconds dial. Key wound and set. Three-quarter plate engraved Walker, Princes Street, Soho London 9744. 7 jewels. Fusee and chain, maintaining power. Pointed tooth scape wheel, tangential pallets. Oversprung steel balance with round rim and blued balance spring. Small triangular plain balance cock with the cranked index over the balance rim to graduations on top plate.
Key size 5. Diameter 44.2mm

John Walker, 40 Princes Street, Leicester Square, listed in Pigot's Directory for 1836-42, at 48 Princes Street, Soho in the PO Directory for 1852 and at both 48 Princes Street, Soho and 68 Cornhill, EC in PO Directory for 1860. Inventor and manufacturer of the crystal case watch Prize medals 1862, 1867 and Railway Guards watch 1875 ref Britten 9th ed.

COLOUR PLATES 6 and 7 (page 7)
Thomas Prest, Chigwell, Essex

Silver open face single bottom case hallmarked London **1845**. Gold hinges. Casemaker FM (Frederick Matthews, 14 Roseman Street, Clerkenwell). Cream enamel dial, black roman numerals, gilt spade hands. Inset seconds dial. Prest's keyless winding. Key to set hands on cannon pinion square. Three-quarter plate. 9 jewels. English lever movement going barrel. Pointed tooth scape wheel. Type 2 Massey escapement. Oversprung three arm gold balance with blue steel balance spring. Plain balance cock with blued index on to top plate engraved Fast Slow.
Diameter 45.5mm

Thomas Prest (1770-1852) was the son of Edward Prest, a skilled watchmaker from Prescot, Lancashire who was apprenticed on 5 January 1784 at the age of fourteen to John Arnold, the watchmaker and chronometer maker of considerable repute. The indenture was for seven years for one penny. John Arnold lived at Wellhall House, Eltham in Kent and it seems likely that Prest worked there as his wife Mary came from those parts. Shortly after getting married he lived in Hackney before moving to Marching Gravel Lane in Chigwell where he became the foreman in John Roger Arnold's workshops. On 20 October 1820 Thomas Prest was to register patent No 4501 for 'A New and additional movement applied to a watch to enable it to be wound up by the pendant knob without any detached key or winder'. This patent could only be applied to going barrel watches. The winding wheel on the barrel arbor had 28 teeth (later examples 34 teeth). The intermediate winding wheel also had 28 teeth horizontally and a similar set of teeth vertically to engage with the winding pinion. Initially there was no provision for hand setting and these watches have robust hands and a screw in the tail of the minute hand to push round to set the hands. The fusee watch was still maintaining its dominance in the English watchmaking industry. There was therefore only a limited commercial value for this invention, but Arnold continued to use it, even when Prest had ceased working for him and become established in his own right (listed Pigot's Directory 1832). It was not long however before other makers sought to improve on the concept. For example in 1838 Louis Audemar of Le Brassus in the Valley of Joux, Switzerland, invented the first keyless watches made in quantity, wound and set through the pendant with the mechanism under the dial and winding to the left. Adolphe Nicole took out a patent in 1844 (UK Patent 10348) for pendant winding utilising a bezel wheel. From 1846 onwards Dent claimed to have sole rights to the design of Adrien Philippe, conceived in Paris in 1842 but rejected by the French watchmakers until his work received recognition at the Exhibition of 1844 when he was awarded a medal. He subsequently joined Compte de Patek and became a partner of the firm in Geneva. A whole book could be written on keyless mechanism! Readers are advised to consult the excellent classification in the series of articles by Dr Vaudrey Mercer that appeared in the Horological Journal *September and November 1984, February and November 1985, June 1986 and January 1987.*

Lever Pocket Watches – English

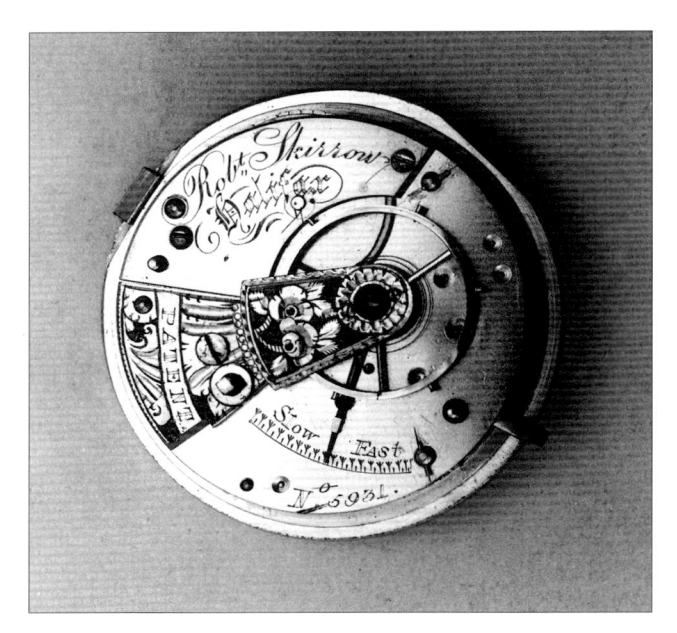

PLATE 15

PLATE 15
Robert Skirrow, 70 Woolshops, Halifax

Silver open face double bottomed engine-turned case hallmarked London **1846.** Casemaker IH (John Hammon, 26½ Sekford Street, Clerkenwell). Dustcap stamped WS 5931. Cream dial, narrow black roman numerals, small minute markings, gold fleur-de-lis hands, inset seconds dial. Key wound and set. Full plate engraved Robt Skirrow Halifax No 5931. Fusee and chain. 7 jewels. English lever movement with maintaining power, pointed tooth scape wheel, tangential pallet. Steel undersprung balance of square cross section with six steel poising screws equally spaced. Blued steel balance spring. Bosley type regulator to an engraved index marked SLOW FAST with Liverpool arrow graduations. The florally engraved triangular balance cock engraved PATENT on foot.
Key size 8. Diameter 50mm

Listed in White's 1853 Directory for Halifax.

PLATE 16A

PLATE 16B

PLATE 16
Arnold, London

White enamel dial, narrow roman numerals with narrow blued steel spade hands. Key wound and set. Fully engraved half plate movement inscribed DETACHED AND VISIBLE LEVER 13 JEWELS No 74410 COMPENSATION BALANCE Arnold London fitted with spring dust ring and half cap released by catch at 13 minute position. Going barrel movement with club tooth scape wheel, fancy tail to pallet and bimetallic compensated balance.
Diameter 42mm

John Roger Arnold, son of John Arnold, died 1843. Business taken over by Charles Frodsham FBHI with the Arnold name being retained. John Arnold and Son *by Dr Vaudrey Mercer.*

Lever Pocket Watches – English

PLATE 17A

PLATE 17
Hatfield & Hall, Manchester

White enamel dial, black roman numerals, blued steel hands, inset sunk seconds. Half plate movement engraved Hatfield & Hall Manchester 8631. Sprung dust ring and half cover. Fusee and chain, maintaining power. Gold three arm balance. Pointed tooth scape wheel. Jewelled scape and balance arbors.
Diameter 44.5mm

Maker recorded Loomes Watch and Clockmakers of the World *as working 1834.*

PLATE 17B

Lever Pocket Watches – English

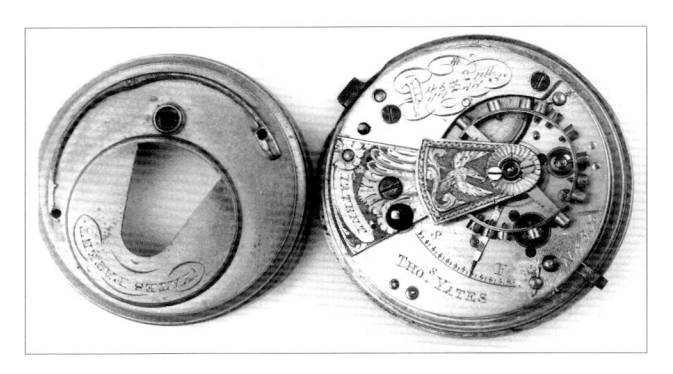

PLATE 18

PLATE 18
Thomas Yates, 159 Friargate, Preston, Lancashire

Silver open face double bottomed case hallmarked London **1884.** Casemaker JJ (James Jackson & Son, 36 Helmet Row, St Lukes, London, EC, silver case maker. This watch was recased in 1884 from a gold case). Dustcap engraved YATES PATENT. Cream enamel dial with narrow black roman numerals, gold spade hands, inset seconds dial. Key wound and key set from cannon pinion square. Inscription on top of full plate fusee lever movement with maintaining power reads THOs Yates Preston NO 806. 11 jewel Liverpool style jewelling to pallet arbor and fourth wheel arbors with small diamond endstone to balance staff. Pointed tooth scape wheel and tangential pallet staff. Bimetallic cut undersprung balance with Bosley style index to index engraved side of balance rim with typical Liverpool arrows and PATENT on balance cock foot in Liverpool style.
Key size 4. Diameter 49mm

Thomas Yates was born in Goosnaugh, fifteen miles east of Blackpool and five miles north of Preston in 1813. By the time he was thirty (1843) he had established himself in business at 159 Friargate, Preston from whence he took out UK Patent No 11443 AD Nov 1846 for a new half seconds dead-beat watch. Like many watchmakers of his generation, he had an experimental outlook and was searching for technical improvement in the lever escapement. He believed that by increasing the weight of the balance but reducing its periodicity to give a train of 7,200 instead of the usual 14,400 it would have to do less work. In practice it was found that the watch was more easily affected by jolting. Unfortunately this was at a time when most of the people who could afford a watch travelled on horseback. Between 1848 and 1858 (159 Friargate and 12 Friargate Preston) he used a 7,200 train; by 1862 he had increased the train to 10,000 with further experimenting in 1871 with a train of 11,200. In 1878 he reverted to 14400. He died on 28 February 1890 at the age of seventy-eight leaving a daughter, Hannah Maria Yates. Probate was granted on 8 May 1891 on an estate of £1,845 13 3d. Antiquarian Horology Vol IX No 3 June 1975 p317/319; X No 6 Spring 1978 p741; XI No 5 Autumn 1979 p521; XV No 2 December 1984 p144/146. The Fusee Lever Watch by Dr Robert Kemp p67-71.

PLATE 19A

PLATE 19B

PLATE 19
Thomas Yates, 159 Friargate, Preston, Lancashire

Open face double bottomed silver case with gold hinges hallmarked Chester **1865,** stamped SQ (Samuel Quillam, Liverpool). Silver pendant hallmarked Birmingham **1865.** Black enamel dial with white roman numerals. Polished steel moon hands. Key wound with maintaining power. Full plate engraved Thos Yates Preston No 3965. 11 jewel English fusee lever movement. Pointed tooth scape wheel, tangential pallets. Bimetallic cut compensated undersprung balance. Bosley style index to Liverpool style arrow engraving side of balance on top plate.
Key size 7. Diameter 48.8mm

Lever Pocket Watches – English

PLATE 20A

PLATE 20
Joseph Slack (Junior), Ipstones, Staffordshire

Silver hunter double bottomed engine-turned case hallmarked London **1849**. Casemaker IH (John Hammon, 11 Sekford Street, Clerkenwell). Dustcap stamped WS (?William Smelt, 2 Berkley Court, Clerkenwell in 1852 Directory) and engraved John Austin D **1854**. Off white dial, narrow black roman numerals, gold spade hands, inset seconds dial. Key wound and set. Full plate movement. 9 jewel English lever movement, fusee and chain, maintaining power, pointed tooth scape with tangential pallet. Undersprung three arm round rimmed steel balance. Bosley type regulator. Floral engraving to cock with diamond endstone. The index on the top plate engraved Josh Slack, Ipstones, No 43930.

Key size 7. Diameter 52.3mm

Joseph Slack Senior is listed 1780-1818 as farmer and clockmaker. Joseph Slack Junior is listed 1845-1854 as clock repairer.

PLATE 20B

Lever Pocket Watches – English

PLATE 21

PLATE 22A

PLATE 21
James Ritchie & Son, 25 Leith Street and 131 Princes Street, Edinburgh

Silver open face engine-turned double bottomed case hallmarked London **1859**. Casemaker RHJ (Robert Henry Jones, 51 Wyngate Street, Goswell Road, London EC). Dustcap stamped inside WH. White enamel dial, black roman numerals, blued steel moon hands. Key wound and set. Full plate engraved Jas Ritchie & Son, Edinburgh, 2851 and 08340. 9 jewel English fusee lever movement with diamond endstone balance staff and jewelled 4th wheel arbor. Pointed tooth scape wheel and tangential pallet with round steel undersprung balance. Regulator index from partially engraved cock over balance to silver quadrant.
Key size 6. Diameter 44mm

PLATE 22
Charles Frodsham, 84 Strand, London

Silver hunter double bottom case hallmarked London **1862**. Casemaker TH (Thomas Holliday). Silver pendant hallmarked London **1861**. White enamel dial, narrow black roman numerals, gold spade hands. Key wound and set. Full plate 11 jewel fusee lever movement engraved on top plate Chas Frodsham. 84 Strand. London 12860. Pointed tooth scape wheel with tangential pallets. Round steel three arm undersprung balance. Pointer index from cock over balance to silver quadrant with Coventry star below the centre point of quadrant.
Key size 4. Diameter 47.2mm

PLATE 22B

Thomas Holliday first recorded 8 Thatched House Row, Lower Road, Islington (1853), then removed to 108 Ratcliffe Terrace, Goswell Road, Clerkenwell EC where listed as gold, silver and Turkey watch case maker. Street renamed (1856), hence Holliday listed at 304 Goswell Road EC until 1866. Thereafter recorded with Edward Holliday at 7 Upper Charles Street, Clerkenwell EC. Directory of Gold and Silversmiths by John Culme. The Frodshams – The Story of a Family of Chronometer Makers by Dr Vaudrey Mercer.

PLATE 23B

PLATE 23
Richard Hillaby, 229 Glossop Road, Sheffield

Silver open face double bottom hallmarked London **1864.** Casemaker BK. Silver pendant hallmarked Birmingham **1864,** maker JJ. Dustcap stamped W and AO, possibly Oxley Bros, 52 Spon End, Coventry. White enamel dial with black roman numerals, blued steel spade hands. Winding up and down dial and inset sunk seconds dial. Key wind and set. Fusee full plate English lever 15 jewel movement with pointed tooth scape wheel and tangential pallet. Oversprung with overcoil and cut bimetallic compensation balance, the balance spring stud being visible on the side of balance cock through the dustcap. The index regulator on the balance cock to an engraved index scale incorporating Coventry fleur-de-lis mark. Plate engraved R Hillaby No 7141 229 Glossop Road Sheffield.
Key size 6. Diameter 51.5mm

PLATE 23B

Lever Pocket Watches – English

PLATE 24A

PLATE 24B

PLATE 24
Robert Roskell, 21 Church Street, Liverpool

Open face double bottom silver case with gold hinges, hallmarked Chester **1864**. Casemaker WR. Dustcap pierced for balance cock. White enamel dial inscribed Robt Roskell Liverpool in small capitals at top of dial, black roman numerals, gilt spade hands, inset seconds dial. Full plate key wound and set. English lever movement engraved on top plate Robt Roskell Liverpool No 67032. Fusee and chain, maintaining power. 9 jewel with Liverpool type jewelling to 4th wheel and diamond endstone to balance arbor. Pointed tooth scape wheel and tangential pallets. Polished steel round edged undersprung balance with Bosley type index to engraved graduations on top plate incorporating Liverpool arrow.
Key size 5. Diameter 50mm

PLATE 25
Edwin Pridham, 4 Bridge Terrace, Harrow Road, Paddington, London

Silver open faced double bottom case hallmarked London **1864**. Casemaker JO (James Oliver, 4 Kings Square, Goswell Road, London EC). Dustcap. White enamel dial, black roman numerals, blued steel spade hands. Key wound and set. Gilt full plate. 7 jewels English lever movement engraved E Pridham 4 Bridge Terrace Paddington. Fusee and chain, maintaining power, pointed tooth scape wheel and tangential pallet. Five arm gold flat rimmed undersprung balance with blued balance spring. Bosley regulator to engraved index scale with SLOW FAST and the Coventry star between. Cock engraved stylised foliage and 134. *The five arm gold balance is an unusual feature.*
Key size 5. Diameter 44.5mm

PLATE 25

Lever Pocket Watches – English

PLATE 26

PLATE 27

COLOUR PLATES 4 and 5 (page 6)
James McCabe, Royal Exchange, London

18 carat gold full hunter engine-turned case hallmarked London **1864**. Casemaker AS (Alfred Stram, 12 Lower Ashby Street, London EC). Dust ring spring loaded. Catch to release at 4 o'clock. Push piece on pendant for full hunter. White enamel dial, roman numerals, gold fleur-de-lis hands. Inset seconds. Three-quarter plate. Key wound and set. Fusee and chain, maintaining power. 9 jewel English lever movement. Pointed tooth scape tangential pallet. Oversprung three arm gold balance with blued steel hairspring. Polished index on small triangular balance cock engraved S F. Top plate engraved Js McCabe, 04315, Royal Exchange, London.
Key size 1. Diameter 43mm

Antiquarian Horology *Volume X Summer 1977 pp308-316*.

PLATE 26
Abraham Levy, Hastings

Silver open face case, dome and bottom hinged, clip-on bezel, hallmarked London **1865**. Casemaker TH (Thomas Holliday, 304 Goswell Road, Clerkenwell, London EC1). White enamel dial, black roman numerals, gilt spade hands. Key wound from back (setting square piercing balance cock foot). Full plate engraved Abraham Levy, Hastings, 20238. Fusee. 13 jewel movement numbered 20238 hinged from front and retained by pin at 6 o'clock position under bezel lip. Diamond endstone to balance with endstones to scape arbor. Jewel fourth wheel arbor. Pointed tooth scape wheel with tangential pallets. Three arm bimetallic oversprung balance. Blued steel Bosley style index.
Key size 7. Diameter 52mm

PLATE 27
George Moore, St Johns Square, London

Silver open face double bottom engine-turned case hallmarked London **1866**. Casemaker TH (Thomas Holliday, 304 Goswell Road, London EC1). Silver pendant hallmarked London **1865**. White enamel dial, black roman numerals, blued steel fleur-de-lis hands, inset sunk seconds dial. Key wound and set. Full plate. Fusee. Engraved on barrel bridge Geo Moore, St Johns Square, London 7/7304. 11 jewel English lever movement with pointed tooth scape wheel and tangential pallets with round steel oversprung balance with index pointer over balance to silver quadrant with Coventry star below.
Key size 6. Diameter 48.50mm

Lever Pocket Watches – English

PLATE 28A

PLATE 28
The London Lever

Silver engraved open face case with hinged back dome and bezel hallmarked London **1867**. Casemaker W. Pendant hallmarked **1867**. White enamel dial inscribed THE LONDON LEVER REGISTERED below trade mark, narrow black roman numerals, narrow blued steel spade hands. Three-quarter plate. Key wound and set. 15 jewel lever movement engraved on top plate THE LONDON LEVER REGISTERED 56393. Club tooth scape tangential pallets, oversprung cut bimetallic balance with blued steel balance spring. Regulator index on balance cock engraved S F with arrow between.
Key size 4. Diameter 40mm

PLATE 28B

PLATE 29

PLATE 29
Wladyslaw Spiridion, 29 Duke Street, Cardiff

Silver open face double bottom engine-turned case hallmarked London **1867**. Casemaker DL (David Lark, 8 Percival Street, Clerkenwell). Dustcap stamped JH. White enamel dial, black roman numerals, gold spade hands, inset sunk seconds dial. Key wound and set. Full plate. Maintaining power, fusee and chain. English lever movement. Top plate engraved W Spiridion Cardiff No 12555. 11 jewel pointed tooth scape wheel with tangential pallet. Cut bimetallic oversprung balance with compensation screws. Regulator index from florally engraved cock over balance to engraved silver quadrant at side of balance with Coventry star adjacent. *An interesting feature is that the click and ratchet wheel for setting up the barrel has been made a decorative feature and is visible through pierced dust cover.*
Key size 5. Diameter 50.4mm

According to Clock and Watchmakers in Wales *3rd ed by Peate, Wladyslaw Spiridion Kliszcewski was a Polish refugee from the 1830-1831 Russo-Polish war. He was interned at Trieste as a political prisoner before being released to travel to London in 1838. Through the good offices of Lord Dudley Coutts-Stuart he was apprenticed to a London watchmaker. He was employed in 1844 by Henry Grant in Castle Street and they moved to Herbert Street, Duke Street in Cardiff in 1852. Spiridion bought the business in 1855 when Grant became a nautical optician at the docks before emigrating to Canada. Spiridion died in 1891.*

PLATE 30

PLATE 30
Thomas Thorpe, Pateley Bridge, West Riding, Yorkshire

Silver open face double bottom engine-turned case hallmarked London **1870**. Casemaker EW (Ebenezer White, 42 Spon Street, Coventry – mark first registered London Assay Office 28 April 1869). Dustcap. White enamel dial, black roman numerals, gold spade hands, inset sunk seconds dial. Key wound and set with maintaining power. Full plate. 7 jewel English lever movement engraved Thos Thorpe, Pateley Bridge, 9078. Pointed tooth scape wheel with tangential pallets. Polished steel round rimmed undersprung balance. Blued steel balance spring. Bosley regulator to engraved graduations SLOW FAST with Coventry star between.
Key size 8. Diameter 52mm

Lever Pocket Watches – English

PLATE 31A

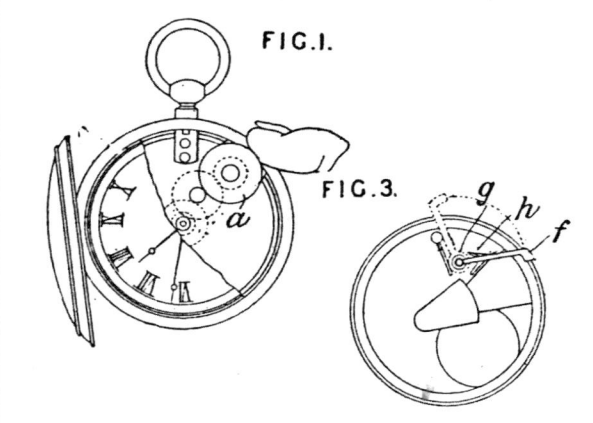

2286. Burdess, A. July 28. 1869
Keyless mechanism; hands, setting by special mechanism.—A ratchet-wheel *g*, Fig. 3, on the barrel or fusee arbor is actuated by the lever *f*, which carries a pawl *h*. The lever passes through a slot in the case, and terminates in a thumb-piece. For setting the hands, a train of wheels is employed, as shown in Fig. 1; the actuating-wheel *a* is milled on its edges, and passes through a slot in the case so that it can be turned by the finger.

PLATE 31B

PLATE 31C

PLATE 31
Adam Burdess, Watch Manufacturer and Tricycle Manufacturer, Dover Street, Holyhead Road, Coventry

Silver open face double bottom case with gold hinges and joints (a sign of quality) hallmarked London **1871**. Casemaker IH (?John Hammon, 11 Sekford Street, Clerkenwell, London EC). Silver pendant also hallmarked London **1871**. Dustcap and slide to movement. Silver engine-turned dial (maker HW) with gold roman numerals, blued steel spade hands and foliate decorative centre with gold pattern outside chapter ring. Full plate fusee keyless 4 jewel movement engraved on top plate

92

Lever Pocket Watches – English

PLATE 32

PATENT A Burdess Coventry 4549. Hand set by knurled disc under edge of dial at 7 minutes past hour. Pointed tooth scape wheel with tangential pallets. Bimetallic cut compensation undersprung balance Bosley type regulator.
Diameter 49.2mm

Adam Burdess took out a patent (No 2286) in July 1869 for a keyless mechanism with the hands being set by a special mechanism. 'A ratchet wheel on the barrel or fusee arbor is actuated by the lever which carries a pawl. The lever passes through a slot in the case and terminates in a thumb piece. For setting the hands, a train of wheels is employed. The actuating wheel is milled on its edges and passes through a slot in the case so that it can be turned by the finger.' A second patent (provisional protection only) was taken out in 1881 (24 December 1881). No Burdess watches include this mechanism.

The association between watch manufacture and watch finishing was not uncommon. As the watch trade dwindled, more and more watchmakers migrated to the thriving cycle making and other precision engineering works in Coventry. An excellent account of the social conditions of the Coventry watch trade can be found in Mary Monte's booklet Brown Boots, *so called because the watchmakers could afford a second pair of boots (brown!) for Sundays. Dr Kemp expounds on his hypothesis for the identification of Coventry finished watches in* The English Fusee Lever *with special reference to the mark (star) engraved at mid point of the Bosley regulating scale.*

**PLATE 32
Peter Gordon, 4 Duke Street, Huntly, Aberdeenshire**

Silver pair-case pocket watch hallmarked London **1872**. Casemaker HB. Dustcap stamped RJ. White enamel dial, black roman numerals, gold spade hands, inset sunk seconds. Key wound and set. Full plate. 7 jewel fusee English lever. Engraved on top plate P Gordon, Huntly, No 1492. Steel three arm round rimmed undersprung balance with Bosley type regulator, Coventry star at mid position of regulator scale.
Key size 6. Diameter 53mm

The PO Directory for 1880 lists Gordon as working.

Lever Pocket Watches – English

PLATE 33A

PLATE 33B

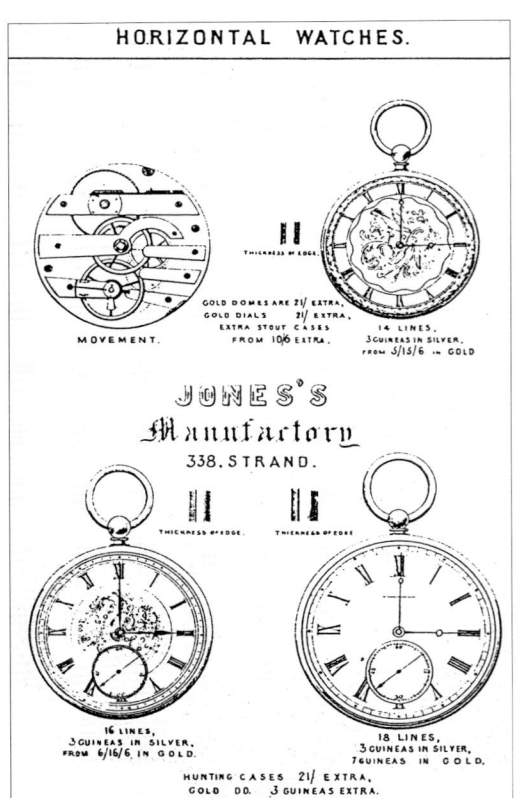

PLATE 33C

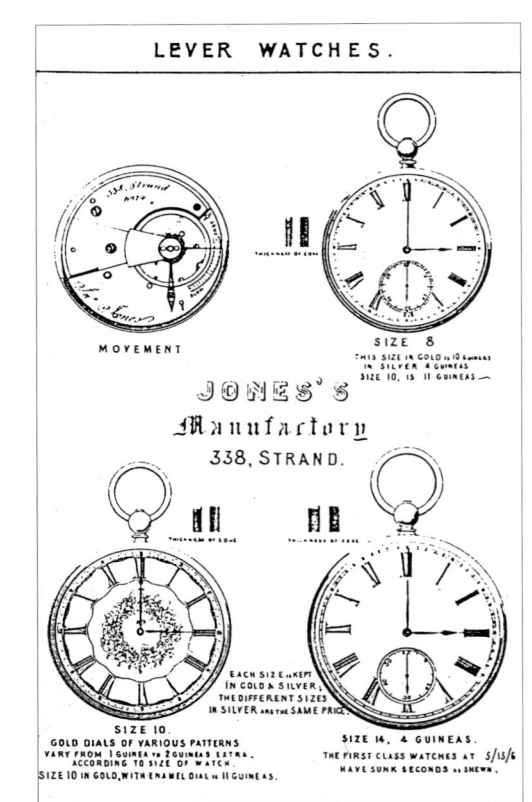

PLATE 33D

Lever Pocket Watches – English

PLATE 33
John Jones, The Watch Manufactury (opposite Somerset House), 338 Strand, London

Silver open face double bottom engine-turned case hallmarked London **1872.** Casemaker AT (Alfred Thickbroom, 6 Spencer Street, Goswell Road, London EC). White enamel dial, black roman numerals and gold moon hands, inset sunk seconds dial. Key wound and set with maintaining power, fusee and chain. Full plate. English lever movement engraved on the top plate Jn Jones, 338 Strand. Pointed tooth scape wheel with tangential pallets. Jewelled pallet arbor and balance staff with endstones. Round steel three arm balance oversprung with plain triangular cock. The index from cock over the balance to a silver quadrant.
Key size 5. Diameter 47.8mm

John Jones was not a founder member of the British Horological Institute but was active in placing it on a sound basis following initial problems. His efforts included the re-establishment of the practical educational classes and constant support in both the

PLATE 33E

financing and planning of the building of the new prestigious premises in Northampton Square, Clerkenwell (cost about £2,000 in 1878). However, he was not particularly popular with the

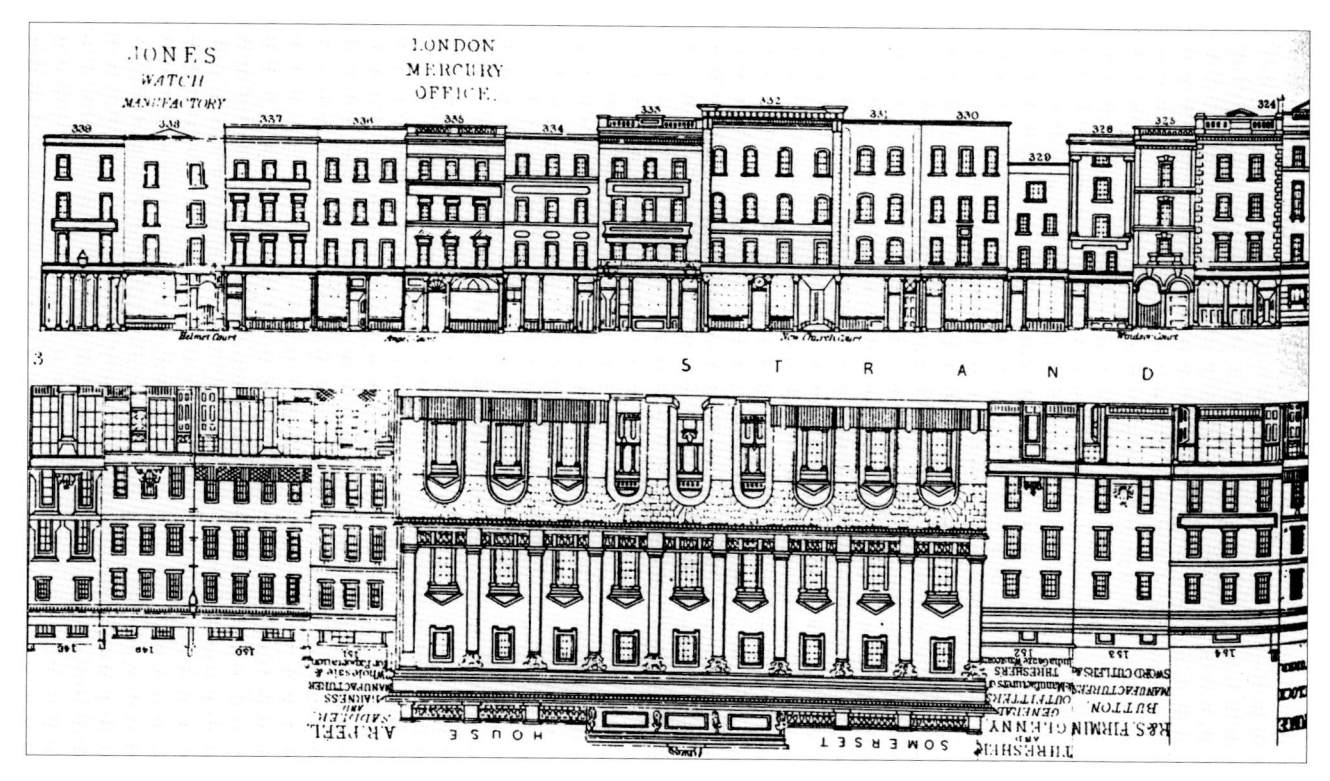

PLATE 33F

Lever Pocket Watches – English

<pre>
          Class 15.   No. 3273.

          CONTENTS OF CASE
                EXHIBITED FROM
     JONES' WATCH MANUFACTORY,
              338, STRAND.

                 FIRST ROW
  Contains the finest specimens of the Lever Escapement the trade can produce, in Ladies'
  Gold ¾-Plate Lever Watches, with Ornamental Gold Dials.   Price 18 Guineas each.
     Some of the Dials show a new application of Jewels to note the divisions of hours on the dial.
                 SECOND ROW
  Contains GOLD HUNTING LEVER WATCHES, suitable for Tropical Climates.   Price
  13 Guineas each, if with Plain Dials.
     Some of these have Dials Illuminated with Gold and Color, a new and hitherto unattained
  process, rendering White Enamel Dials more appropriate for Ladies.
                 THIRD ROW
  Contains the Perfection of RAILWAY WATCHES in Silver, with Compensation Balances
  and Isochronal Springs, suitable as Presents to their Employés by Railway Establishments;
  also, TWO-DAY WATCHES, suitable for Travellers.   Price 9 Guineas each.
                 FOURTH ROW
  Contains GOLD ¾-Plate HUNTING LEVERS, fully Jewelled with the finest Lever
  Escapements, approaching mathematical precision—Price 21 Guineas each.   Also, some
  Whole-Plate Levers with Spring Caps, a new and convenient arrangement, effecting
  increased flatness above the ordinary cap—Price 14 Guineas each.
                 FIFTH ROW
  Contains beautiful LEVER WATCHES FOR LADIES, with decorated Gold Dials, of
  great mechanical excellence.   Price 11 Guineas and 12 Guineas each.
                 SIXTH ROW
  Contains SILVER LEVER WATCHES, named "THE CHALLENGE WATCH."
  Price 3 Guineas each.

     For some years past the English Manufacture has suffered from the cheaper products of
  the Swiss Mountains,—the wages of labour are so small in Switzerland, compared with
  England, that the Swiss Manufacturers have been gaining large profits while the English
  Manufacturers have barely stood their ground.   The superior merits of English work are
  confessed on all hands, but cheapness and showiness are qualities more easily estimated than
  geometric precision.   To meet this competition J. Jones has, by new methods of manufacture,
  produced a good Lever Watch at 3 Guineas, the first announcement of which has been reserved
  for this Exhibition, and he trusts the production will have a national result.
</pre>

PLATE 33G

Clerkenwell watchmakers as he brought out a cheap fusee watch at 4gns and also stocked a Swiss going barrel for 3gns 'manufactured expressly for this house in Switzerland and examined by our own workmen here'. His catalogue description of the former reads 'Elegant, accurate and durable Silver double bottomed Watches, LEVER Escapement of four beats to the second, and hand to mark the same; manufactured especially to compete with the Swiss, without reducing the accuracy of performance, and claiming particularly the notice of those who can appreciate the merits of the British workman, and are anxious to encourage native manufacture.' Variations included Caps for an extra 10/6d; Hunting cover in silver an extra 10/6d and engraved back in English work an extra 21/-. He retired to his farm at Send Green near Woking and died in 1909 in his ninety-fifth year.

Horological Journal *November 1903 p32-35, October 1909. Sketch of the History and Principles of Watchwork by J Jones.*

PLATE 34A

PLATE 34B

PLATE 34
John Jones, 338 Strand, London

White enamel dial, narrow black roman numerals, blued steel spade hands, sunk seconds dial. Dustcap cut away to allow access to 'winding handle' (engraved WATKINS'S PATENT). Full plate. Engraved Jn Jones 338 Strand

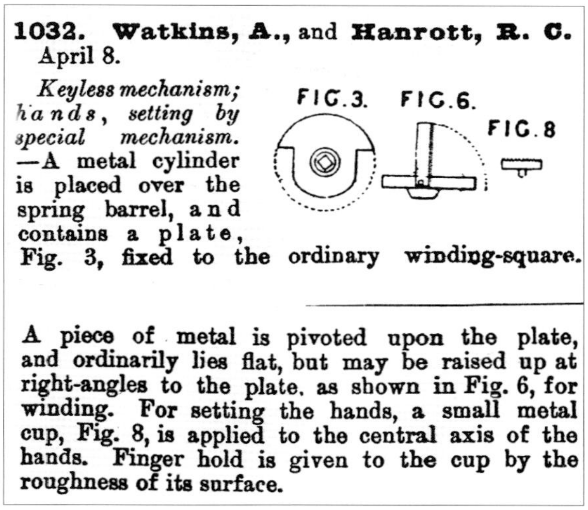

Lever Pocket Watches – English

1032. Watkins, A., and Hanrott, R. C.
April 8.

Keyless mechanism; hands, setting by special mechanism. —A metal cylinder is placed over the spring barrel, and contains a plate, Fig. 3, fixed to the ordinary winding-square.

A piece of metal is pivoted upon the plate, and ordinarily lies flat, but may be raised up at right-angles to the plate, as shown in Fig. 6, for winding. For setting the hands, a small metal cup, Fig. 8, is applied to the central axis of the hands. Finger hold is given to the cup by the roughness of its surface.

PLATE 34C

12259. 11 jewel English fusee movement with diamond endstone, balance staff and endstones fitted to scape wheel arbor. Steel three arm undersprung balance Movement fitted with Watkins's patent keyless winding.
Diameter 40.5mm

Watkins A and Hanrott RC UK Patent April 8 1870 No 1032, keyless mechanism.

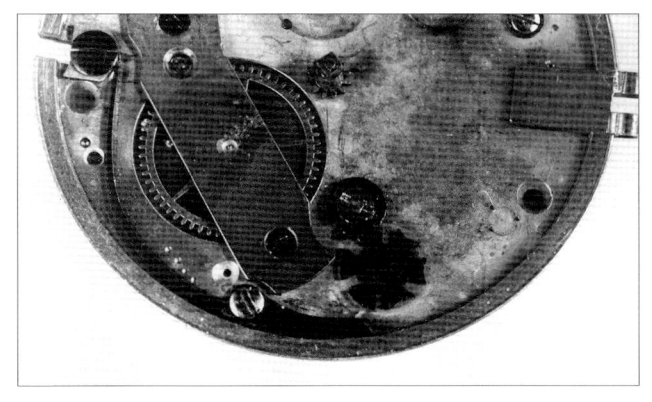

PLATE 35

PLATE 35
John Jones, 338 Strand, London

English lever full plate movement with fusee and chain. Top plate engraved Jno Jones. The pillar plate stamped 7202 124. Cylinder pillars jewelled pallet, scape and seconds arbor.

An unusual feature is the stop work fitted to fusee arbor under dial.

PLATE 36

PLATE 36
Anon

Full plate English lever watch movement and key wound and set originally fitted. Dustcap. Typical Liverpool jewelling to fusee arbor. 17 jewel with diamond endstone. Bimetallic cut compensated undersprung balance with Bosley regulator and Liverpool arrows to regulate index. Balance cock fully engraved with engraved barrel plate inscribed JEWELLED IN 12 HOLES. Movement stamped on dial plate TR 20906, probably Thomas Russell, Liverpool.
Diameter 45.6mm

Lever Pocket Watches – English

PLATE 37A

PLATE 37B

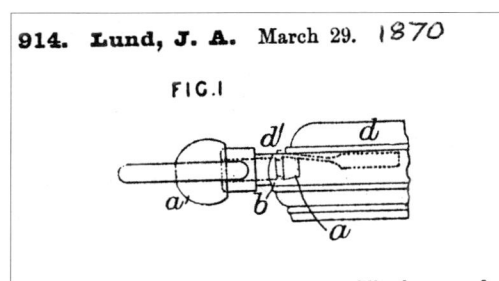

> **914. Lund, J. A.** March 29. 1870
>
> FIG.1
>
> *Watch keys; pendants; fusees.*—The key *a* of a watch is made in such a way as to fit within the pendant as shown, and is secured by a spring *d* having a nose-piece *d¹*, which is caused to enter a groove *b* in the key by the act of shutting the case. In some cases, in order to reduce the size of the pendant, the key is made solid. The fusee is then made with a socket, which is in one with the cap, the bottom pivot being in one with the ratchet-wheel.

PLATE 37C

PLATE 37D

PLATE 37
Lund Bros, 41 Cornhill, London

Silver open face double bottom engine-turned case hallmarked London **1874**. Casemaker PW (Philip Woodman, 10 Great Sutton Street, Clerkenwell, London EC). Dustcap. White enamel dial inscribed Lund Bros, 41 Cornhill, London, black roman numerals, gold spade hands, inset sunk seconds dial. Key wound and set – the key in the form of a removable winding button, subject of a patent (No 914). Full plate fusee and chain with maintaining power. 11 jewel English lever movement. Pointed tooth scape wheel and tangential pallets.

Lever Pocket Watches – English

PLATE 38A

Oversprung bimetallic compensated balance and blued steel balance spring. Top plate engraved LUND BROS 41 Cornhill London 9906. Floral engraving and grotesque mask on triangular balance cock with regulator index over the balance rim to silver quadrant and Coventry star engraved between Slow Fast on top plate.
Key size 8. Diameter 50.3mm

Lund Bros was the brand marketed from the premises of Barraud & Lunds. Cedric Jagger, in his book on Paul Philip Barraud, says that hundreds of these watches were made in silver cases up until about 1883. They are, however, not so readily found today. It is worth remembering that when the key is in situ the watch had all the appearance of a more conventional keyless watch. UK Patent No 914 20 March 1870 JA Lund. Paul Philip Barraud by Cedric Jagger.

PLATE 38
John Bennett, 65 and 64 Cheapside, London

Silver full hunter double bottom case numbered 9013. Case and pendant hallmarked London **1874**. Casemaker PW (Philip Woodman, 10 Great Sutton Street, Clerkenwell, London EC). Dome engraved Maker to the Royal Observatory. Dustcap stamped W & AO (?Oxley Bros, 51 Spon End, Coventry). White enamel dial, black roman numerals, gold spade hands, inset seconds dial. Key wound and set. Full plate. Engraved John Bennett, 65 & 64 Cheapside, LONDON 23428. Fusee. 7 jewel. Pointed tooth scape wheel, tangential pallet. Undersprung three arm gold balance with Bosley regulator.
Key size 6. Diameter 52.5mm

Sir John Bennett was notable figure both in the City and the horological world. Born in 1814, son of a watchmaker (John

PLATE 38B

Bennett of Greenwich), he joined his mother in the family business after his father's death. Around 1847 he set up his own business at 65 Cheapside as goldsmith and watchmaker. An exhibition at the Guildhall in 1986 quite correctly hailed him as Sir John Bennett Clockmaker, Showman and Sheriff. He became free of the City of London in 1849 and entered City politics as Common Councilman for the ward of Cheap in 1862, a position he held until 1889. In 1871-2 he served as Sheriff of London and Middlesex. He was knighted on 14 March 1872. In June 1877 he was elected Alderman of Cheap but was rejected by the Court of Aldermen. After he had been elected and rejected twice more the Aldermen elected another candidate in his place. The business became a limited company in 1889 and shortly afterwards he retired from the business altogether. He died in 1897. The firm continued to trade as Sir John Bennett Ltd until 1940 when the premises were destroyed by enemy action. Closure had already been decided upon and therefore alternative premises were not sought. Sir John was not always popular with

PLATE 38C

the horological trade as he was openly critical of the lack of organisation among the craftsmen of the day and subsequent higher prices. He saw no point in necessarily supporting the English watchmakers if he could obtain more reasonably priced watches from Switzerland. He was a pioneer of mass-marketing. He had a flamboyant personality which he utilised to its full potential in the marketing of his own products. His advertising campaigns make present-day efforts pale and insignificant. Perusal of the pages of the Horological Journal *of this period finds many accounts of the altercations between Sir John and, in particular, the watchmakers in Clerkenwell.*

PLATE 39
Josh Sewill, 61 South Castle Street, Liverpool

Silver open face double bottom engine-turned case hallmarked London **1874**. Casemaker IT. Dust ring secured by two dog screws. White enamel dial, roman numerals, gold spade hands, inset seconds dial. Key wound and set.

PLATE 39A

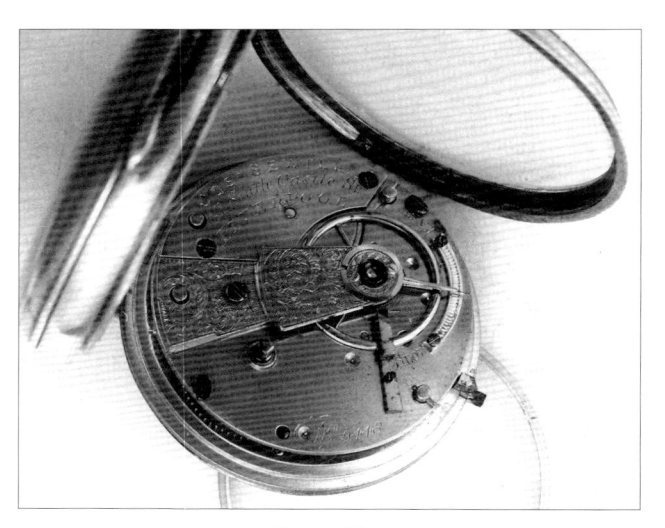

PLATE 39B

Full plate 7 jewel ENGLISH movement with diamond endstone to balance. Pointed tooth scape wheel, tangential pallets. Oversprung but bimetallic balance. Engraved on top plate Josh SEWILL, 61 South Castle St, Liverpool, No 9448. Balance cock with foliate engraving and regulator index over the balance to quadrant on top plate.

Key size 5. Diameter 42.5mm

Lever Pocket Watches – English

PLATE 40A

PLATE 40
William Potts & Sons – retailer's name on dial,
12-13 Guildford Street, Leeds
John Barr, Bridge Street, Earlstown,
Newton-le-Willows, Lancashire

Silver open face double bottom case with gold hinges, hallmarked Chester **1875**. Casemaker HG (Hugh Green, 59 Mount Pleasant, Liverpool). Dust cap stamped W. Pendant maker JJ. Cream enamel dial signed with name of retailer William Potts & Sons, Leeds, black roman numerals, blued steel hands, inset sunk seconds dial. Full plate 9 jewel lever movement with fusee and chain. Maintaining power. Pointed tooth scape wheel and tangential pallets. Top plate engraved John Barr, Earlstown No 89. Liverpool window jewelling 4th arbor. Undersprung three arm gold balance with blued steel balance spring. Bosley type regulator. Triangular balance cock engraved scrolls and index markings engraved incorporating Liverpool arrows.
Diameter 51.4mm

It has not been possible to trace John Barr as a finisher and it is possible that William Potts and Sons accepted for sale themselves a watch previously commissioned by John Barr.

PLATE 40B

Lever Pocket Watches – English

PLATE 41A

PLATE 41B

PLATE 41
Thomas Morison, Ayton, Near Gateshead, Durham

Silver pair-case hallmarked Birmingham **1875**. Casemaker I. Dustcap stamped AC. White enamel dial, broad black roman numerals, gold spade hands, inset sunk seconds dial. Key wound and set. Full plate. 7 jewel. Top plate engraved Thos Morison. Ayton. 30993. English lever movement. Fusee and chain and maintaining power. Pointed tooth scape wheel with tangential pallet. Oversprung three arm gold balance with blued steel balance spring. Regulator index from a triangular cock with floral and foliate engraving over balance rim to silver quadrant.
Key size 6. Diameter 52mm

Morison listed in Clock and Watchmakers of the World *Vol 2 2nd ed by Brian Loomes as watchmaker working in 1875.*

PLATE 42
John Poole, 57 Fenchurch Street, London

Silver open face case hallmarked London **1875**. The engine-turned case stamped PW (Philip Woodman succeeded Christopher Rowlands in 1870 at 56 Great Sutton Street, Clerkenwell; retired 1906; died 30 March 1908). Pendant hallmarked London **1875**. White enamel dial signed John Poole, 57 Fenchurch St, London, 5086, black roman numerals, gold spade hands, inset sunk seconds dial. Movement and case numbered 5086 with dustcap. Full plate fusee. Top plate engraved John Poole, 57 Fenchurch Street, LONDON 5086. 11 jewels. English lever movement with jewelled scape arbor including end jewel. Pointed tooth scape wheel with tangential pallet. Round steel oversprung three arm balance with blued steel balance spring (size 5). Regulator index over balance to engraved quadrant.
Diameter 48.7mm

PLATE 42A

102

Lever Pocket Watches – English

PLATE 42B

John Poole was born in 1840. His father was the well-known marine chronometer maker John Poole (born 1817, died 1867), inventor of auxiliary compensation for use in marine chronometers. The Horological Journal 1 August 1865 lists John Poole junior at 14 Middleton Street, Middleton Square – chronometer maker.

**PLATE 43
John Neve Masters, Rye, Sussex**

COASTGUARD WATCH
Heavy silver open faced double bottomed engine-turned case hallmarked Birmingham **1876**. Casemaker's mark WE (William Ehrhardt Lt, 72 Great Hampton Street, Birmingham – first registered Birmingham Assay Office 14 November 1867). Pendant CH (Charles Harrold) hallmarked Chester **1876**. Dustcap engraved COASTGUARD JN MASTERS RYE. White enamel dial with inset seconds signed COASTGUARD WATCH JN Masters Rye and above seconds dial Made in England, heavy black roman numerals, gold spade hands. Key wound and set. Full plate. 9 jewel English lever movement number 960612. Going barrel with club tooth scape and tangential pallets. Undersprung cut bimetallic balance. Blued balance spring with Bosley type regulator to engraved index on top plate. Triangular engraved cock of foliate design. Rare unusual silver cock (hallmarked Birmingham **1876**).
Key size 7. Diameter 51mm

According to listing in The Clockmakers of Sussex *by EJ Tyler, Masters was born in Tenterden in 1846 and died in 1928. He worked for Whitehead of Sevenoaks and took over the business of CF Lewns, High Street, Rye in 1869. Obviously a stalwart public figure as he became Mayor of Rye in 1893. See also VERACITY WATCH BY JN Masters made by William Ehrhardt hallmarked 1897 (Plate 70).*

PLATE 43A

PLATE 43B

103

Lever Pocket Watches – English

PLATE 44A

PLATE 44B

PLATE 44C

PLATE 44
Thomas Russell & Son, Church Street, Liverpool

Silver case hallmarked Chester **1877**. Casemaker's mark TR Number 101235. White enamel dial with sunk seconds dial inscribed Thos Russell & Son, black roman numerals, gold spade hands. Key wound. Inscribed top plate THOS RUSSELL & SON Machine Made Lever Russells Patented Click. 15 jewel going barrel lever movement with bimetallic uncut balance. Blued oversprung balance spring.
Key size 7. Diameter 51.4mm

Thomas Russell is listed as a watch manufacturer in 1848 with premises at 20 Slater Street, Liverpool, later moving to 30 Slater Street. See illustration of factory where the Russell Time O'Day watches were manufactured. After 1859 there was a change of name (Thomas Russell & Son). It appears that around this date Thomas Russell had relinquished the reins of his business to Thomas Robert Russell and Alfred Holgate Russell. By 1870 they were listed as makers to the Queen. They continued in partnership until 1878 first at Slater Street and Holborn, London and by 1877 Church Street, Liverpool. After 1878 TR

104

Lever Pocket Watches – English

PLATE 44D

PLATE 44E

Russell continued as proprietor of the Russell Watch and Chronometer Manufactory at Cathedral Works, 18 Church Street, Liverpool with addresses at Piccadilly, London and Toronto, Canada. This became Russell's Ltd in 1894. From this date it seems that they were a retail jewellers with branches in Liverpool and by the early 1900s at Manchester and Llandudno. Alfred Russell continued under the name of Thomas Russell & Son (Alfred Russell & Co). They were listed in 1880 as watch and chronometer manufacturers and machine made keyless lever and jewellery merchants by appointment to Her Majesty the Queen and HRM the Duke of Edinburgh and The Admiralty. By the following year the listing included their being importers of Swiss watches, musical boxes etc. First at Sandon Buildings, Post Office Place, subsequent moves took them to 12 Church Street, Liverpool and Holborn, London. Eventually the firm was converted into a limited liability company and by 1938 had become Thomas Russell & Son Watch Co Ltd.

The Official Trade Mark Journal lists registration 12510 Thomas Russell and Son, Church Street, Liverpool and the words 'Machine Made Lever'. The advertisement shown of their successors, Alfred Russell & Co (Plate 44E), gave the following details concerning the trade mark – 'First class keyless and ordinary watches' carried the registered trade mark of 'Thos Russell & Son, Makers to the Queen, Liverpool and London'. Second quality watches merely had 'RUSSELL'. The trade mark in the centre bottom of the advertisement with the name Alfred Russell & Co Liverpool indicated the 'Finest quality keyless watches'. The trade mark of one hand encircled with the words 'Russell's Patent and Machine Made Lever' referred to machine made watches. Finally horizontal watches carried the trade mark of a pair of shaking hands in an oval with the words 'T R & S REGISTERED'.

This watch and the following two examples have been included in this section as an interesting example of how the wholesale manufacturers also imported Swiss movements for casing and hallmarking in this country and either sold under their own name or, as with Parkinson & Frodsham, sold on to another retail outlet. The circular distributed in 1881 (Plate 44D) demonstrates their close association with Longines at this date. The movement shown has the words THOS RUSSELL & SON LONGINES around the edge and PATENT STEM WINDER on plate.

Lever Pocket Watches – English

PLATE 45
Thomas Russell & Son, Church Street, Liverpool

Silver hunter case with gold hinges hallmarked Chester **1878.** Casemaker's stamp TR. White enamel dial, Thos Russell & Son 15001, black narrow roman numerals, narrow steel spade hands, inset sunk seconds dial. Keyless wound and set by lever at edge of bezel at 4 o'clock position. Half plate going barrel. 15 jewel movement inscribed on barrel bridge RUSSELL'S MACHINE MADE LEVER. Club tooth scape wheel with tangential pallets with ring at tail of pallet. Uncut oversprung bimetallic balance with blued steel balance spring. Index on balance cock which is engraved with the Longines Trade Mark (winged hourglass registered May 1880). The identical movement is illustrated under Longines Première Periode in *Le Livre D'Or de Horlogerie* p35.
Diameter 42mm

PLATE 45A

PLATE 45B

106

Lever Pocket Watches – English

PLATE 46A

PLATE 46B

PLATE 46
Parkinson & Frodsham, 4 Change Alley, Cornhill, London

Silver open face case hallmarked Chester **1880.** Snap-on bezel, hinged back and dome, gilt hinges. Casemaker's mark TR above star. White enamel dial inscribed PARKINSON & FRODSHAM LONDON, broad black roman numerals, blued steel spade hands. Keyless wound with push piece to set hands at one o'clock position. Half plate movement engraved Russell's Machine Made Keyless Lever, 18 Church Street, Liverpool. Going barrel. 15 jewels. Club tooth lever with tangential pallet. The long body of the pallet counterbalanced at the tail with a ring. Small index on to balance cock which is engraved PATENTED with registered mark of Longines.
Diameter 53.8mm

The name Parkinson & Frodsham is usually associated with pocket and marine chronometers. Founded in 1801, the firm continued in business until their premises were bombed in 1944 for the third time. They ceased to exist in 1947. Kelly Post Office 1880 and 1890 Directory advertisement pages for London show 'Speciality two guineas keyless watch'.

PLATE 47
Thos Russell & Son, Church Street, Liverpool

White enamel dial with ⅕ seconds markings, sunk centre dial, black roman numerals, gold spade hands with blued steel narrow spade centre seconds hands. Keyless wound. Inscribed on dial and movement CHRONOGRAPH CENTRE SECONDS Thos Russell & Son Makers to the Queen Liverpool 101432. Top plate also engraved with trade mark of winged wheel in clouds and words TEMPUS FUGIT. *Trade Mark Journal* lists this trade mark (No 15965 19 September 1883) Alfred Russell & Co, 24/25 Sandon Building, P O Place, Liverpool. Going barrel. 15 jewel English lever pointed tooth scape with cut bimetallic balance and blued steel Breguet overcoil spring with regulator index on balance cock. *Note typical high class English jewel setting.*
Diameter 47.2mm

PLATE 47

107

Lever Pocket Watches – English

PLATE 48B

PLATE 48A

PLATE 48
John Nelson, Haymarket Street, Bury

Silver open face double bottom case hallmarked London **1882.** Casemaker JN (besides the possibility of John Nelson, mark could be James Newman, 92 Cloudesley Road, N London or James Thomas Newman, 42 Spencer Street, Clerkenwell). Dustcap stamped W & AO (?Oxley Bros., 51 Spon End, Coventry). White enamel dial, John Nelson, Bury. Black roman numerals, gold spade hands, up and down dial and inset sunk seconds dial. Full plate. Fusee and chain. English lever movement with pointed tooth scape wheel and tangential pallets. Bimetallic cut balance with compensation screws and blued steel balance spring. Index on triangular cock of engraved foliate design. The index is therefore adjustable without removing dust cover. The plate engraved John Nelson Bury 22253.
Diameter 53mm

Jeweller and Metalworker, 15 Dec 1949, p1158, will of Albert, son of John Nelson.

PLATE 49
Elias Wolfe, Sunderland

Silver open face double bottomed engine-turned case with milled edge to body, hallmarked Chester **1883.** Casemaker CH (Charles Harris, 6 Norfolk Street, Holyhead Rd, Coventry). Pendant J. White enamel dial inscribed IMPROVED PATENT, broad black roman numerals, blued steel spade hands, inset sunk seconds. Key wound and set. Full plate engraved Elias Wolfe. Sunderland. No 7620. Fusee and chain with maintaining power. 7 jewels. English lever movement. pointed tooth scape wheel, tangential pallet. Undersprung uncut imitation bimetallic balance with blue steel balance spring. Triangular balance cock engraved in scroll pattern with the Bosley type regulator to index markings engraved on top plate incorporating SLOW FAST between Coventry star.
Key size 8. Diameter 53.5mm

Kelly Directory lists Joseph Wolfe, Watchmaker, Borough Road, Sunderland and Keith Bates in Clock and Watchmakers of Northumberland and Durham *lists Joseph Wolfe as clock dealer at 10 Sussex Street, Sunderland in 1884.*

Lever Pocket Watches – English

PLATE 49A

PLATE 49B

PLATE 50
Camerer Kuss & Co, 56 New Oxford Street, London

Silver open face case hallmarked London **1884.** Snap-on bezel, hinged back and dome. Casemaker Joseph Walton, 7 Upper Charles Street, Clerkenwell. White enamel dial inscribed with retailer's name – Camerer Kuss & Co, 56 New Oxford St, London, black roman numerals, blued steel narrow spade hands, inset sunk seconds. Keyless wound and set with push piece at 3 minutes past position. Three-quarter plate engraved Camerer Kuss & Co 56 NEW BOND ST LONDON 32503. 7 jewelled English lever movement. Pointed tooth scape wheel, tangential pallet. Bimetallic cut oversprung balance with blued steel balance spring. Index regulator to graduations on small triangular cock incorporating Coventry star in mid line.
Diameter 52.7mm

The firm Camerer Cuss & Co was established in 1788 to import and sell Black Forest clocks. The original spelling of Kuss was changed at the outset of the First World War to 'Cuss'. Originally at 2 Broad Street. London, they acquired 522 (later renumbered 56) New Oxford Street in 1868. They left these premises for their present establishment in Ryder Street in 1983. Camerer Cuss & Co 1788-1988 – The Bicentenary.

PLATE 50

109

Lever Pocket Watches – English

PLATE 51A

PLATE 51
James Pyott, 74 West India Dock Road, London

18 carat gold engraved open face double bottomed case hallmarked London **1884.** Casemaker JM (John Martin, 32A Colebrooke Row N). Gold engine-turned and engraved dial, black roman numerals with fleur-de-lis hands. Key wound and set. Full plate. 9 jewel English lever movement with fusee and chain and maintaining power. Oversprung bimetallic compensated balance. Triangular cock engraved foliage with index over balance to quadrant with adjacent Coventry star in mid line position. Top plate engraved James Pyott, 74 West India Dk Rd, London, No 919.
Key size 4. Diameter 38mm

James Pyott was for many years co-editor of the Horological Journal *He was an important maker of marine chronometers. Originally at Spencer Street, Clerkenwell (1864-1865), he worked at 74 West India Dock Road between 1876-c1900. He died in 1918.*

PLATE 51B

Lever Pocket Watches – English

PLATE 52A

PLATE 52B

PLATE 52
John Gaydon, 99 High Street, Barnstaple, Devon

Silver open face double bottomed case hallmarked London **1885**. Casemaker Charles Harris, 6 Norfolk Street, Coventry. Dustcap stamped TW (Thomas Whittaker, 39 Hill Street, Coventry). Dustcap not pierced to show balance cock. White enamel dial inscribed Jno GAYDON BARNSTAPLE, black roman numerals, gilt spade hands. Key wound and set. Full plate Jno Gaydon Barnstaple 41512. Fusee and chain with maintaining power. 7 jewel English lever movement. Polished round rimmed steel undersprung balance with blued balance spring. Bosley type regulator to engraved index graduations on the top plate incorporating Coventry star. Triangular balance cock engraved with floral and scroll pattern.
Key size 7. Diameter 50.3mm

COLOUR PLATE 8 (page 7) and PLATE 53
George Wilson, 29 Middlegate, Penrith

Silver open face double bottomed engine-turned back, snap-on bezel, hallmarked London **1885**. Casemaker RS. Pendant hallmarked London **1885**. White enamel dial, black roman numerals, blued steel spade hands, inset sunk seconds dial. Key wound and set. Dial opens 4 to 10, apparently a much safer way of opening to avoid injury to seconds hand. Full plate going barrel, English lever movement. 7 jewels. Pointed tooth scape wheel with tangential pallets. Oversprung three arm gold balance recessed into top plate with special adjustable balance spring stud. Regulator index on balance cock engraved with Coventry star. Barrel is removable. The two seals screwed to inside of the dome close over the winding holes.
Key size 6. Diameter 51.1mm

The dome is subject of UK Patent 1 Sept 1884, No 11852; UK Patent 9 Sept 1884, No 12199 taken out by Reuben Squire, 37 Myddelton Square, London EC. Prize Medal at Inventors Exhibition, South Kensington, 1885. Trade Mark Journal 19 September 1883 25513 lists Reuben Squire, 35 Myddelton Square, as having trade mark 'A star surrounded by rays', the words the 'London Watch Company', 'Standard' and the initials 'LWC'.

PLATE 53

111

PLATE 54A

PLATE 54B

PLATE 54
James William Benson, 58-60 Ludgate Hill, London

Heavy brass open face double bottomed case. Casemaker GJT (George James Thickbrook, 6 Spencer Street, London EC). GWR engraved on back. Dustcap. Movement and case stamped 31798. White enamel dial JW BENSON LONDON, black roman numerals, broad spade blued steel hands, inset sunk seconds dial. Key wound and set. Full plate. English lever movement engraved on top plate JW Benson 58 & 60 Ludgate Hill London TO HRH THE PRINCE OF WALES & HIM THE EMPEROR OF RUSSIA No 31798. Movement with fusee and chain maintaining power. 7 jewels with ruby endstone to balance staff. Pointed tooth scape wheel, tangential pallets. Cut bimetallic oversprung balance with compensation screws. Blued steel balance spring and plain triangular cock. Bosley type regulator to engraved graduations and Slow Fast with Coventry star between.
Key size 9. Diameter 57.5mm

Great Western Railway incorporated 31 August 1835, Brunel having been appointed engineer to the project in 1833. SS Benson & JW Benson are listed between 1847-1855 at Cornhill and Ludgate Hill as watchmakers, gold and silversmiths. Later they became JW Benson with premises between 1855-1897 at 33 Ludgate Hill. The renumbering of Ludgate Hill and Circus in the mid-1860s resulted in a change to 58-60, their watch factory at this date being at 4 & 5 Horseshoe Court at the rear of the Ludgate premises. There was a move to rebuilt and enlarged premises (62-64) in the 1880s with their new steam workshops now nearby at 38 La Belle Sauvage Yard. Their watches were well finished with three main trade names, The Ludgate, The Bank and The Field. They also retailed watches brought in from other smaller manufacturers in this country and, as demand grew, Switzerland. Benson claimed his workshops 'contained an efficient staff of workmen...employed not only in the manufacture, but in the repair of Watches'. This is an excellent example of how the approximate date of an unhallmarked watch can be established by reference to dates of address changes, in this instance the incorporation of the GWR, ie after 1835 but prior to the mid-1860s.

Lever Pocket Watches – English

PLATE 55A

PLATE 55B

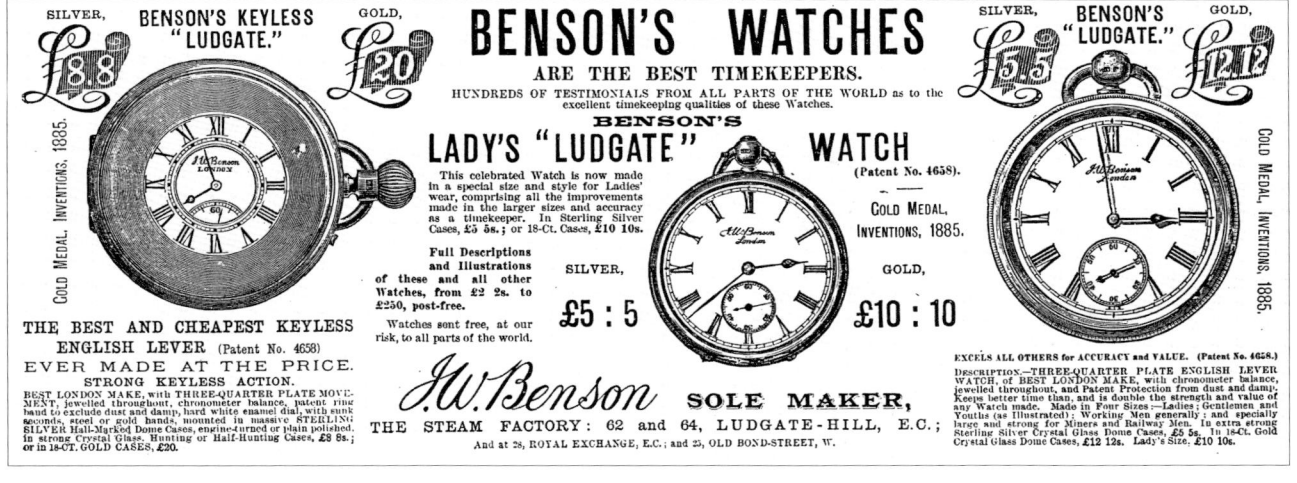

PLATE 55C

PLATE 55
John William Benson, Ludgate Hill, London

Silver open face case hallmarked London **1886.** Snap-on bezel, hinged back and dome. Casemaker JWB. White enamel dial inscribed JW Benson, London, black roman numerals, gold spade hands, inset sunk seconds dial. Key wound and set. Three-quarter plate engraved JW Benson Ludgate Hill London The LUDGATE WATCH BEST LONDON MAKE BY WARRANT TO HM QUEEN Patent No 4658 No 27416. English lever. 13 jewels movement with pointed tooth scape wheel, tangential pallets. Bimetallic cut compensated oversprung balance. Blued steel balance spring and polished index on top of small triangular balance. Cock engraved FS.
Key size 5. Diameter 54.2mm

113

Lever Pocket Watches – English

BENSON'S WATCHES. ENGLISH WARRANTED

Guaranteed for Accuracy, Durability, and Strength, at Maker's Cash Prices.

The "BEST" LONDON WATCH made

In Silver Cases, £5.5 In 18-ct. Gold Cases, £12.12

at the price is **BENSON'S 'LUDGATE'**

STRONGEST and CHEAPEST THREE-QUARTER PLATE English Lever Watch, with Chronometer Balance, Patent Dust-proof Ring Band, ever made at the price.
Made in Three Sizes at one price, £5 5s.
In massive 18-ct. Gold Cases, with Crystal Glass, Gentlemen's, £12 12s.; Ladies', £10 10s.

The Watch to suit all Classes.

Can be obtained on "**The Times**" System of Purchase by **MONTHLY PAYMENTS**. Particulars Post Free.

Old Watches and Jewellery taken in Exchange.

Before purchasing elsewhere, write for Guide Book, post free.

J. W. BENSON, LTD.,
Steam Factory:
LUDGATE HILL, LONDON, E.C.

(1)

PLATE 55D

UK Patent No 4658, 25 Oct 1881, Samuel Morgan, Kingsland, Middlesex, 'To protect the watch from dust and afford extra space for the barrel or balance without increasing the diameter of the case'.

PLATE 56
Faller Bros, 31 Union Street, Inverness, Scotland

Silver pair-case hallmarked Chester **1886**. Casemaker CH (Charles Harris, 6 Norfolk Street, Coventry). Dustcap. Antique pinned bow. White enamel dial, narrow black roman numerals, gold spade hands, inset seconds dial. Key wound and set. Full plate engraved Faller Brothers Inverness 15916. Fusee and chain maintaining power. 9 jewel English lever movement. Pointed tooth scape with tangential pallets. Oversprung three arm gold balance with a triangular balance cock engraved in scroll design. The regulator index over the rim of the balance to silver quadrant.
Key size 5. Diameter 52mm

Kelly Directory 1880 lists Faller Brothers, 31 Union Street, Inverness as watch and clock makers. Compare this watch with that by Thomas Morison, Ayton (Plate 41).

PLATE 56

114

Lever Pocket Watches – English

PLATE 57A

PLATE 57B

PLATE 57
Emily J Gee, London

Silver open face double bottomed engine-turned case hallmarked London **1887**. Casemaker PW (Philip Woodman & Son, 33 Smith Street, Clerkenwell). Gilt movement and dust ring. White enamel dial, black roman numerals, blued steel spade hands. Key wound and set with maintaining power. Full plate fusee and chain. English lever movement. Engraved on top plate Emily J Gee London 1046. Pointed tooth scape wheel, tangential pallets. Three arm gold oversprung balance with blued steel spring. Polished index over balance to silver quadrant SLOW FAST engraved on top plate with Coventry star between. In original presentation case – maroon leather and silk and velvet lined to match.
Key size 4. Diameter 38.5mm

It has not yet been possible to trace this Emily Gee – a similar watch is known by David Gee hallmarked London 1885. It was of course common practice for widows to continue their husbands' businesses. Adam Gee 1853-64 and Susannah Gee and Standley 1867-75 could be antecedents.

PLATE 58
William Gibson, Donegall Place and Castle Place, Belfast

Silver open face double bottom silver case hallmarked London **1888**. Casemaker JJ (James Jackson, 36 Helmet Row, St Lukes, London EC). Dustcap stamped W & AO (Oxley Bros). Movement and case numbered 15499. White enamel dial, black roman numerals, gold spade hands, inset sunk seconds dial. Coventry key wound and set. Full plate fusee. 1 jewelled movement. Top plate engraved Wm Gibson & Co Belfast 15499. Pointed tooth scape wheel with tangential pallets. Gold three arm undersprung balance with Bosley type regulator to index scale incorporating Coventry 4 arrow star. The triangular cock engraved with shamrock.
Key size 8. Diameter 48.5mm

PLATE 58

115

Lever Pocket Watches – English

PLATE 59

PLATE 59
Alfred Henry Drinkwater, Jedburgh House, Butts, Coventry

Silver open face double bottomed case hallmarked Chester **1888**. Casemaker RJP (RJ Pike, Earlsdon Terrace, Earlsdon, Coventry). Dustcap stamped WO (?W J Oxley, watch cap maker, 4 Dover Street, Holyhead Road, Coventry). Off white enamel dial with sunk centre, black roman numerals, gold fleur-de-lis hands, inset sunk seconds dial. Key wound and set. Full plate. English lever. 11 jewelled movement engraved on top plate A H Drinkwater Coventry 16575 Manufacturer. Fusee and chain maintaining power. Pointed tooth scape wheel with tangential pallets. Undersprung and bimetallic balance with compensation screws and blued steel balance spring. Polished Bosley type regulator to engraved graduations with Slow Faft with Coventry star engraved between. The triangular balance cock engraved with scrolls and foliate design.

Key size 8. Diameter 54.3mm

English lever full-plate movement.

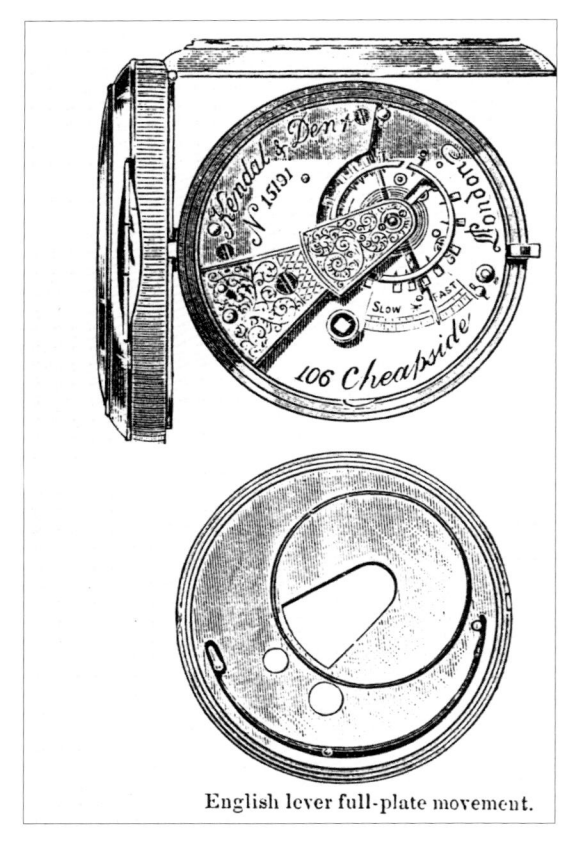

PLATE 60A

English lever three-quarter-plate movement.

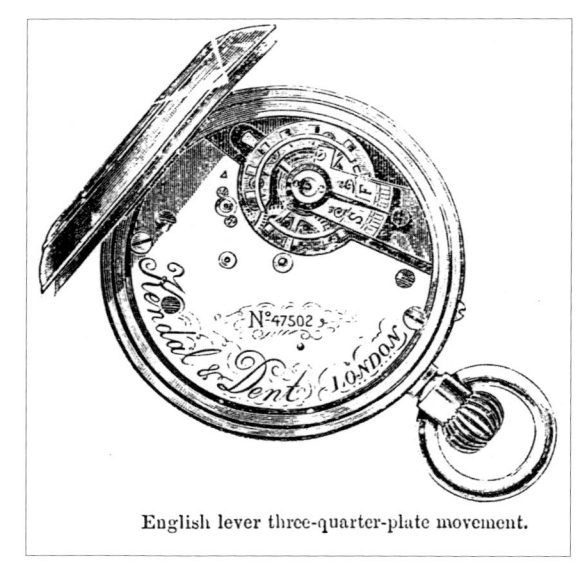

PLATE 60B

PLATE 60
James Kendal & John G Dent, 106 Cheapside, London EC

Silver open face double bottomed engine-turned case hallmarked Birmingham **1889**. Casemaker stamp K & D, dustcap stamped inside 118847, pendant CH (Charles

Lever Pocket Watches – English

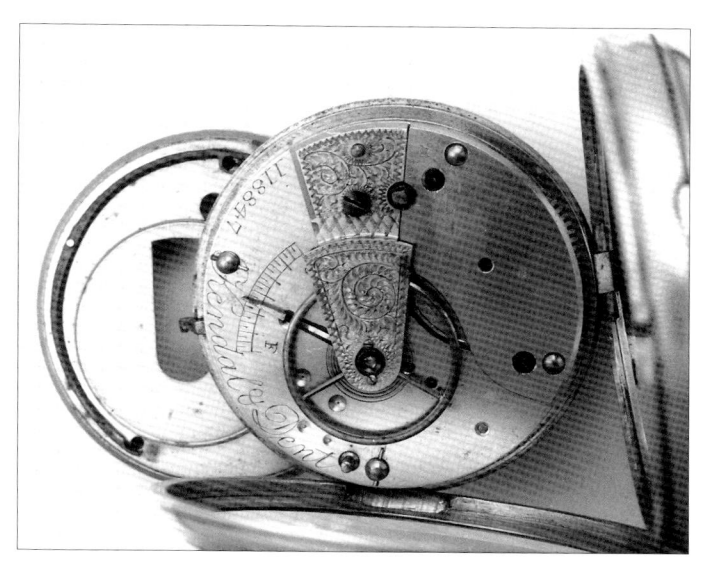

PLATE 60C

Harrold). White enamel dial inscribed KENDAL & DENT, black roman numerals, blued steel spade hands, inset sunk seconds dial. Key wound and set with going barrel. Full plate. English lever movement. 7 jewels movement. Pointed tooth scape wheel with tangential pallets. Undersprung three arm brass balance with blued steel balance spring and Bosley type regulator to engraved graduations on top plate which is inscribed Kendal & Dent 118847. Balance cock engraved with scrolled pattern.
Key size 6. Diameter 52.6mm

James Francis Kendal wrote in 1892

> Some English lever watches are still made with full plate movements, but the greater number are of the 3/4 plate construction, in which a portion of the top plate, or the one which is visible on opening the watch, is cut away, so that the upper escapement pivots are carried by the brackets, technically termed cocks. Full plate watches of the better kind have the movements covered with a cap, affording an immunity from dirt which is not secured with an open movement. Full plate watches are therefore to be recommended for wearers engaged in dusty vocations.
>
> There are two advantages of the 3/4 plate over its predecessor; firstly, the escapement can be more readily removed and replaced, and secondly, the balance being sunk below the level of the plate, a thinner watch is possible. Occasionally the 4th wheel pivot is also carried in a cock in order to get in a slightly larger balance than could otherwise be admitted. This necessities cutting away the plate still further and the watch is then spoken of as a half-plate; but very few are made this way.

History of Watches and Other Timepieces *by Kendal* published 1892. *In* Edward John Dent and His Successors *Dr Vaudrey Mercer refers to the advertisements in the* London Gazette *18 November 1883 on the association between James Kendal and John Gilder (otherwise Dent) as having no connection with E Dent & Co.*

PLATE 60D

117

Lever Pocket Watches – English

PLATE 61A

PLATE 61
James Moore French, 9 Royal Exchange, London

Silver open face case with hinged dome and bottom hallmarked London **1890**. Casemaker JW (?James Woodhouse, 31 Wynyatt Street, Goswell Rd, EC or John Woodman, Gt Sutton Street, Clerkenwell EC). Silver pendant hallmarked Birmingham **1889**. White enamel dial inscribed FRENCH Royal Exchange LONDON, black roman numerals, narrow blued spade hands, inset sunk seconds dial. Key wound and set. Three-quarter plate engraved J M French Royal Exchange London No 5047. 15 jewelled English fusee lever movement with pointed tooth scape wheel and tangential pallets. Oversprung bimetallic cut compensated balance with index regulator on balance cock.

Key size 4. Diameter 45.7mm

James Moore French, 1808-1842, chronometer maker of repute. The firm continued to use his name.

PLATE 61B

Lever Pocket Watches – English

PLATE 62A

PLATE 62
Harriet Samuel, 97 Market Street, Manchester

Silver open face engine-turned double bottomed case hallmarked Birmingham **1890.** Casemaker HS. Dustcap stamped inside 128873. White enamel dial inscribed H SAMUEL MARKET ST MANCHESTER, black roman numerals, blued steel spade hands, inset sunk seconds dial. Full plate key wound and set signed H Samuel MANCHESTER No 128873. 7 jewel going barrel movement with pointed tooth scape wheel and tangential pallets. Undersprung gold three arm balance with Bosley type regulator to engraved graduations on top plate. The triangular balance cock covered with engraving.
Key size 7. Diameter 52.4mm

1897 Directory lists H Samuel also at 34 Yorkshire Street, Rochdale and 168 Friargate, Preston. This is the precursor of H Samuel the famous High Street jewellers.

PLATE 63
Army & Navy Co-operative Society Ltd, 105 Victoria Street, London SW

Silver full hunter case with gold hinges to dome bottom and front cover, hallmarked London **1892.** Casemaker FT (F Thoms, 25 Spencer Street, Clerkenwell). Pendant maker CH (Charles Harrold, 2 & 3 St Pauls Square, Birmingham). White enamel dial, black roman numerals, blued steel spade hands, inset sunk seconds dial. Keyless movement, push piece to set hands at 17 minutes past position. Three-quarter plate engraved Army & Navy Co-operative Society Ltd, 105 Victoria St, London SW, No 6322. Going barrel. 19 jewel movement with screwed in jewel setting endstones to pallet staff and scape arbors. Oversprung bimetallic compensation cut balance. Index on balance staff with Coventry fleur-de-lis mark above index scale.
Diameter 50.3mm

PLATE 62B

PLATE 63

Lever Pocket Watches – English

PLATE 64A

PLATE 64B

PLATE 64
Henry John Norris, 30 Union Street, Coventry

Silver open face double bottomed case with engine-turned back and milled edge to body, hallmarked London **1892.** Casemaker HJN. Dustcap stamped inside TW O8 (Thomas Whittaker, Gas Street, Coventry – Kelly Directory 1888). White enamel dial signed H J Norris Coventry immediately above inset sunk seconds dial with arabic numerals narrow black roman numerals, broad spade pale gold hands, inset seconds dial. Key wound and set. Full plate. 7 jewel going barrel. English lever movement engraved on the top plate H J Norris Coventry 39608. Pointed tooth scape wheel and tangential pallets. Undersprung bimetallic balance with compensation screws. Bosley type index to engraved graduations and S R on the top plate. Plain triangular balance cock.
Key size 5. Diameter 46.5mm

PLATE 65A

PLATE 65
Coventry Co-operative Watch Manufacturing Society, 32 Bishop Street, Coventry

Silver open face engraved double bottomed case (lady's fob watch) hallmarked Chester **1892.** Casemaker WB (William Bird, 61 Craven Street, Chapel Fields, Coventry). Dust ring secured by two dog screws. Pendant CH (Charles Harrold, 2 & 3 St Pauls Sq, Birmingham). White enamel dial, black roman numerals, gold spade hands. Key wound and set cannon pinion square. Full plate. Going barrel. 7 jewel English lever movement with

120

Lever Pocket Watches – English

PLATE 65B

PLATE 65C

pointed tooth scape wheel and tangential pallets. Undersprung three arm steel balance with round rim, blued steel balance spring. Bosley type regulator to engraved graduations on top plate numbered 4. The triangular balance cock engraved with 238, the trade mark of the Society – registered 1883.
Key size 6. Diameter 42.5mm

Henry Gannay interested a number of watchmakers to form a co-operative in order better to market their watches. Formed in 1876, they were relatively successful for about ten years but were gradually forced to expand to deal in foreign watches. Their main downfall was a reluctance to introduce the use of machine tools and preferring to continue in the traditional methods. It is doubtful if any watches were actually finished by them after the turn of the century although the registration of the trade mark continued until after the First World War.

PLATE 66
Charles T Cowell, Douglas, Isle of Man

Silver open face double bottomed case hallmarked London **1893**. Casemaker CH (Charles Harries, 3 Hertford Place, Butts, Coventry). Dustcap stamped TW (Thomas Whittaker. 39 Hill. Coventry). White enamel dial, black roman numerals and gold spade hands, inset sunk seconds dial. Key wound and set. Full plate engraved Chas T Cowell DOUGLAS ISLE OF MAN 25922. Fusee and chain. English lever movement. Pointed tooth scape wheel with tangential pallets. Gold three arm undersprung balance with Bosley regulator. Index calibrations engraved on top plate with Coventry star in mid position. Triangular balance cock stamped with floral design.
Key size 6. Diameter 49.2mm

PLATE 66

121

Lever Pocket Watches – English

PLATE 67A

PLATE 67B

PLATE 67C

PLATE 67
Frodsham, 84 Strand & 115 New Bond Street, London

Silver demi-hunter case, hinged dome and back, hallmarked London **1895.** Casemaker's mark HMF (Harrison Mill Frodsham). White enamel dial signed CHAS FRODSHAM LONDON 08061, black roman numerals, blued steel double swell spade hands, inset seconds dial. Keyless wound and push piece to set hands at 19 minute past position. Three-quarter plate. 15 jewel English lever going barrel movement with pointed tooth scape wheel and tangential pallets. Oversprung with overcoil cut bimetallic compensated balance. Blued steel balance spring. Polished index on balance cock engraved with index scale and F S. Top plate engraved Chas Frodsham By Appointment to the Queen 84 Strand & 115 New Bond Street London No 08061. The raised area of top plate over barrel engraved with scroll pattern.
Diameter 50mm

Lever Pocket Watches – English

PLATE 68A

PLATE 68
Charles Frodsham & Co, 115 New Bond Street, London

Silver demi-hunter hallmarked London **1896**. Casemaker's mark HMF 201 (Harrison Mill Frodsham). White enamel dial signed C Frodsham & Co 08548, black roman numerals, blued steel double spade hunter hands, inset sunk seconds dial. Keyless wound. Three-quarter plate. 17 jewels Rotherham movement number 08548 with push piece to set hands. Top plate signed Chas Frodsham 115 New Bond St 08548 ADJUSTED. Pointed tooth scape wheel. Tangential pallets and oversprung and overcoil bimetallic cut compensation balance with gold screws.
Diameter 50mm

PLATE 68B

PLATE 69
D Robertson, 18 Arcade, Glasgow

Silver open face double bottomed case hallmarked Chester **1896**. Casemaker IM. White enamel dial, black roman numerals, gold spade hands. Key wound and set. Three-quarter plate. Fusee and chain. 11 jewel movement with maintaining power. Top plate engraved D Robertson & Co Glasgow 50119. Pointed tooth scape wheel, tangential pallets. Large three arm gold oversprung balance with blued balance spring. Small triangular cock with diamond endstone engraved FAST SLOW.
Key size 5. Diameter 53.8mm

PLATE 69

Lever Pocket Watches – English

PLATE 70
John Neve Masters, Hope House, 21 High Street, Rye, Sussex

VERACITY WATCH

Silver open face double bottomed case hallmarked Birmingham **1897**. Casemaker WE (William Erhardt, Time Works, Barr Street, West Birmingham). Silver dust cover with blued steel slide and engine-turned decoration. White enamel dial inscribed Veracity WATCH JN Masters Rye Made in England, broad black roman numerals, gold spade hands. Key wound and set. Full plate going barrel. 11 jewel English lever movement. Club tooth scape wheel, tangential pallets. Cut bimetallic undersprung balance with Bosley type regulator to engraved index on top plate. Top plate engraved 282449.

Key size 7. Diameter 50mm

PLATE 70A

PLATE 70B

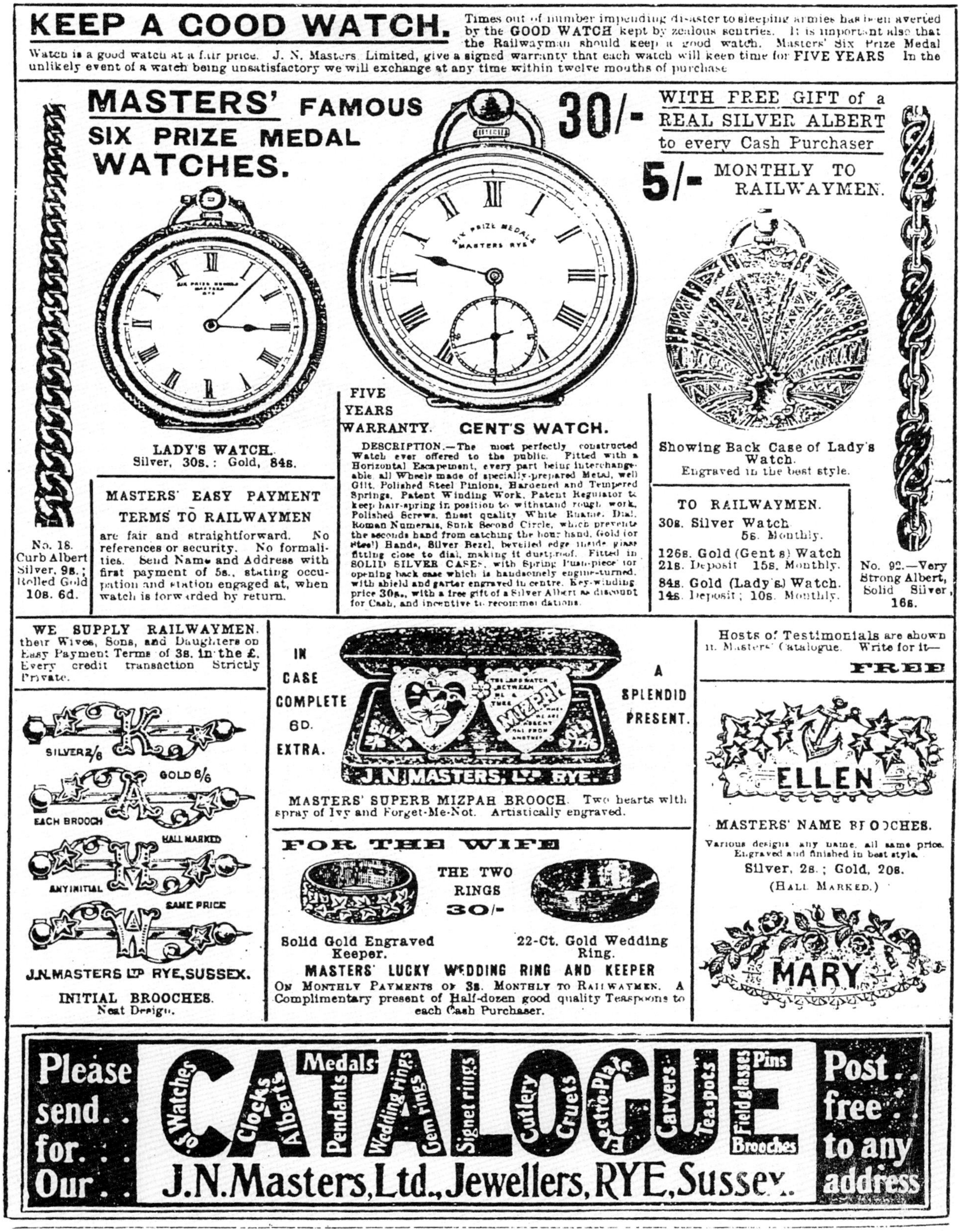

KEEP A GOOD WATCH. Times out of number impending disaster to sleeping armies has been averted by the GOOD WATCH kept by zealous sentries. It is important also that the Railwayman should keep a good watch. Masters' Six Prize Medal Watch is a good watch at a fair price. J. N. Masters, Limited, give a signed warranty that each watch will keep time for **FIVE YEARS**. In the unlikely event of a watch being unsatisfactory we will exchange at any time within twelve months of purchase.

MASTERS' FAMOUS SIX PRIZE MEDAL WATCHES.

30/-

WITH FREE GIFT of a REAL SILVER ALBERT to every Cash Purchaser

5/- MONTHLY TO RAILWAYMEN.

LADY'S WATCH.
Silver, 30s.; Gold, 84s.

FIVE YEARS WARRANTY. GENT'S WATCH.
DESCRIPTION.—The most perfectly constructed Watch ever offered to the public. Fitted with a Horizontal Escapement, every part being interchangeable, all Wheels made of specially-prepared Metal, well Gilt, Polished Steel Pinions, Hardened and Tempered Springs, Patent Winding Work, Patent Regulator to keep hair-spring in position to withstand rough work, Polished Screws, finest quality White Enamel Dial, Roman Numerals, Sunk Second Circle, which prevents the seconds hand from catching the hour hand, Gold (or Steel) Hands, Silver Bezel, bevelled edge inside glass fitting close to dial, making it dust-proof. Fitted in SOLID SILVER CASES, with Spring Dust-piece for opening back case which is handsomely engine-turned, with shield and garter engraved in centre. Key-winding price 30s., with a free gift of a Silver Albert as discount for Cash, and incentive to recommendations.

Showing Back Case of Lady's Watch.
Engraved in the best style.

MASTERS' EASY PAYMENT TERMS TO RAILWAYMEN
are fair and straightforward. No references or security. No formalities. Send Name and Address with first payment of 5s., stating occupation and station engaged at, when watch is forwarded by return.

TO RAILWAYMEN.
30s. Silver Watch
5s. Monthly.
126s. Gold (Gent's) Watch
21s. Deposit 15s. Monthly.
84s. Gold (Lady's) Watch.
14s. Deposit; 10s. Monthly.

No. 18. Curb Albert Silver, 9s.; Rolled Gold 10s. 6d.

No. 92.—Very Strong Albert, Solid Silver, 16s.

WE SUPPLY RAILWAYMEN
their Wives, Sons, and Daughters on Easy Payment Terms of 3s. in the £. Every credit transaction Strictly Private.

SILVER 2/6
GOLD 6/6
EACH BROOCH
HALL MARKED
ANY INITIAL
SAME PRICE

J.N.MASTERS LTD RYE, SUSSEX.

INITIAL BROOCHES.
Neat Design.

IN CASE COMPLETE 6D. EXTRA.

A SPLENDID PRESENT.

J.N. MASTERS, LTD, RYE.

MASTERS' SUPERB MIZPAH BROOCH. Two hearts with spray of Ivy and Forget-Me-Not. Artistically engraved.

FOR THE WIFE
THE TWO RINGS 30/-

Solid Gold Engraved Keeper.
22-Ct. Gold Wedding Ring.

MASTERS' LUCKY WEDDING RING AND KEEPER
On Monthly Payments of 3s. Monthly to Railwaymen. A Complimentary present of Half-dozen good quality Teaspoons to each Cash Purchaser.

Hosts of Testimonials are shown in Masters' Catalogue. Write for it—
FREE

ELLEN

MASTERS' NAME BROOCHES.
Various designs any name, all same price. Engraved and finished in best style.
Silver, 2s.; Gold, 20s.
(HALL MARKED.)

MARY

Please send for Our **CATALOGUE** of Watches, Clocks, Alberts, Medals, Pendants, Wedding rings, Gem rings, Signet rings, Cutlery, Cruets, Electro-Plate, Carvers, Teapots, Field glasses, Pins, Brooches. Post free to any address

J.N. Masters, Ltd., Jewellers, RYE, Sussex.

PLATE 70C

Lever Pocket Watches – English

PLATE 71A

PLATE 71B

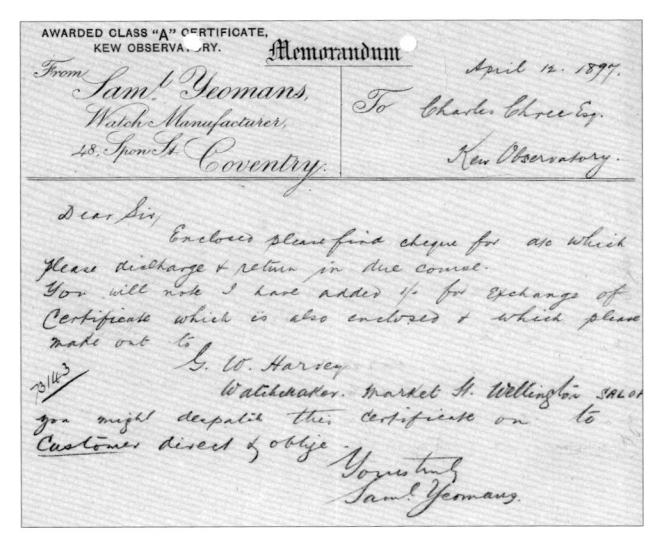

PLATE 71C

PLATE 71
W Potts & Sons, Leeds & Newcastle

Open face double bottomed silver case hallmarked Chester **1897**. Casemaker's stamp SY (Samuel Yeomans, 48 Spon Street, Coventry). Dustcap stamped RCH. White enamel dial W POTTS & SONS LEEDS AND NEWCASTLE, black roman numerals, gold spade hands, inset sunk seconds dial. Key wound and set. Full plate. Signed on top plate W Potts & Sons LEEDS AND NEWCASTLE UPON TYNE 75944. Going barrel. English lever 7 jewelled movement with pointed tooth scape wheel and tangential pallets. Oversprung cut bimetallic balance with blued steel balance spring. Regulator index pointer on triangular cock visible through pierced dust cover.

Key size 7. Diameter 53mm

Certificates were issued by Kew Observatory for watches that performed within certain specified rates at various degrees of temperature and various positions. Class A certificates were for watches with a high rated performance over five days for eight periods. They were held in high regard. Yeomans submitted watches made (finished) by him at 48 Spon Street for trial to the Observatory and his name often appeared in the published lists of successful applicants. The attached letter indicates that he had sold a watch subsequent to it receiving a Certificate to a retailer (G W Harvey, Market Street, Wellington, Salop) who now wished to have the certificate reissued in his name – no comment! The fee charged by the Observatory for this service was 1/-.

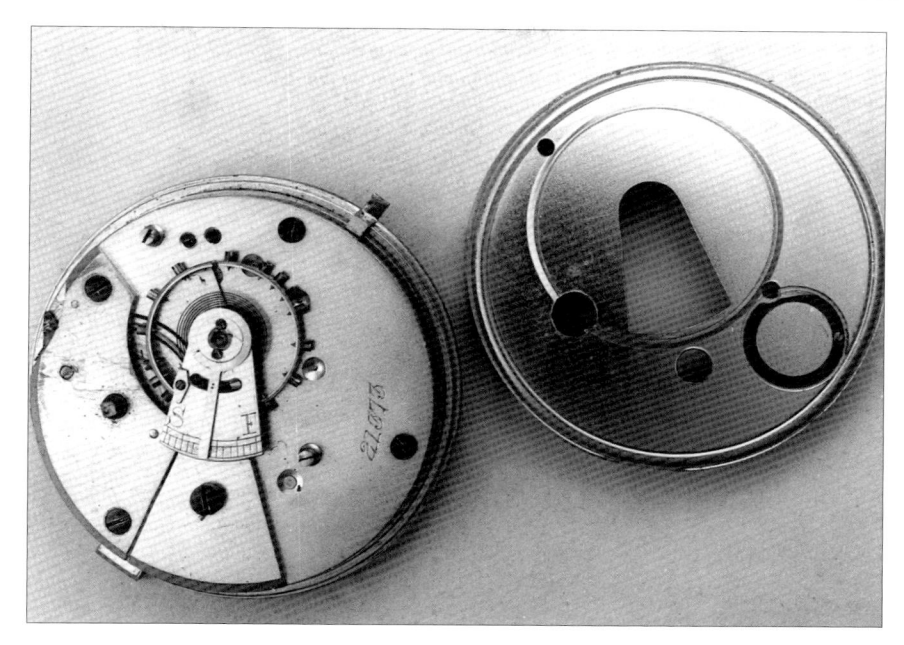

PLATE 72

PLATE 72
Charles Hutton Errington, 8 Spon Street, Coventry

Silver open face double bottomed engine-turned case hallmarked Birmingham **1897.** Casemaker CHE (Charles Hutton Errington, first registered Birmingham Assay Office 14 Dec 1886). Dustcap secured by small circular slide and stamped inside CHE PATENT 235. White enamel dial, black roman numerals, gold spade hands, inset sunk seconds dial. Key wound and set. Full plate. Top plate engraved 21373. 7 jewel going barrel movement with dummy wheel. Pointed tooth escapement with tangential pallets. Oversprung cut bimetallic balance with blued balance spring. Index regulator on balance cock with the unusual feature of the index curb passing through the cock.
Key size 7. Diameter 51.7mm

Charles Hutton Errington had been trained originally as an engraver but subsequently (around 1875) started a watch manufacturing concern finishing watches for the trade – Errington Watch Company. He had been highly successful in the production of a cheap watch. He took out several patents. One in 1881 (No 1433) was for a method of setting hands from the pendant and (No 5636) was for a centre second stop watch. Another in 1891 (No 6617) was for a device for raising the barrel ratchet click and readily letting down the mainspring. No 10356 in 1892 was for a watch bolt and spring in one piece while No 18766 in the same year described an arrangement to prevent the barrel ratchet when removed from damaging the centre wheel. His company was taken over by Messrs Williamson in 1895.

Lever Pocket Watches – English

PLATE 73A

PLATE 73B

**PLATE 73
Anon**

Lady's silver open face double bottomed fob watch, case and pendant hallmarked **1898.** Case stamped JR within a diamond (? John Rotherham ? James Richardson ? James Reid, all of Coventry). White enamel dial, black roman numerals, blued steel spade hands. Dust ring to full plate going barrel movement numbered 25157. 7 jewels. Pointed tooth scape wheel with tangential pallets. Gold three arm undersprung balance. Plain triangular balance cock with Bosley type regulator.
Diameter 41.5mm

**PLATE 74
D Kissenisky, Edinburgh**

Silver case (finely decorative – Scottish thistle around gold wreath) hallmarked Birmingham **1898.** Casemaker's mark

PLATE 74A

PLATE 74B

Lever Pocket Watches – English

PLATE 74C

WGM. White enamel dial signed D KISSENISKY EDINBURGH, black roman numerals, blued steel spade hands, inset sunk seconds dial. Key wound and hand set. Blued steel spring. Full plate fusee English lever. Top plate signed D Kissenisky EDINBURGH 8001. 11 jewels. Bimetallic and oversprung balance. Regulator index over balance to silver index plate at side of balance.
Key size 8. Diameter 56.5mm

PLATE 75
Lowe & Sons, 6 Bridge Street Row, Chester

Lady's silver open faced engraved double bottomed case hallmarked Chester **1899.** Casemaker CH (Charles Harris, 6 Norfolk Street, Coventry). Dust ring secured by two oval headed screws. White enamel dial, narrow black roman numerals, gold spade hands, inset sunk seconds dial. Key wound and set. Full plate. Going barrel. 7 jewels English lever movement. Pointed tooth scape wheel with tangential pallets. Undersprung bimetallic balance with blued steel balance spring and Bosley type regulator. Index scale engraved on top plate. Engraved scrolls on balance cock. Top plate engraved Lowe & Sons Chester 33019.
Key size 2. Diameter 40mm

Lowe & Sons were a famous firm – the proprietor was F Lowe who died at the age of seventy-four. His brothers GB and James F, Assay Masters, both died in 1911. Chester Clocks and Clockmakers by Nicholas Moore.

PLATE 75A

PLATE 75B

129

Lever Pocket Watches – English

PLATE 76B

PLATE 76
Joseph Harris & Sons, 4 St Ann's Square, Manchester

Silver open face double bottomed case with milled edge to body hallmarked Chester **1899**. Casemaker's mark JH & S. Dustcap. Pendant hallmarked Birmingham **1899** CH (Charles Harrold). White enamel dial, black roman numerals, gold spade hands, inset sunk seconds dial. Key wound and set. Full plate. 9 jewel. English lever fusee movement with maintaining power. Pointed tooth scape wheel with tangential pallets. Undersprung three arm gold balance with blue steel balance spring. Bosley type regulator with the index scale engraved on top plate SLOW FAST. Engraved balance cock of scroll design. Top plate engraved J Harris & Sons LONDON AND MANCHESTER 13523.
Key size 8. Diameter 52.6mm

P O Directory 1887 lists Joseph Harris & Sons at 104 Corporation Street, Manchester and the works at Coventry with the P O Directory 1897 giving 4 St Ann's Sq, Manchester.

PLATE 76A

PLATE 77

PLATE 77
Coventry Co-operative Watch Manufacturing Society

Silver open face double bottomed case hallmarked Chester **1899.** Casemaker's stamp W & F. Dustcap. White enamel dial, black roman numerals, gold spade hands, inset sunk seconds dial. Full plate going barrel. 7 jewels English lever movement with pointed tooth scape wheel and tangential pallets. Brass undersprung three arm balance with Bosley type regulator to graduations engraved on top plate. A triangular balance cock engraved with the monogram of a cross bearing the letters CCWMS. This was the trade mark of the Coventry Co-operative Watch Manufacturing Society. Top plate engraved 6/402. *Note the fractional numbering system on top plate.*
Diameter 53.2mm

These initials were originally tentatively decoded as William M Smith of 1 Craven Street, Coventry – an excellent example of drawing the wrong conclusion through insufficient data.

Lever Pocket Watches – English

PLATE 78A

PLATE 78B

PLATE 78C

PLATE 78
Rotherham & Sons, 27 Spon Street, Coventry

Silver open face engine-turned case with snap-on bezel, hinged back and dome, hallmarked Birmingham **1900.** Casemaker's mark JR (John Rotherham, first registered mark 20 Feb 1877). White enamel dial, black roman numerals, blued steel spade hands, inset sunk seconds dial. Keyless wound and set, push piece at 4 minutes past position. Three-quarter plate. Top plate engraved 145642 1A. 19 jewels English lever movement of high quality. Pointed tooth scape wheel and tangential pallets with oversprung and overcoil to cut bimetallic balance. Blued steel balance spring. Regulator index on balance cock engraved with index scale S F. *A very fine watch.*
Diameter 52.2mm

Samuel Vale founded the firm in 1747, the changing styles being Vale, Howden & Carr 1754-1790; Vale & Rotherham 1790-1840; Richard Kevitt Rotherham c.1840; Rotherham & Sons 1851-1912; Rotherhams & Sons 1912-recently From their early days they embarked upon a factory system although the first attempts were no more than efficient organisation of local labour. They introduced the use of machinery – some imported from the States – and it has been claimed they made the first English machine-made watch. They exhibited widely and in 1889 at the Paris Exhibition they were showing 'genuine English watches'. They owe their longevity to their diversification into cycle parts and precision engineering. There is an excellent account of a visit to their factory in the Horological Journal *for June 1887, and another in the* Watchmaker, Jeweller and Silversmith *for August 1897.*

PLATE 79A

PLATE 79
THE 'EXPRESS' ENGLISH LEVER
John George Graves (Retail Jeweller), Sheffield

Silver open face double bottom case hallmarked Chester **1901**. Casemaker's mark JGS. Dustcap with blued steel slide. White enamel dial signed THE 'EXPRESS' ENGLISH LEVER JG GRAVES SHEFFIELD, black roman numerals, gold spade hands, inset sunk seconds dial. Key wound. Full plate. Going barrel. English lever movement with reversing pinion and club tooth escapement. Bimetallic cut oversprung balance with compensation screws. The regulator index over the balance to engraved quadrant. Top plate engraved REVERSING PINION No 566733.
Key size 7. Diameter 52.5mm

JG Graves was born in Horncastle, Lincolnshire in 1866. He was apprenticed at the age of fourteen to W Wichman in Sheffield. He eventually started up on his own as a repairer while at the same time he travelled peddling watches. His business rapidly expanded. Realising that many people could not afford to purchase a watch outright he introduced one of the first deferred payment mail order businesses. The Lancashire Watch Company was his main supplier but, as he gradually added a wider range of

PLATE 79B

PLATE 79C

goods and eventually his mail order catalogues were household reading, imported watches and clocks were naturally included. He was renowned for his advertising campaigns – the most successful being the wide scale use of the back page of the Saturday edition of the London papers. A fascinating character – a man of his times – he was also a generous benefactor to many local causes.

Lever Pocket Watches – English

PLATE 80A

PLATE 80B

PLATE 81A

PLATE 81B

PLATE 82A

PLATE 82B

gold spade hands, inset sunk seconds dial. Key wound and set. Full plate. Top plate engraved Jas Reid & Sons WATCH MANUFACTURERS CHESTER ST COVENTRY 17738. Going barrel. English lever 7 jewel movement with pointed tooth scape wheel with tangential pallets. Undersprung cut bimetallic balance with blued steel balance spring with Bosley type regulator. Triangular engraved cock of foliate design.
Key size 11. Diameter 52mm

PLATE 80
THE 'EXPRESS' ENGLISH LEVER
JG Graves, Sheffield

Silver open face engine-turned double bottomed case hallmarked Chester **1905**. Casemaker's stamp JGG in shield with Js (J G Graves). Dustcap with blued spring slides. White enamel dial signed THE 'EXPRESS' ENGLISH LEVER JG Graves Sheffield, black roman numerals, gold spade hands and inset sunk seconds dial. Key wound and set. Full plate. 7 jewels English lever movement engraved on top plate JG Graves SHEFFIELD 774418 REVERSING PINION. Club tooth scape wheel with tangential pallets. Oversprung cut bimetallic balance with blued balance spring. The regulator index from engraved triangular balance cock over the balance rim to silver quadrant.
Key size 7. Diameter 54.2mm

PLATE 81
THE RENOWNED ENGLISH LEVER
James Reid, Chester Street, Coventry

Silver open face double bottomed case, gold hinges, milled edge to body, hallmarked Chester **1901**. Casemaker JR & Bs (J Reid & Bros, Chester Street, Coventry). Dust cover stamped W & AO (? Oxley Bros, 51 Spon End, Coventry). Pendant CH (Charles Harrold 2 & 3 St Pauls Square) hallmarked Birmingham **1901**. White enamel dial signed THE RENOWNED ENGLISH LEVER 17738 JAs REID & BROs CHESTER ST COVENTRY, black roman numerals,

PLATE 82
Fattorini & Sons, 21 Kirkgate & 27 Westgate, Bradford

Silver open face double bottomed engine-turned case hallmarked Chester **1901**. Dustcap. White enamel dial inscribed FATTORINI & SONS BRADFORD, black roman numerals, gold spade hands, inset sunk seconds dial. Key wound and set. Full plate. English lever movement with going barrel and reversing pinion inscribed on the top plate Fattorini & Sons BRADFORD REVERSING PINION No 487742. 7 jewels. Club tooth scape with tangential pallets. Oversprung cut bimetallic balance with blued steel balance spring. The index over the balance rim to a silver quadrant. Movement probably manufactured by Lancashire Watch Co, Prescot.
Key size 8. Diameter 52.2mm

Founder of the dynasty was Antonio Fattorini who settled in England after the Battle of Waterloo in 1815. By 1831, however, two separate businesses had emerged with one son settling in Kirkdale, Bradford and the other (Thomas F) in Skipton, Leeds. Antomio Fattorini was born in Italy in 1797. He established jewellers' shops in Leeds and Harrogate. Opened in Kirkgate, Bradford, trading as A. Fattorini & Sons but following a family split c.1859 became Fattorini & Sons with a further change in 1909 to Fattorini & Sons Ltd. Advertised in 1884 as 'The cheapest house in the trade for American Waltham Watches…'. Thomas Fattorini & Sons was founded in 1859 and is recorded as working in Skipton under this title in 1902.

Lever Pocket Watches – English

PLATE 83A

PLATE 83
Rotherham & Sons, 27 Spon Street, Coventry

Silver open face double bottomed engine-turned case hallmarked Birmingham **1903**. Casemaker JR. Dustcap stamped inside 13668 PAT 7204. White enamel dial with trade mark (eight pointed star), black roman numerals, gold spade hands, inset sunk seconds dial. Key wound and set. Full plate. 1 jewel English lever. Going barrel movement with pointed tooth scape wheel and tangential pallets. Oversprung cut bimetallic balance with blued steel balance spring. Regulator index over the balance rim from a plain triangular cock to an engraved silver quadrant. Top plate engraved 213668.
Key size 7. Diameter 53.4mm

PLATE 84
GWR (Great Western Railway), Rotherham & Sons Ltd
27 Spon Street, Coventry
1 Holborn Circus, London EC1

Heavy nickel case open face case, snap-on bezel, hinged back and dome. White enamel dial signed GWR, broad black roman numerals and broad blue spade hands, inset sunk seconds dial. Keyless wound, push piece to set hands at 3½ minute past position. Full plate. Going barrel. 13 jewel English lever movement with tangential pallets, endstones to 4th wheel arbor and balance staff.

PLATE 83B

Oversprung cut bimetallic balance with blued steel balance spring. Index from plain triangular balance cock over balance rim to silver quadrant. Top plate engraved Rotherhams LONDON 216295. Great Western Railway Watch No 1640 engraved on back cover.
Diameter 57.7mm

PLATE 85
John Hawley & Sons, Reliable Works, 35, 36 and 37 Cow Lane, Coventry

Gold 9 ct open face engraved double bottomed case hallmarked Birmingham **1904**. Dustcap engraved John Hawley Coventry 65207. Pendant hallmarked 9.375, maker CH (Charles Harrold 2 & 3 St Pauls Square, Birmingham). Gilt engine-turned dial engraved John Hawley Coventry over trade mark of conifer, black roman numerals and blued steel spade hands, inset seconds dial. Key wound and set. Full plate. 1 jewel English lever movement with fusee and chain, maintaining power and pointed tooth scape wheel and tangential pallets. Oversprung cut bimetallic balance with compensation screws. Blued steel balance spring and regulator index and graduation incorporating SLOW FAST with Coventry star between visible through pierced dust cover. Top plate engraved Gold and Silver Medals Awarded 65207 Established 1857.
Key size 9. Diameter 54.9mm

The firm of John Hawley & Son was founded in 1857 by John Hawley senior. His son joined him in 1879. By 1900 they were working in a three storey building (Reliable Works) in Cow Lane, Coventry. As well as being efficiently equipped with automatic machinery for the making of the various parts, they also claimed to

Lever Pocket Watches – English

PLATE 84A

PLATE 84B

PLATE 85A

PLATE 85B

137

Lever Pocket Watches – English

PLATE 85C

make their own gold and silver cases. They were one of the first makers of movements with going barrels. One of their specialities was the Collier's Friend, a movement adapted for rough wear and with a good sturdy case. They exhibited in the 1891-2 Tasmania Exhibition and supplied both home and overseas markets.

PLATE 86
STANDARD TIMEKEEPER
James W Morris & Co, 5 Market Street, Faversham, Kent

Silver open faced engine-turned double bottomed case hallmarked Chester **1904.** Casemaker CHE (Charles Hutton Errington, Holyhead Road, Coventry). Dustcap. Pendant hallmarked Birmingham **1900**, maker CH (Charles Harrold, 2 & 3 St Pauls Square, Birmingham). White enamel dial signed J W MORRIS & CO STANDARD TIMEKEEPER FAVERSHAM, broad black roman numerals, broad spade blued steel hands. Key wound and set. Full plate. 9 jewel English lever movement. No pillars. Going barrel with club tooth scape wheel and tangential pallets with brass body. Oversprung cut bimetallic balance with compensation screws. Blued steel balance spring. Regulator index and scale on plain triangular cock visible through pierced dustcap. Top plate signed J W Morris & Co FAVERSHAM 10519.

Key size 8. Diameter 53.5mm

PLATE 86B

PLATE 86A

138

Lever Pocket Watches – English

PLATE 87A

PLATE 87B

PLATE 87
THE FIELD WATCH
James William Benson, 62 & 64 Ludgate Hill, London

18ct gold-demi hunter case hallmarked London **1906**. Casemaker's stamp JWB. Patent dust ring. Pendant stamped PW (?Philip Woodman & Sons). White enamel dial signed J W Benson London upper centre, black roman numerals, blued steel double spade hands, inset sunk seconds dial. Keyless wound with push piece at 3 o'clock position to set hands. Gilt three-quarter plate. English lever movement. Top plate engraved THE FIELD WATCH BEST LONDON MAKE BY WARRANT TO HM THE LATE QUEEN VICTORIA J W Benson 62 & 64 Ludgate Hill No 111609. 15 jewel movement with pointed teeth to scape wheel with tangential pallets. Bimetallic cut compensated balance with oversprung blued balance spring with Breguet overcoil. Polished index on cock engraved F S with Coventry star between.
Diameter 52mm

Lever Pocket Watches – English

BENSON'S "FIELD" WATCH

In 18-ct. Gold Cases, £25

In Silver Cases, £15

BEST LONDON MADE HIGH CLASS WATCH
at a Moderate Price.   Half-Chronometer.   Breguet Sprung.

In Hunting, Half-Hunting, or Crystal Glass 18-ct. Gold Cases, £25; or Silver Cases, £15.

Hundreds of Testimonials from wearers in all parts of the World.

NOTE.—This and all other **WATCHES, CLOCKS, CHAINS, JEWELS, PLATE, FITTED BAGS, &c.,** can be obtained on "*THE TIMES*" System of Purchase by **MONTHLY PAYMENTS.**

GUIDE BOOK for Purchasers sent post free.

J. W. BENSON, LTD.,
STEAM FACTORY:
LUDGATE HILL, LONDON, E.C.

PLATE 87C

---

524 — THE GRAPHIC — OCTOBER 26, 1889

BENSON'S "BANK" SILVER CASES £5 / £5

BENSON'S WATCHES
Guaranteed for Strength, Accuracy, Durability, and Value.

BENSON'S LADY'S KEYLESS LEVER WATCH.

SILVER CASES Is fitted with a ¾-Plate LEVER Movement, Compound Balance, Jewelled throughout, and Strong KEYLESS Action.

The Cases are of 18-Carat Gold, Strong, and Well Made, either Hunting, Half-Hunting, or Crystal Glass, Richly Engraved all over, or Plain Polished, with Monogram Engraved Free.

PRICE £10.   Or in Silver Cases, £5.

LADY'S GOLD ALBERT CHAINS TO MATCH, FROM £1 15s.

These Watches sent Free and Safe, at our risk, to all parts of the World, on Receipt of Draft, Cash, or Post Office Order, Payable at General Post Office.

For further Particulars see Benson's New ILLUSTRATED BOOK containing Illustrations of Watches from £2. 10s. to £500, Jewellery, Clocks, Presentation and Domestic Plate. The Largest and most Complete Catalogue published, Free on Application to

THE STEAM FACTORY—
J. W. BENSON, 62 & 64, LUDGATE HILL, E.C.;
and at 28, ROYAL EXCHANGE, E.C.; and 25, OLD BOND STREET, W.

SILVER KEYLESS ENGLISH LEVER WATCH
THREE-QUARTER PLATE Movement, Compensation Balance, Jewelled in Rubies, in Strong, Sterling Silver, Crystal Glass Cases, £5.

BENSON'S "FIELD." SILVER CASES £15 / GOLD CASES £25 / £10

ENGLISH LEVER HALF-CHRONOMETER
Best London make, for Rough Wear, with Breguet Spring to prevent variation when worn on horseback, &c. Specially adapted for Hunting Men, Colonists, Travellers, and Soldiers; from whom HUNDREDS of TESTIMONIALS have been received. In Hunting, Half-Hunting, or Crystal Glass Cases, 18-carat Gold, £25; or Silver, £15.

PLATE 87D

Plate 88
Dent, 61 Strand and 4 Royal Exchange, London

Silver open face case, snap-on bezel, hinged back and dome, hallmarked London **1911**. Casemaker's stamp JW & Co (Joseph Walton & Co, 25/26 Spencer Street, Clerkenwell). White enamel dial with Dent trade mark (DENT within triangle with LONDON below), black roman numerals, blued steel spade hands, inset sunk seconds dial. Keyless wound and push piece at 4 minutes past position to set. Three-quarter plate. 19 jewels. Fine English lever movement with pointed tooth scape wheel and tangential pallets. Bimetallic cut compensated balance oversprung with Breguet overcoil of blued steel balance spring. Small triangular cock with index scale and polished index finger. Top plate engraved with trade mark Watchmaker to the King 61 Strand and 4 Royal Exchange No 57959.
Diameter 50.8mm

Edward John Dent was a renowned marine chronometer maker and subsequently won the contract for the making of the Great Clock of Westminster ('Big Ben'). The triangular trade mark was obtained in 1876 and was applied to all E Dent & Co watches, clocks etc. Dents moved to 4 Royal Exchange from 34/35 Royal Exchange December 1884. The Dent factory was at 4 Hanway Place, London. This could be a Rotherham movement. Edward John Dent and His Successors by Dr Vaudrey Mercer.

Plate 88a

Plate 88b

Lever Pocket Watches – English

PLATE 89A

PLATE 89B

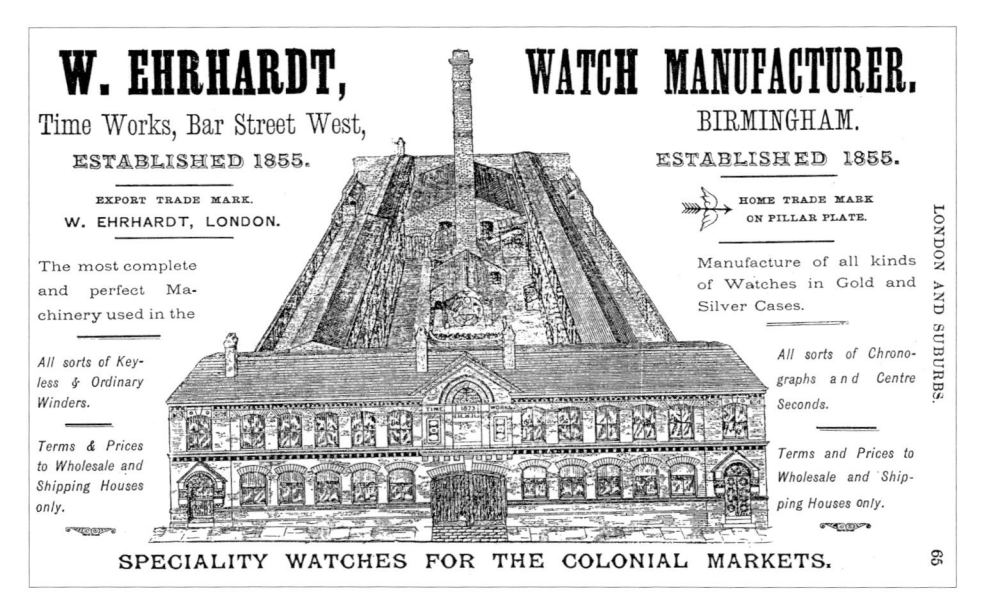

PLATE 89C. *Kelly's Directory,* 1880

PLATE 89
William Ehrhardt, Time Works, Barr Street, Birmingham

Open face Stylic silver case with hinged bottom and bezel, hallmarked Chester **1920**. Casemaker's mark SMCo Ltd.

White enamel dial signed W Ehrhardt LONDON with arabic luminous numerals and fenestrated luminous spade hands, inset sunk seconds dial. Keyless wound with push piece at 4 minutes past position to set hands. Full plate. Going barrel. 7 jewel movement. Unusual feature of movement is balance cock on rotating platform for

PLATE 89D

adjustment of poise of bimetallic cut compensated balance with Coventry star mark. Top plate engraved D 2149.
Diameter 50.2mm

The firm Smith, Ewen & Stylic Ltd, 42 Frederick Street, Birmingham still appear in the Watch and Clock Yearbook *for 1958. William Ehrhardt was born in 1831 in Dargun, Mecklenburg-Schwerin, Germany. At the age of fifteen he was apprenticed to a watchmaker in the local town and at nineteen he came to London to the Great Exhibition of 1851, a visit that was to give him inspiration and ideas for the future. He was subsequently employed at Upjohn and Brights in King William Street, just off the Strand, and attended the Royal School of Mines in Jermyn Street. Not sure where to settle he visited Coventry and Birmingham. Although he wished to settle in an area to his liking, he also recognised the fact that he was unlikely to succeed with a new manufactory in Coventry where the industry was well established. He went in 1855 to Birmingham where he married the following year. He was now able to start a*

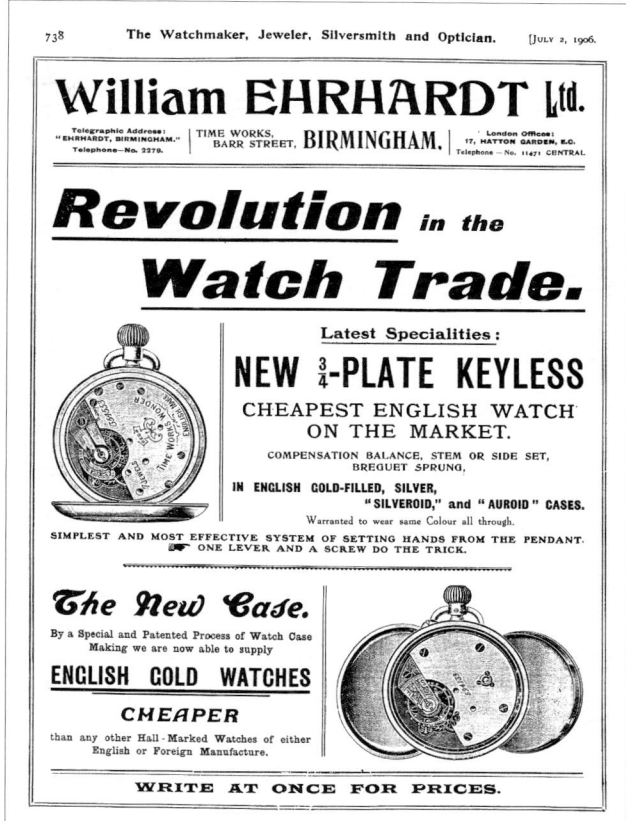

PLATE 89E

small workshop at 30 Paradise Street, Birmingham where he made use of Lancashire movements and outworkers. By 1862 he was in new premises at 26 Augusta Street, Birmingham and exhibiting 'Various kinds of watches and instruments with the manufacture thereof' – Class XV, North East Gallery, No 3247 at the International Inventions Exhibition in London. Until 1865 he experienced difficulty in obtaining skilled employees but this was the year that the American Civil War came to an end and he was then able to move to 72 Gt Hampden Street and install more machinery. Rapid development enabled him to build the Time Works in 1873 and to open at Barr Street, Hockley, Birmingham the following year. William Ehrhardt claimed to be the first successful manufacturer of machine-made watches in England. By 1880 the firm was producing 500 watches per week – some by machinery and some by hand (ref Ironmongers 15 March 1880). He died on 20 November 1897 leaving a widow, six sons and two daughters. The two eldest sons continued his business with William Ehrhardt junior becoming the director of the new limited company. In 1900 a further company was registered, The British Watch Materials Manufacturing Co Ltd.

Lever Pocket Watches – English

PLATE 90A

PLATE 90
H Williamson Ltd, London

Nickel open face case Dennison No 150747, screw back and bezel. Black enamel dial H WILLIAMSON LTD LONDON 98664F, luminous arabic numerals and spade hands, inset seconds dial with smaller arabic numerals at 3 o'clock position. Keyless wound, pull to set. Top plate inscribed 7 JEWELS WARRANTED ENGLISH 302027. Going barrel. Straight line lever. Club tooth scape wheel. Cut bimetallic balance with blued steel oversprung and overcoil balance spring. Regulator index on balance cock. *This watch was issued for military use.*

Diameter 58mm

H. Williamson Ltd 1892-1931. This was originally a wholesale business (jet jewellery). In 1895 they absorbed the Errington Watch Co of Coventry and in 1899 they purchased the business of Suter & Co Swiss (Buren) watchmakers, as well as the watch factory of Albert Mondandon at Chaux de Fonds. They were accused of contravening the Merchandise Marks Act by utilising Swiss-made parts in their 'English made' watches. They lost the case, details of which make feisty reading in the Horological Journals between 1899 and 1890! Three of their trade marks were the Acme, the Sphinx (first used 1895) and the Astral (first introduced 1910). A detailed account of the watches made by the company as well as their chequered history can be found in Watches 1850-1980 by M Cutmore.

PLATE 90B

Lever Pocket Watches — English

PLATE 91A

PLATE 91B

PLATE 91
Anon

Silver open face engine-turned case, hinged dome and bottom with hinged bezel, hallmarked Birmingham 1920. Casemaker's stamp ALD (Dennison Watch Case Co Ltd) No 397318. White enamel dial, black roman numerals, broad spade blued hands. Keyless movement 300752 stamped with S within five pointed star together with 7 JEWELS WARRANTED ENGLISH. Half plate. Club tooth scape wheel. Straight line pallets. Compensated cut oversprung overcoil bimetallic balance.
Diameter 50.7mm

Corresponds to 16 size material by Williamson/Errington in Hirst Fourth Wide Awake Catalogue p273.

PLATE 92
George Hutchinson, High Street, Clapham, London

Silver open face case, snap-on bezel and hinged dome, hallmarked London **1922**. Casemaker's stamp FT (F Thoms, 25 Spencer Street, Clerkenwell). White enamel dial, black roman numerals, blued steel spade hands, inset seconds at 3 o'clock position. Keyless wound and set with push piece at 4 minutes past position. Three-quarter plate. 13 jewels. English lever movement with top plate engraved G Hutchinson Clapham No 1277. Pointed tooth scape

PLATE 92

wheel with tangential pallets. Oversprung cut bimetallic balance with blued steel balance spring. The regulator index on finely engraved triangular balance cock.
Diameter 52.2mm

George Hutchinson, High Street, Clapham listed as watch, clock and chronometer maker in Britten 9th ed.

145

Lever Pocket Watches – English

PLATE 93

PLATE 93
British Watch Co Ltd

Silver open face case by Dennison Watch Case Co Ltd stamped ALD 435998/D, hallmarked Birmingham **1923**. White enamel dial, short black roman numerals with blued steel spade hands, inset sunk seconds dial. Keyless. Three-quarter plate. 7 jewels. English lever top plate engraved Safety Pinion 803801 BRITISH WATCH CO LTD LONDON. Club tooth scape wheel with straight line lever escapement. Bimetallic cut compensation balance with Breguet overcoil. Oversprung blued balance spring with triangular balance spring stud.
Diameter 49.4mm

'We have been asked by several readers for particulars as to the British Watch Company, Limited, of St. John's-square, Clerkenwell, E.C. The Company was registered on the 30th of November, with a capital of £2,000, in £1 shares, with the object of acquiring the business carried on by B. Nichols, J.C. Sellars, and I.B. Hogan as the British Watch Company. The directors are B. Nichols, J.C. Sellers, E.H. Funkley, and Ethel M. Whibley. Their qualification is one share, and their remuneration £5 extra for his services. The registered office of the Company is 2, 3, and 4, St. John's-square, E.C. Truth has been making some very scathing remarks concerning this company, and its sending out circulars with signatures of the persons to who its literature is directed, proofs of which have also come into our possession. In one particular case, in Norfolk, considerable apprehension has been expressed by a small farmer as to how these people could possible have got hold of his signature. Doubtless we shall hear more about the British Watch Company Limited.' The Watchmaker, Jeweller, Silversmith and Optician, *1 February 1906* . M Cutmore in Watches 1850-1980 *comments on a firm of identical name listed in Kelly's 1921 Directory at Ehrhardt's London address!*

PLATE 94A

PLATE 94B

PLATE 94
THE BANK WATCH
James W Benson, 62 & 64 Ludgate Hill, London

Silver open face engine-turned case, snap-on bezel, hinged back and dome, hallmarked London 1923. Casemaker's stamp JWB, pendant CH (Charles Harrold, 2 & 3 St Paul's

146

Square, Birmingham). White enamel dial signed J W Benson London, black roman numerals, gold spade hands, inset sunk seconds dial. Keyless wound and set, push piece at 4 minute past position. Three-quarter plate. 9 jewels. English lever. Going barrel movement. Pointed tooth scape wheel with tangential pallets. Cut bimetallic oversprung balance with gold compensation screws. Blued steel balance spring. Polished index on small triangular cock engraved S F with Coventry star between. Top plate engraved THE BANK WATCH BEST LONDON MAKE NO 1098613 JW Benson 62 & 64 Ludgate Hill LONDON.
Diameter 51mm

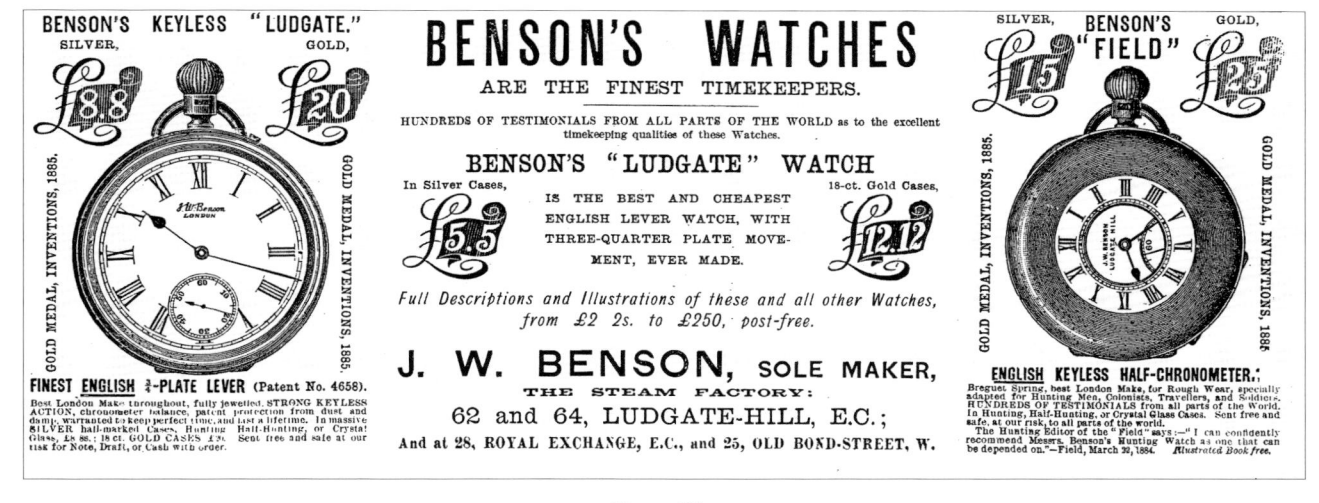

PLATE 94C

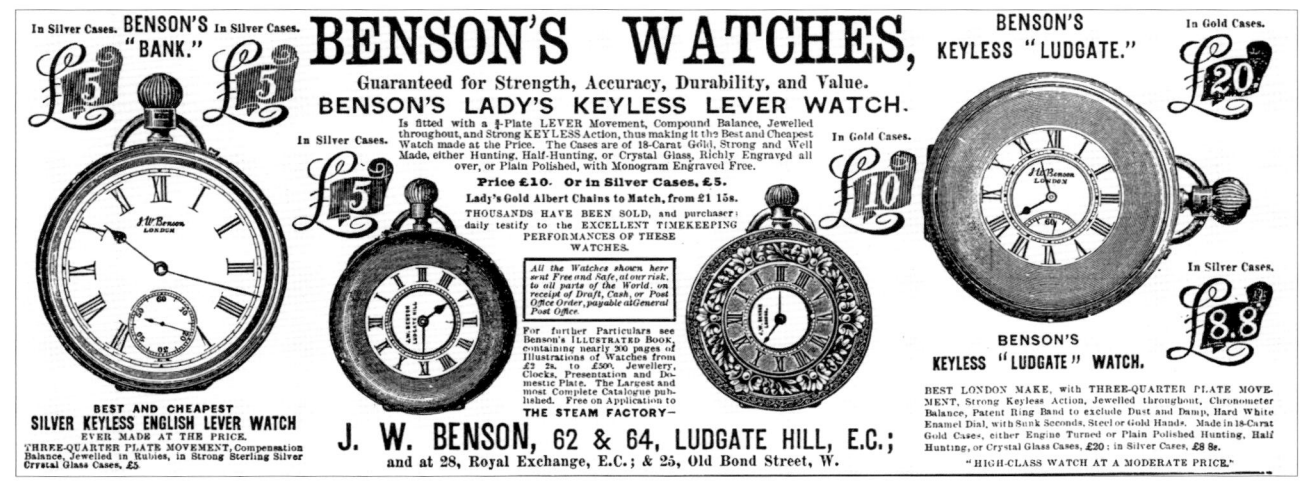

PLATE 94D

PLATE 95

PLATE 95
George & Richard Beesley, 56 Great Cross Hall Street and 4-6 Boundary Street, Liverpool

White enamel dial, upper centre dial signed R & G Beesley Liverpool, black narrow roman numerals, gilt spade hands, inset sunk seconds dial. Unusual keyless wound and set. Winding square for key winding still retained on fusee arbor. Lever to select hand setting is at 5 o'clock position. English fusee lever movement with maintaining power and Liverpool jewelling to train including fusee arbor. Cylindrical pillars. Bimetallic cut compensated and oversprung balance with compensation screws. Single roller. Index from balance cock over balance to quadrant on top plate. Dustcap stamped SW and has unusual feature of a collet screwed to cover around winding hole. It would have been possible thereby to fit a sealing cap to the old winding hole as an option. Barrel click and ratchet wheel and engraved balance cock visible through dust cover. *Appears to be an experimental model.*

THE CENTRE SECONDS CHRONOGRAPH

An interesting anomaly amongst the English lever watches is the so-called Centre Seconds Chronograph These were made almost exclusively in Coventry and Prescot around the turn of the last century, their design and solidarity appealing to the successful provincial Victorian businessman or doctor on whom the expansion of mechanisation generally, the growth of the railroad and an increasing interest in sporting events had no doubt had their influence. They also wished their timepieces to display the latest technology.

The description (chronograph) appearing on the dial, however, does cause confusion. These watches have little in common with the true chronograph, whether it be of English or Swiss origin. These were much more sophisticated mechanisms which had developed following the invention in 1862 of the resetting heart piece by Adolphe Nicole in 1862 (to be described in a separate chapter). The pocket watches under discussion here were good quality, heavy, large three-quarter plate English fusee lever watches with 18,000 trains and invariably key wound. These were superseded towards the end of the nineteenth century by keyless movements with going barrels, a substantial watch in a silver or gold case. The silver cases often had gold hinges. The large white enamel dial had a sunk centre in order to provide extra clearance for the counterpoised centre seconds hand with the seconds chapter ring finely marked 0-300, thus indicating ⅕ second intervals. Time measurement was controlled by a stop/start slide in the band of the case at the 2 o'clock position. This controlled a wire brake acting directly on the roller of the balance, thus stopping and starting the whole mechanism of the watch.

Lever Pocket Watches – English

PLATE 96A

**PLATE 96
John Matheson & Co, 47 & 48 Shore, Leith, Edinburgh**

Silver open face case, hinged bezel back and cuvette, hallmarked London **1885.** White enamel dial inscribed MATHESON & CO LEITH CHRONOMETER MAKER TO HRH THE DUKE OF EDINBURGH & Royal Navy and the Royal Crest Dieu et Mon Droit, black roman numerals, blued steel spade hands and centre sweep seconds hand. Dial calibrated ⅕ seconds (0-300). Keyless wound and push piece to set at 3½ minute past position. Three-quarter plate. 13 jewel. Going barrel. Coventry movement with pointed tooth scape and tangential lever. Oversprung cut compensated balance with regulator index on balance cock engraved FAST SLOW with Coventry star. Top plate engraved Matheson & Co Leith Chronometer Maker to HRH Duke of Edinburgh & Royal Navy 71396. *(NB. The Duke of Edinburgh was later to become King Edward VII.)*
Diameter 53.5mm

PLATE 96B

Lever Pocket Watches – English

PLATE 97A

PLATE 97
Joseph Hargreaves & Co, 17 Norton Street, London Road, Liverpool

Silver open faced case, hinged back and cuvette, hallmarked Chester **1890.** Casemaker's stamp J H & Co. White enamel dial with sunk centre inscribed MAKERS TO THE QUEEN AND HRH THE PRINCE OF WALES J HARGREAVES & CO LIVERPOOL, narrow black roman numerals, gold spade hands. Blued steel centre seconds and ⅕ second markings on dial (0-300). Keyless winding with push piece to set at 3½ minute position. Top plate engraved J Hargreaves & Co Liverpool 51818. Three-quarter plate. Going barrel. 15 jewel pointed tooth and tangential lever movement with oversprung overcoil cut bimetallic balance and blued steel balance spring. Regulator index on balance cock engraved FAST SLOW.
Diameter 54.6mm

PO Directory England, Scotland and Wales 1892 lists Joseph Hargreaves at this address as a watch manufacturer.

PLATE 97B

Lever Pocket Watches – English

PLATE 98A

PLATE 98
Coventry Co-operative Watchmakers Society, 32 Bishop Street, Coventry

Silver open face double bottomed case hallmarked Chester **1900**. Casemaker's stamp W & F, pendant maker CH (Charles Harrold). White enamel dial with sunk centre and signed CENTRE SECONDS CHRONOGRAPH and ⅕ second markings (0-300), black roman numerals, gold spade hands, blued steel spade centre seconds. Key wound and set at back. Three-quarter plate. 15 jewel movement engraved on the top plate 55354 and trade mark CCWMS within stellar cross. Pointed tooth scape and tangential lever with oversprung cut bimetallic balance and blued steel balance spring. Regulator index on cock engraved SLOW FAST. Gold slide at 2 o'clock position.
Key size 8. Diameter 55.5mm

Trademark Journal Application No 9619 CCWMS on Stellar Cross Horological Journal *September 1883 p9.*

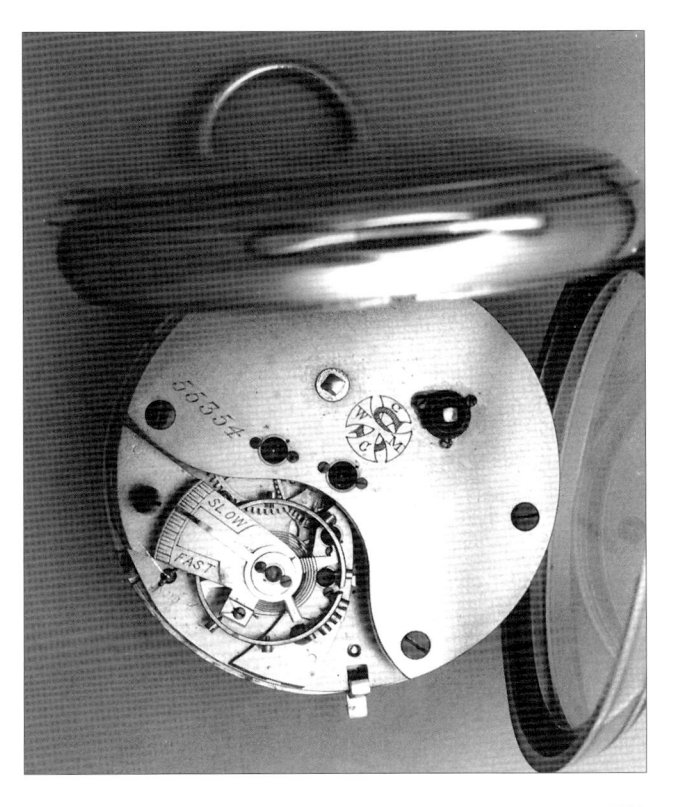

PLATE 98B

151

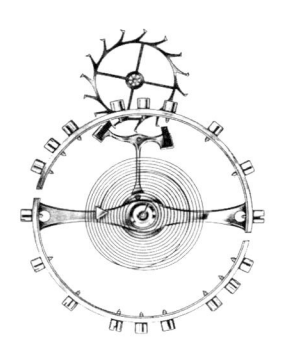

PLATE 1. Swiss lever escapement

Part 2 – Swiss

PLATE 2
Lezard & Sons, Viaduct Chambers,
36 Holborn Viaduct, London EC

Silver open face case hallmarked Birmingham **1883.** Snap-on bezel, hinged back and cuvette. Case mark JL/EL (Joseph Lezard & Edward Lezard registered 21 January 1878), repeated on pendant and bow. White enamel dial, black roman numerals with arabic at five minute intervals outside chapter ring, blued steel spade hands, inset seconds. Keyless wound and push piece to set. Top plate stamped L & S with lizard facing right 215380. 13 jewel half plate club tooth and straight line lever movement. Oversprung uncut bimetallic balance and blued steel balance spring. Regulator index on balance cock engraved FAST SLOW.
Diameter 50.8mm

Joseph Lezard and Edward Lezard traded as Lezard Bros at the above address as watchmakers and importers of watches. J Lezard retired 15 August 1887 with EJ Lezard declared bankrupt in 1888.

PLATE 2A

PLATE 2B

152

PLATE 3B

PLATE 3A

PLATE 3C. *Hirst Brothers Catalogue,* c.1916

PLATE 3
THE LONDON LEVER, Kendal & Dent, 106 Cheapside, London EC

Silver open face case, hinged bezel, back and cuvette, imported hallmark London **1910.** Casemaker WHS (William Henry Sparrow). Dustcap with slide with separate aperture for balance cock, winding and setting. White enamel dial signed KENDAL & DENT MAKERS TO THE ADMIRALTY THE LONDON LEVER SWISS MADE, black roman numerals, blued steel cathedral hands, inset sunk seconds. Key wound and set from back. 11 jewel three-quarter plate. Club tooth scape wheel with straight line pallet with oversprung uncut bimetallic balance. Index on engraved balance cock.

Key size 6. Diameter 51.3mm

See pages 116-117 for further details on Kendal & Dent.

Lever Pocket Watches – Swiss

PLATE 4A

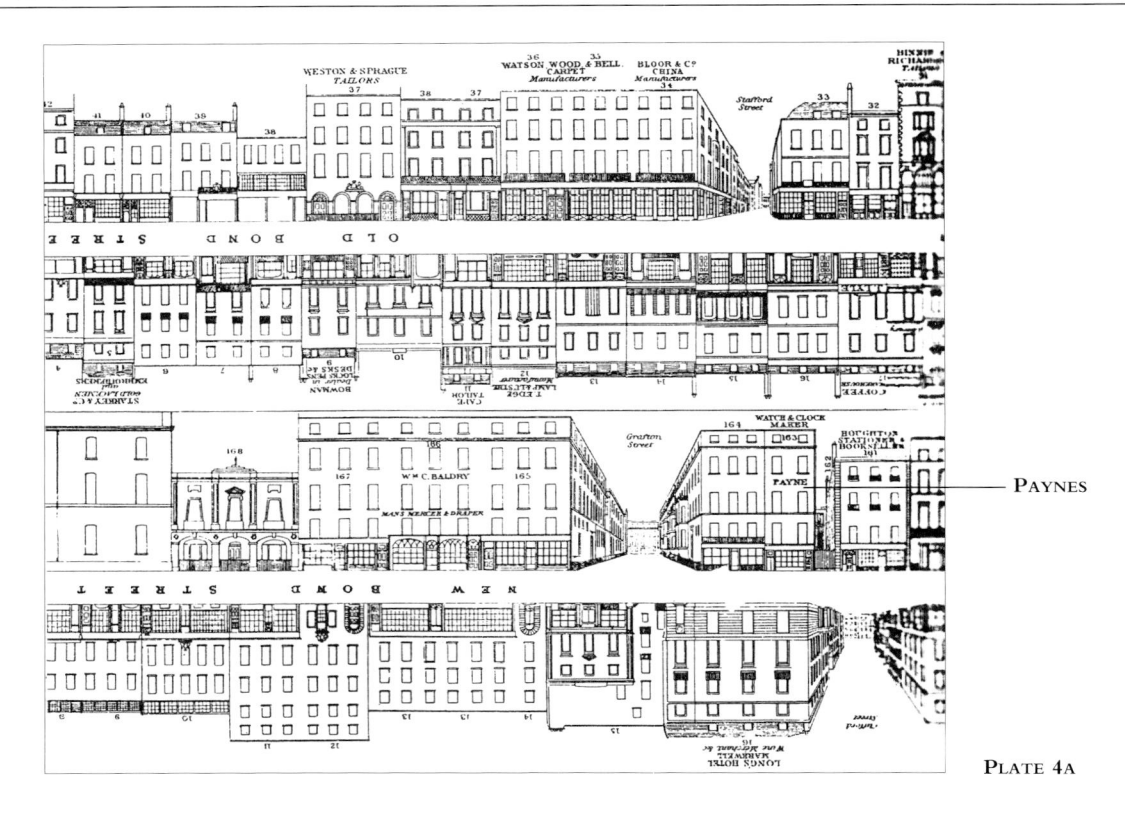

PLATE 4
Payne & Co, 163 New Bond Street, London

Lady's silver (.925) open face fob watch with hinged back and cuvette engraved Payne & Co 163 New Bond St London No 51448. White enamel dial, black roman numerals, blued steel narrow spade hands. Key wound and set. Mercurial gilt Swiss bar movement with finely blued steel screws. 13 jewels. Pointed tooth scape wheel and gold oversprung balance.

Key size 3. Diameter 42mm

PO Directory indicates that the firm William Payne became & Co between 1880 and 1890. However, JB Hawkins in Thomas Cole and Victorian Clockmaking quotes an entry for 1851 as being Wm Payne Watchmaker, 163 New Bond Street with the following year Wm Payne & Co Watch, Clock and Pedometer Maker of the same address. He goes on to say that by the middle of the 19th century this firm was occupying three floors of a large and imposing business building (see Tallis map 1838/1840, Plate 4A). Although listed as watchmakers in later years they no doubt were mainly retailers. This particular watch was obviously imported from Switzerland and merely carried their name as retailer.

PLATE 4B

Lever Pocket Watches – Swiss

PLATE 5

PLATE 6A

PLATE 5
Anon

10 ct gold filled demi-hunter Dennison Moon grade case inscribed ENGLISH MAKE GUARANTEED TO BE MADE OF TWO PLATES OF 10ct GOLD WITH PLATE OF COMPOSITION BETWEEN AND TO WEAR 20 YEARS. White enamel dial, short black roman numerals, blued steel half-hunter hands. Keyless wound and pull to set. 17 jewel bar movement No 103607 SWISS MADE. Club tooth scape wheel and straight line lever escapement. Oversprung with overcoil cut bimetallic balance and blued steel balance spring. Micro-regulator on curved balance cock, the regulator in the form of an 8 pointed star activating pinion and rack adjustment. Engraving on balance cock AF RS and graduations black filled.
Diameter 50.7mm

PLATE 6
Anon

0.935 silver open face case, snap-on bezel, hinged back and cuvette. Case mark GS (George Stockwell – importer), hallmark London **1917**. White enamel dial, luminous arabic numerals and luminous skeleton spade hands, inset seconds dial. Keyless wound and pull to set. Top plate stamped Brevet 65777. Half plate. 13 jewel Swiss club tooth lever

PLATE 6B

movement with straight line lever and two arm mono-metallic oversprung balance with blued steel balance spring. Index on balance cock engraved FS/AR.
Diameter 50.7mm

155

Lever Pocket Watches – Swiss

PLATE 7A

PLATE 7B

PLATE 7C

PLATE 7
NEW "CASTLE" LEVER, Reid & Sons

10 ct gold open face Dennison case, hinged back, cuvette and bezel. Inside cuvette stamped ENGLISH MAKE GUARANTEED TO BE MADE OF TWO PLATES OF 10ct GOLD WITH PLATE OF COMPOSITION BETWEEN AND TO WEAR 20 YEARS. Inside back stamped Moon trade mark AND DENNISON WATCH CASE CO LTD, Dennison bow. White enamel dial signed REID & SONS' NEW "CASTLE" LEVER SWISS MADE, black arabic numerals, blued steel spade hands, inset seconds dial. Keyless wound and pull to set. Half plate. Swiss club tooth straight line lever. Engraved on top plate BUREN 15 jewels and BUREN BUREN on barrel wheel, also PATENTED under balance wheel. Oversprung and overcoil cut bimetallic balance with blue steel balance spring. Snail and whiplash micro-regulator.

Diameter 50.5mm

H Williamson Ltd, 81 Farringdon Road, London claim in their advertisement carried in the Horological Journal for January 1916 that the Buren watch was made throughout by the only Swiss factory of English ownership, ie H Williamson Ltd! Reid & Sons were established in 1788 as manufacturing and retail silversmiths with premises at 12 Dean Street before moving to 14 Grey Street, 41 Grey Street and 48 Granger Street and finally moving to new premises, Gem Buildings, Blackett Street, Newcastle. Reid and Sons became a limited liability company, Reid & Sons Ltd, in March 1930. In 1967 they became a subsidiary of Northern Goldsmiths.

Lever Pocket Watches – Swiss

PLATE 8A

PLATE 8
DREADNOUGHT, Record Watch Co,
Geneva and Tramelan, Switzerland

Silver open face case hallmarked Chester **1932**, snap-on bezel and back, engine-turned barleycorn pattern back, inside back stamped S & E (Smith, Ewen & Stylic Ltd, 42 Frederick Street, Birmingham, ref *W & C Yearbook* 1958). Special split pendant and bow. White enamel dial signed RECORD Dreadnought Swiss Made either side of inset seconds, black arabic numerals, blued steel spade hands. Keyless wound and pull to set. Top plate stamped RECORD W CO 17 JEWELS 4 ADJTS SWISS MADE 164475. Oversprung and overcoil cut bimetallic balance with blued steel balance spring and whip lash micro-regulator.

Diameter 50mm

PLATE 8B

157

Lever Pocket Watches – Swiss

PLATE 8C

PLATE 8C
DREADNOUGHT, Record Watch Co,
Geneva and Tramelan, Switzerland

Silver open face Dennison case hallmarked Birmingham 1917. Snap-on bezel, hinged back and cuvette. White enamel dial, black arabic numerals, blued steel moon hands, inset seconds. Keyless wound and pull to set. Centre bar inscribed TRENTON RECORD DREADNOUGHT WATCH FACTORIES 7 JEWELS. Elsewhere SWISS MADE and 3 ADJUSTMENTS. Oversprung and overcoil cut bimetallic straight line club tooth lever movement with small index on balance cock.

PLATE 8D. *La Livre d'Or de l'Horlogerie*

Lever Pocket Watches – Swiss

| Manufacture des montres | **Record Watch Co S.A.** | GENÈVE TRAMELAN |

Calibres ancre — Anker Werke — Calibros ancora — Lever movements

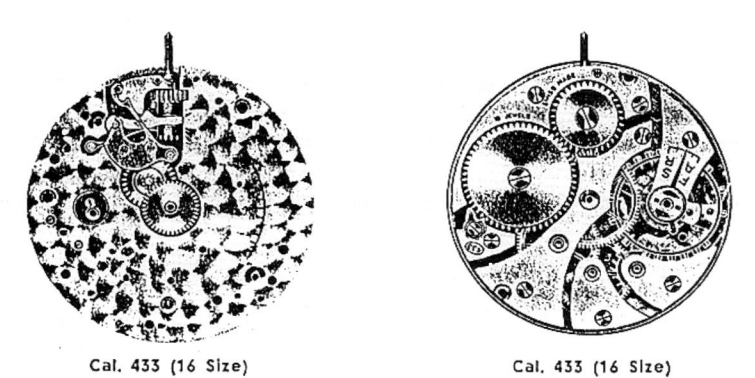

Cal. 433 (16 Size)   Cal. 433 (16 Size)

Pièces de rechange du Calibre 433 (16 Size)

When ordering exchange parts, it is necessary to indicate:
1. The description or the number of the part required.
2. The size of the movement and the number of the caliper.

PLATE 8E. *La Classification Horlogerie* 3rd ed., 1949 by Jobin

Lever Pocket Watches – Swiss

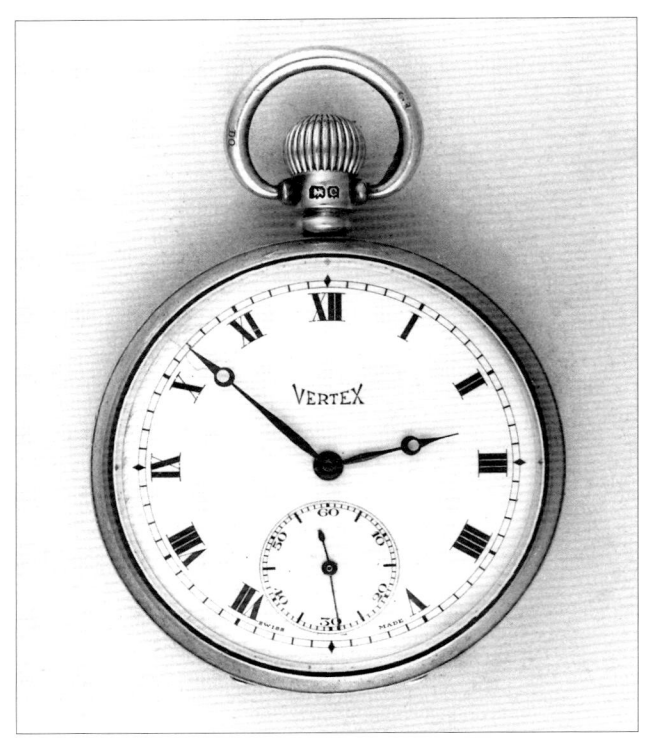

PLATE 9A

PLATE 9B

PLATE 9
Vertex

Silver open face case, hinged back and cuvette with snap-on bezel, hallmarked Birmingham **1927** and stamped ALD DENNISON WATCH CASE CO LTD. Dennison patented bow. White enamel dial, VERTEX upper centre, SWISS MADE below, short black roman numerals, blued steel moon hands, inset seconds dial. Keyless wound and set. Half plate. 15 jewel movement. Top plate stamped VWC (Vertex Watch Co). Club tooth scape wheel. Straight line lever. Oversprung cut bimetallic balance.
Diameter 49.2mm

The Vertex watch was promoted by Claude Lyons, a watch salesman, 28 Hatton Gardens, London – the early movements marketed under this name came from various sources. The Revue watches manufactured in the Thommen factory were marketed in this country under the brand name of Vertex. The watchmaking industry was introduced into the Waldenburg valley in Switzerland to compensate for the loss of employment when the newly built railway robbed the area of its traditional means of livelihood. Around 1853 the Society for Watchmaking had been founded, employing watchmakers who had migrated bringing their skills from the French speaking areas of Switzerland. There were twelve scattered workshops but when the project was taken over in 1859 by Louis Tschopp and Gideon Thommen these workshops were organised into one factory. They thrived and expanded to become one of the major manufacturers. Thommen ran the business alone after 1870 with his four sons taking over between 1891 and 1905. It became a limited company in 1905. They became Revue-Thommens in 1908 and continued to flourish with 1953 seeing the celebration of their centenary.

PLATE 10
Vertex

Silver 202 half-hunter case, hallmarked Birmingham **1927**, stamped ALD DENNISON WATCH CASE CO, hinged bezel back and cuvette. Pendant stamped ALD and hallmarked Birmingham **1926**. White enamel dial with short black roman numerals, blued steel spade hands and inset seconds dial. Keyless movement. Half plate. 21 jewel to centre. Club tooth scape wheel with straight line lever oversprung overcoil cut bimetallic balance and whiplash and snail micro-regulator and five adjustments. Backplate stamped 21 JEWELS SWISS MADE 5 ADJ VW and VERTEX twice with six pointed star on the winding wheel.
Diameter 52.3mm

Lever Pocket Watches – Swiss

PLATE 10

PLATE 12

PLATE 11

PLATE 11
Jaeger Le Coultre SA, Geneva and La Sentier, Valle du Joux, Switzerland

Chrome open face case, snap-on back. White painted dial, black arabic numerals with 3, 9, 12 skeletonised and luminous, luminous skeletonised hands, inset sunk seconds. 15 jewel movement. Top plate engraved Jaeger Le Coultre SWISS MADE (LETTERS S REVERSED). Pillar plate numbered 467/0. Club tooth escapement with straight line lever. Oversprung monometallic balance with compensation screws. Index on balance cock engraved R A S F.
Diameter 52mm

This firm was founded in 1833, Antoine La Coultre with David La Coultre joined by Edmund Jaeger. After 1937 all watches have the brand name of JAEGER LE COULTRE.

PLATE 12
Anon

Rolled gold open face case Moon grade DENNISON WATCH CASE CO LTD and stamped ALD 679001 corresponding to the 1939 period with typical Dennison bow, hinged back, cuvette and bezel. White enamel dial, short blade roman numerals, blued steel spade hands and inset seconds dial. Keyless wound and pull set. Half plate. 17 jewel. Club tooth scape wheel with straight line pallet with oversprung overcoil bimetallic cut compensated balance with whiplash regulator. Engraved 17 JEWELS 4 ADJUSTMENTS SWISS MADE 1271109 and four adjustments.
Diameter 51mm

Lever Pocket Watches – Swiss

PLATE 13

PLATE 13
Weill & Cie, Chaux de Fonds, Switzerland

Silver .935 open face case, hinged back cuvette and bezel with engine-turned back. White enamel dial, black roman numerals, blued steel spade hands, inset seconds dial. Pillar plate stamped W & Cie SWISS MADE. Keyless wound and push piece to set hands. Three-quarter plate. 15 jewel. Polished steel scape wheel and straight line lever. Oversprung overcoil cut bimetallic balance. Regulator under balance cock.
Diameter 46.3mm

PLATE 14
C Jagermann, Nachfe, Berlin

Silver open face case with Swiss woodcock silver mark and German silber Stempel mark No 38027. The cuvette is engraved C Jagermann Nachfe Berlin. White enamel dial, Louis style hands, inset sunk seconds. Keyless wound movement with push piece to set hands at 55 minute position. Three-quarter plate. 15 jewel lever escapement. Oversprung overcoil bimetallic compensated balance. Index on balance cock.
Diameter 51.5mm

PLATE 14

Lever Pocket Watches – Swiss

PLATE 15A

PLATE 15
Jean Perret SA, 1211 Genève 1, Rue Chapponnière 1

Open face, snap-on bezel and back with light blue enamel surround to bezel and back of case with gilt dot and comma decoration. Gilt dial inscribed JEAN PERRET GENEVE, radial bars and gilt pointer hands. Keyless wound and pull to set. Top plate stamped SWISS. 17 Jewels and pillar plate 63. Club tooth straight line lever escapement with oversprung monometallic balance with imitation compensation screws. Regulator index on balance cock stamped + –.
Diameter 39mm

Listed in the 1979 Swiss Industries Fair Catalogue 'Horloges, Models exclusifs, nouveautés et créations en argent, or, acier, métal chrome, rhodie ou plaque.'

PLATE 15B

PLATE 16
Anon

Silver full hunter, hinged bezel back and cuvette. Case numbered 4543 17178, cuvette engraved REMONTOIR AU PENDANT MI CHRONOMETRE. White enamel dial, narrow black roman numerals, narrow blue steel moon hands, inset sunk seconds dial. Rare keyless winding and set hands. Bar movement. 15 jewel. Overcoil blued steel spring. Uncut bimetallic balance. Club tooth lever.
Diameter 50mm

PLATE 16A

PLATE 16B

PLATE 17A

PLATE 17
Vauchay, Vve de Louis Goering, Switzerland

Open face FORTUNE gold filled case, hinged back by AWC Co (American Watch Case Company) made in Canada. Silver dial with engine-turned chapter ring with floral decorative centre signed VAUCHAY SWISS MADE lower edge dial, stylised black arabic numerals, blued steel cathedral hands, blued steel seconds dial. Keyless spotted bar movement stamped VAUCHAY 17 JEWELS SWISS MADE with club tooth scape wheel and straight line pallets. Oversprung overcoil bimetallic cut balance with whiplash regulator.

Diameter 45.3mm

Répertoire des Marques de Fabrique 1945 *gives VAUCHAY as the trade mark of Vve de Louis Goering Montres Elaine et Imperial SA, Switzerland.*

PLATE 17B

Lever Pocket Watches – Swiss

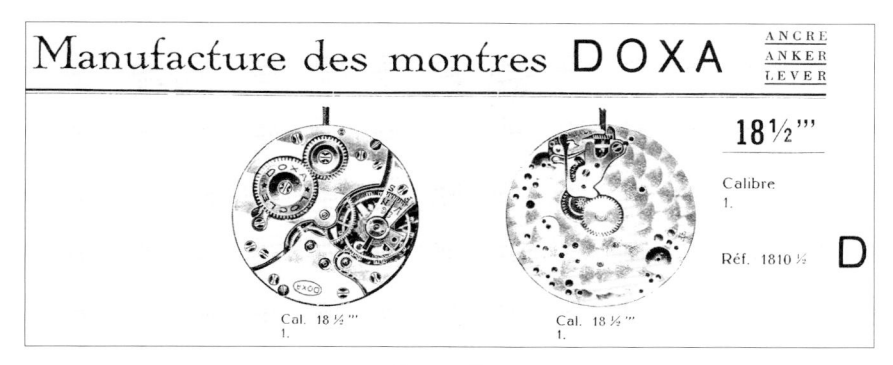

PLATE 18A

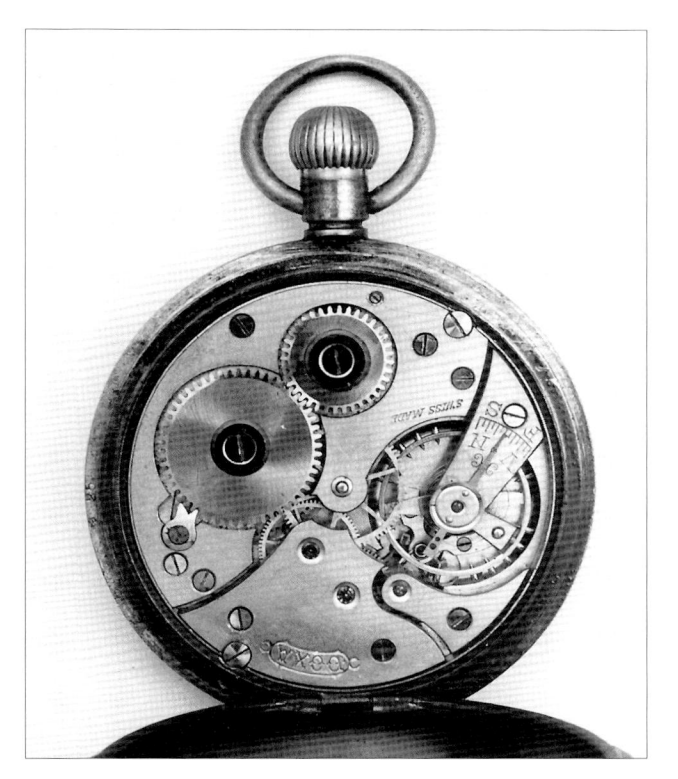

PLATE 18B

PLATE 18C

PLATE 18
DOXA, Fabrique des Montres DOXA SA,
Le Locle, Switzerland

Oxydised steel open face case, snap-on bezel, hinged back and cuvette. White enamel dial, sunk centre, black arabic numerals, blued steel spade hands, inset seconds dial. Keyless wound and pull to set. Top plate stamped DOXA SWISS MADE. 15 jewels. Half plate movement with club tooth scape and straight line lever with oversprung monometallic balance and white balance spring. Index on balance cock FS AN.
Diameter 50.8mm

Doxa was founded by George Ducommun (1868-1936).

165

Lever Pocket Watches – Swiss

PLATE 19A

PLATE 19
Nirvana Watch Co, La Chaux de Fonds, Switzerland

Nickel plated brass open face case with hinged bezel back and cuvette, inside back inscribed NIRVANA INTERNATIONAL CRC EXTRA. White enamel dial signed Nirvana NON MAGNETIC, black roman numerals, blued steel spade hands, inset seconds dial. Key wound and set from back. Top plate stamped 531038 and pillar plate SWISS MADE under rim of balance. 7 jewel. Half plate movement with club tooth scape and straight line lever. Oversprung monometallic balance with white balance spring. Index on balance cock.

Key size 7. Diameter 54.6mm

Kelly Directory 1913 lists Nirvana Watch Co, La Chaux de Fonds and Wholesale Department, 42 Holborn Viaduct, London EC1.

PLATE 19B

PLATE 20A. *Horological Journal* Volume 57 November 1914

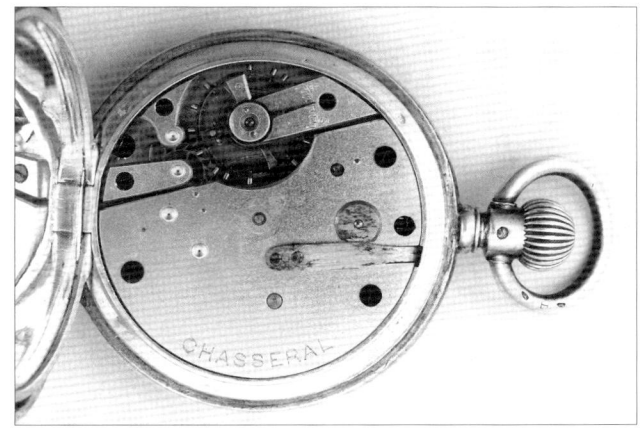

PLATE 20B

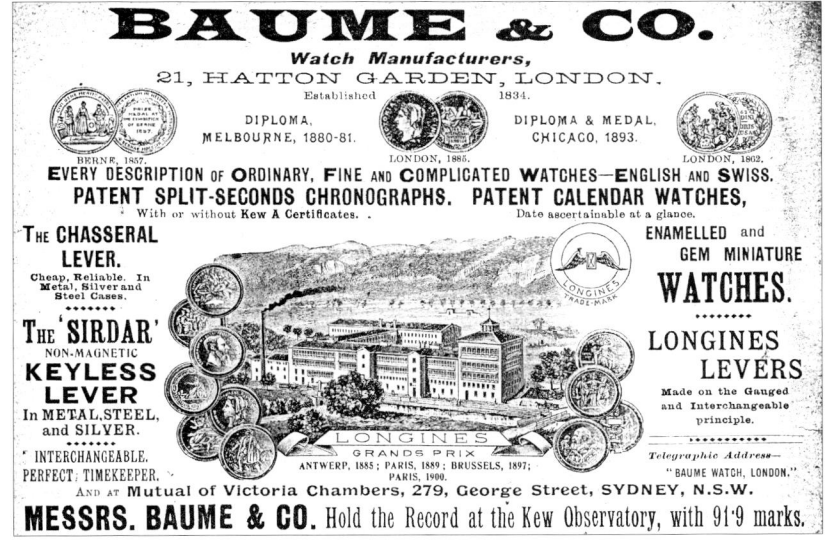

PLATE 20C. *Horological Journal* Volume 44 June 1902, back cover

PLATE 20
CHASSERAL, Ernst Françillon, St Imier, Switzerland

Silver open face case, hinged back and cuvette with snap-on bezel, Swiss hallmark O.935 503178 6 (corresponds to Longines manufacturing period 1888-1893). White enamel dial, black roman numerals, blued spade steel hands, inset seconds. Keyless wound and pull set. Three-quarter plate. 4 jewel. Movement stamped CHASSERAL with oversprung monometallic balance with compensation screws. Club tooth scape wheel with tangential pallet. *Special feature is a very unusual keyless mechanism.*
Diameter 48.3mm

Longines was founded in 1866 by Ernest Françillon assisted by his cousin Jacques David – a young and talented engineer – in the village of St Imier. Their object was to produce a reliable timepiece at a competitive price. To this end they adopted the use of machinery and worked towards ensuring, by mass-production methods, an exactitude of manufacture that allowed a total interchangeability of parts. In 1867 they had only twenty workers but nevertheless won an award in the Paris Exposition. By 1914 they had over 1,000 workers and had become a private company under the directorship of the family of the late founders, E Gagnebin, B Savoye and Arthur Baume. Baume & Co marketed the watches in this country. The name Longines referred to the strips of cultivated land along the banks of the River Suze. The name Chasseral alluded to the site of the factory which was on the edge of the Forêt de l'Envers under the Crest of Chasseral known as 'the pride of the Vallon'. For a list of Longines' movement numbers and dates see W H Samelius, Dean of American Watchmakers compiled by O R Hagans, p370.

Lever Pocket Watches – Swiss

PLATE 21
Ernest Françillon et Cie, St Imier, Switzerland

Nickel case, dome stamped CHASSERAL SWISS with dome and cuvette stamped 779096 (corresponds to Longines period c.1894). White enamel dial signed WW, black roman numerals, blued steel spade hands, inset seconds. Top plate stamped SWISS MADE CHASSERAL + 3846. The unusual keyless wound and set movement with half plate 4 jewelled club tooth lever movement with tangential pallet and oversprung monometallic brass balance.
Diameter 50mm

Patent Swiss 3846, Class 64, 23 Juillet 1891, Ernest Françillon & Cie. Mechanism de Remontoir avec mise à l'heure par le pendant.

PLATE 21A

PLATE 21B

168

Plate 22
Longines, St Imier, Switzerland

Silver open face case with hinged back and cuvette stamped JO within lozenge, AU NECRE 19 BOULD ST DENIS PARIS on cuvette and numbered 737566 corresponding to manufacturing date 1892. White enamel dial, black roman hour and arabic minute numerals, blued steel spade hands, inset sunk seconds dial. Keyless wound and set movement with push piece at 5 minutes past position. Half plate. 15 jewels. Stamped LONGINES MODÈLE DÉPOSÉ. Club tooth scape with straight line pallet. Overcoil with uncut bimetallic Longines registered trade mark on balance cock under index regulator.

Diameter 41mm

Plate 22a

Plate 22b

Lever Pocket Watches – Swiss

PLATE 23A

PLATE 23B

PLATE 23
CAMPAIGN, Marcks & Co Ltd,
Bombay & Poona, India

Silver .935 open face case with snap-on bezel, hinged back and cuvette, inside back engraved CAMPAIGN MEDAL SWISS MADE 667131 – (corresponds to Longines manufacturing period 1890). White enamel dial signed MARCKS & Co Ltd BOMBAY & POONA Swiss Made, broad roman numerals, blued steel spade hands, inset sunk seconds dial. Keyless wound and push piece to set at 5 minutes past position. Half plate 15 jewelled Swiss club tooth and tangential lever movement with uncut bimetallic oversprung balance and blued steel balance spring. Regular index on balance cock engraved SF.
Diameter 48.8mm

A very similar but not absolutely identical calibre (Calibre 15.47 1904 model) is illustrated Hirst Bros 4th Wide Awake Catalogue 1915 p271.

PLATE 24
Longines, St Imier, Switzerland

Silver .935 open face engraved case hallmarked with Swiss Bear Number 989667, snap-on bezel, hinged back and cuvette. White enamel dial, black roman numerals, blued steel spade hands, centre seconds. Keyless wound and push piece to set 5 minute past position. 13 jewel one plate club tooth and straight line lever movement with oversprung cut bimetallic balance and blued steel balance spring. Regulator index on balance cock engraved FAST SLOW. Serial number corresponds to Longines manufacturing period 1898.
Diameter 35.3mm

Ref Smith and Son Guide to the Purchase of a Watch p45 for illustration of similar watch listed as a nurse's watch (centre seconds hand for counting the beat of the pulse).

PLATE 25
Longines, St Imier, Switzerland

Silver .935 open face case with imported hallmark London 1913 stamped AB, snap-on bezel, hinged back and cuvette. White enamel dial signed J W BENSON LONDON SWISS MADE lower edge dial, black roman numerals, gold spade hands, inset seconds dial. Keyless wound and push piece to set hands. 15 jewel Swiss lever movement 3026421 15 JEWELS SWISS MADE. Club tooth scape wheel with straight line pallet and oversprung overcoil bimetallic balance with index on balance cock.
Diameter 48.8mm

PLATE 24A

PLATE 24B

PLATE 25A

PLATE 25B

Lever Pocket Watches – Swiss

PLATE 26A

PLATE 26B

PLATE 26C

PLATE 26
Longines, St Imier, Switzerland

Silver open face niello case with hinged back and cuvette, inside cuvette stamped PD & LX. White enamel dial inscribed Longines, stylised arabic numerals 1-12 with smaller arabic numerals 13-23 with 0 at 24 hour position on outside chapter ring, inset seconds dial. Keyless wound and set. Half plate. 13 jewel Swiss lever movement numbered 3398231 (corresponds to 1916 production period) LONGINES. Club tooth scape wheel and straight line pallet. Cut bimetallic oversprung balance with blued steel balance spring and index on balance cock.

Diameter 45.2mm

Niello is a form of decoration in which the design is deeply engraved into the silver with an undercut edge and then filled with the niello composition, a mixture of approximately 50% lead, 25% silver and 25% copper with a small amount of borax and sulphur. After heating in a furnace to fuse the mixture to the metal, the sulphides remain black, presenting a pleasing contrast between the white silver and the black infill.

Lever Pocket Watches – Swiss

PLATE 27

PLATE 28

PLATE 27
Omega, Louis Brandt, Bienne, Switzerland

Silver .925 case with imported hallmark London **1908,** open face, hinged bezel back and cuvette. White enamel dial, short black roman numerals, gold spade hands, inset seconds dial. Keyless wound, pull to set. 15 jewels. Omega trade mark on top plate (Omega in circle). Club tooth scape and straight line lever escapement, highly polished. Oversprung and overcoil cut bimetallic balance with blued steel balance spring. Micro-regulator in the form of 12 point star with internal snail adjustment on balance cock engraved F S.
Diameter 51mm

PLATE 28
LABRADOR, Omega, Louis Brandt Frères, Bienne, Switzerland

Open face, hinged bezel, back and cuvette. Stamped inside back with emblem of Aesculapius and inside cuvette WARRANTED TO WEAR 20 YEARS FAHYS MONTAUK (Joseph Fahys, Sag Harbor, Long Island, New York, Watch Case Co 1857 and on into the 1900s at New York City and Carlsladt, New Jersey). White enamel dial, black roman numerals, blued steel spade hands. Inset seconds dial. Half plate keyless 15 jewel movement engraved on top plate Labrador REG MARK SWISS. Club tooth scape wheel, straight line pallet. Oversprung overcoil bimetallic cut balance with whiplash micro-regulator.
Diameter 50.5 mm

173

Lever Pocket Watches – Swiss

PLATE 29A

PLATE 29B

PLATE 30A

PLATE 30B

Lever Pocket Watches – Swiss

PLATE 31A

PLATE 31B

PLATE 29
Omega, Louis Brandt Frères, Switzerland

Nickel on brass open face case with snap-on bezel, inside back cover stamped OMEGA FAB SUISSE SWISS MADE 1440. White enamel dial inscribed OMEGA with trade mark, luminous skeletonised arabic numerals at 3, 6, 9, 12 with plain black at intermediate positions, skeletonised luminous spade hands, inset seconds dial. Keyless wound, pull to set hands. Half plate. Club tooth scape wheel, tangential pallets. Monometallic oversprung overcoil balance with index on balance cock. Stop work recoil click. Backplate stamped OMEGA SWISS 15 JEWELS No 9980758.
Diameter 52mm

PLATE 30
Omega, Louis Brandt Frères, Switzerland

Silver .800 open face case, Swiss woodcock hallmark and Deutsch (German) silber Stempel, hinged bezel back and cuvette. White enamel dial signed OMEGA, black arabic numerals, gilt Louis spade hands, inset seconds. Keyless wound, pull to set. Half plate. 15 jewel. Club tooth and straight line lever movement. Oversprung cut bimetallic balance to blued steel balance spring. The regulator index on balance cock engraved with Omega trade mark and A R F S.
Diameter 47.8mm

PLATE 31
ZENITH RAILWAY POCKET WATCH,
Zenith Watch Co, Le Locle, Switzerland

Heavy nickel open face case, snap-on bezel and hinged back and cuvette. Back engraved with railway engine and monogram AC intertwined, cuvette engraved GRAND PRIX 1900 ZENITH. White enamel dial signed ZENITH, short black roman numerals I to XII and red arabic 13 to 24, broad spade blued steel hands, inset seconds. Keyless wound and push piece to set. Top plate inscribed ZENITH 2877141. Half plate. 15 jewel. Club tooth straight line lever movement with oversprung and overcoil cut bimetallic balance with blued steel balance spring and Zenith micro-regulator on balance cock.
Diameter 56.5mm

The Zenith company was founded in 1865 by George Favre Jacot with the trade mark of BILLODES being registered in 1884, 1885 and 1896. They won a Gold Medal at the National Swiss Exhibition in Geneva in 1896 and the Grand Prix at the Paris Exhibition in 1900 followed by yearly prizes. Examples of their pocket watches in the rating trials at Kew in 1926 and 1929 were recorded as 97.2 and 97.3 respectively. Their London address was 119 High Holborn. As shown in the accompanying advertisement, the Zenith Standard movement was supplied in all sizes from 9 to 21 lines as well as American sizes 000000 0, 6, 12, 14, 16 and 18. The range of cases included nickel, gunmetal, silver, gold filled and gold cases.

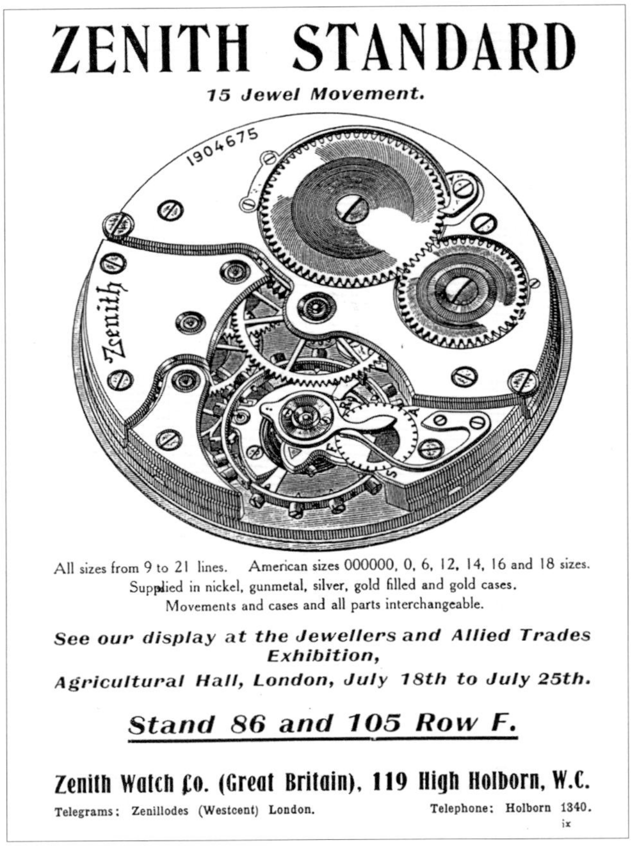

Lever Pocket Watches – Swiss

PLATE 31C. *Horological Journal*, 1914

PLATE 31D. *Horological Journal*, 1914

PLATE 32A

PLATE 32B

| 1883] | ABRIDGMENT CLASS WATCHES &c. | [1883 |

2624. Spence, W. H., [*Droz et fils, A.*]. May 26.

Cases for keyless watches are so constructed as to be hermetically sealed against damp and foreign matter. The rim B, which carries the glass cemented to it by an insoluble and infusible composition, is made to screw or otherwise fit upon the flange A, which is made higher and squarer than usual. By slightly greasing the parts an airtight fastening is secured. There is no opening at the back of the case.

Keyless mechanism.—The pendant C has a left-handed screw-thread for receiving the stem E of the button F, which, when screwed down, rests externally on the shoulder G and internally on the top end of the pendant, a double closing being thus obtained. To effect the winding, the button F is given about a turn to the right, so as to loosen it, after which it is moved alternately from left to right and *vice versâ* in the ordinary manner. The winding operation may be continued without limit, the completion of the winding being known by the sound of the mainspring passing from one notch to another in the barrel. To set the hands, the stem E of the button is liberated from the screwed part of the pendant, which is done by making three or four turns to the right, and then drawing the button into the position shown by dotted lines in Fig. 1, and in the raised position shown in Fig. 2; which action raises the arm L and causes the rocking lever M connected therewith to drive a second rocking-lever N, and drop into a notch O therein. This second lever, which carries a stud *p* inserted in the groove R of the sliding pinion S, will move the latter into gear with the wheel T connected with the minute hand; then, by turning the button, the watch can be set as required.

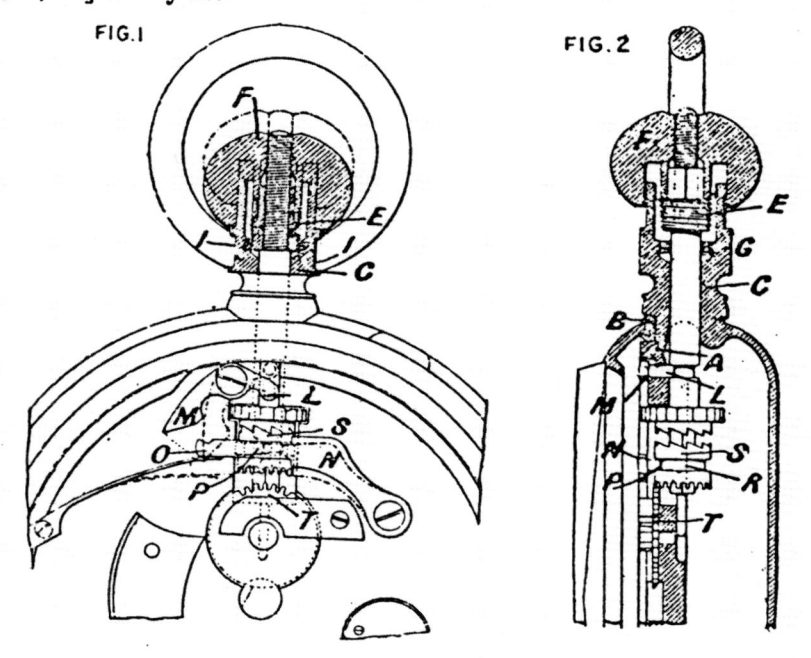

Bows and pendants.—The pendant is formed as shown at I, I to receive the bow.

PLATE 32C

PLATE 32
WATERPROOF, Fabrique d'Horlogerie, Alcide Droz & Fils, St Imier, Suisse

Nickel open face case, snap-on bezel, back of case stamped PATENTED IN ALL COUNTRIES and trade mark. White enamel dial inscribed WATER-PROOF 29805 with the index regulator cranked around balance to operate at the side of the dial, black roman numerals, blued steel spade hands. Index scale FS on the dial outside the chapter ring at 16-23 minute position, *a most interesting and unusual feature.* Inset sunk seconds. Keyless wound, pull to set hands. Swiss club tooth lever escapement, jewelled balance and pallet staff and scape arbor. Watch movement number on front plate under dial 3746C3. To gain access to the dial the snap-on front bezel is removed and two small dog screws are turned to release dial which, after removal, enables access to larger dog screws holding the movement in the case.

Diameter 49.3mm

Ref UK Patent No 2624 Spence WH (the actual patentees being Droz et Fils A), 26 May 1883. Marque de Fabrique et de commerce Imperméable No 968 & 969, 22 Jan 1884. Emblem of eagle carrying watch and AD & F was the mark registered No 1112 Alcide Droz & Fils Fabricant, St Imier, 12 Mars 1884.

Plate 33
PERSEO, Cortébert Watch Co
(Juillard & Cie founded 1790), **Bureaux de Ventre, Chaux de Fonds, 25 Rue du Parc, Switzerland**

Nickel open face case with hinged bezel and back. White enamel dial, black roman numerals, blued steel spade hands, inset seconds. Keyless wound, push piece to set hands at 11 o clock position. Top plate engraved PERSEO. Half plate. 15 jewel. Club tooth straight line lever with oversprung overcoil cut bimetallic balance and blued steel balance spring. Index on balance cock engraved A R F S.
Diameter 52.4m

PLATE 33

PLATE 34A

PLATE 34B

Plate 34
Anon

Gilt full hunter case with florid engraving front and back and centre of back cover engraved with rustic scene and cottage. White enamel dial, black roman hours and arabic numerals at 5 minute intervals outside chapter ring, Louis style gilt hands, inset sunk seconds. Keyless wound and setting lever at 5 o'clock position at edge of bezel. Barrel bridge stamped 928356 and side of dial plate stamped SWISS. 17 jewels. Full plate club tooth Swiss lever escapement with oversprung monometallic balance with blued steel balance spring. Regulator index from cock over the balance to silver quadrant.
Diameter 55.5mm

The finely blued screws adjacent to the jewel settings are non-functional, ie for decorative and sales promotion purposes only. The watch appears to have been intended for the American market.

Lever Pocket Watches – Swiss

PLATE 35A

**PLATE 35
Anon**

Heavy nickel open face case with snap-on bezel and hinged back. 46mm white enamel dial with chapter ring of 30mm, black roman numerals and blued steel spade hands. Keyless winding and push piece at 3 minutes past position. 12 jewel bar movement with club tooth scape and straight line lever. The lever has forked tail embracing the scape arbor. Oversprung three arm monometallic balance with blued steel balance spring. Index on balance cock engraved FS. In spite of the keyless mechanism the watch can still be wound and set with key from the back of the movement, the winding and setting squares having been retained.
Diameter 55.8mm

PLATE 35B

179

Lever Pocket Watches – Swiss

PLATE 36A

PLATE 36B

PLATE 36
Marvin Watch Co, Springfield

Gilt brass full hunter case, hinged front back and cuvette. White dial with very light blue chapter ring inscribed M W Co Springfield, thin black roman numerals with gold diamond markers between, narrow spade blued steel hands, inset seconds dial. Keyless wound with lever at edge of bezel at 5 o' clock position to set hands. Full plate 17 jewelled movement engraved MARVIN WATCH CO SPRINGFIELD 148972 on the top plate and the *side* of the pillar plate stamped SWISS. Oversprung uncut bimetallic balance with blued steel balance spring. Long index over balance rim to silver quadrant on top plate.
Diameter 55.5mm

This is an example of a Swiss watch made and named to give the impression that it was an American watch from Springfield, Massachusetts, USA – see National Association of Watch and Clock Collectors Bulletin Whole Number 201 August 1979, Vol XXI, No 4, p426-434, Fig 7 and catalogue No 22, 1885, SF Myers, 179, Broadway Illustration number 92703.

PLATE 37
K Serkisoff & Co (retailer), Constantinople

Silver .800 full hunter case, Swiss woodcock hallmark. Front cover stamped Billodes. Registered Mark (Antique Bow). White enamel dial with name of retailer K SERKISOFF & CO CONSTANTINOPLE, Turkish numerals, blued steel spade hands, inset sunk seconds dial. Key wound and set at back. BILLODES engraved on bridge over barrel ratchet and Billodes trade mark. Half plate. Swiss club tooth straight line lever movement with oversprung uncut bimetallic balance and blued steel balance spring. Regulator index on balance cock.
Key size 5. Diameter 52mm

PLATE 38
Bravingtons Jewellers (retailer), 296 298 Kings Cross, London N1, General Watch Co (manufacturer), Helvetia, Bienne, Switzerland

Chrome open face case, screw bezel and back. Inside back stamped ACIER INOXYDABLE SWISS 99324, outside back engraved BRAVINGTON. White enamel dial, black roman hour markings, 5 minute marking in small arabic numerals between roman numerals and chapter ring, centre seconds. Keyless wound and pull to set hands. Bar movement fully jewelled to centre. Club tooth straight line lever escapement with oversprung and overcoil cut bimetallic balance and blued steel balance spring. Regulator index on balance cock engraved A R F S.
Diameter 62.5mm

PLATE 37A

PLATE 37B

PLATE 38A

PLATE 38B

Lever Pocket Watches – Swiss

PLATE 39

PLATE 40A

PLATE 40B

Plate 39
Moeris, St Imier, Switzerland

Nickel open face case, screw bezel and back. Back engraved GSTP 141252 with Government Property broad arrow mark. Black enamel dial with MOERIS in white top centre and arabic numerals 3, 9 and 12, luminous dots for 5 minute markings, luminous pointer hands, white seconds ring. Keyless wound and pull to set. Top plate engraved MOERIS 15 JEWELS ADJUSTED 19H and pillar plate stamped SWISS MADE. Half plate movement with club tooth straight line lever escapement. Oversprung overcoil monometallic balance with poising screws. Blued steel balance spring and index on balance cock engraved FS RA.
Diameter 52mm

Compare with Moeris pin pallet GSTP General Service Timepiece Military issue page 230. Moeris Marque Deposé No 17816, 1904 18 Oct, Fritz Moeris, St Imier.

Plate 40
A McMillan, Ottawa, Tavannes Watch Co (manufacturer), Tavannes, Switzerland

Gold filled FORTUNE open face with milled edge to body by AWC Co (American Watch Case Co), screw bezel and back. White enamel dial, A McMillan Ottawa upper centre, short black roman numerals, narrow spade blued steel hands, inset sunk seconds dial. Keyless wound and pull to set. Top plate engraved SWISS US PAT 24 MAY 1904 21 JEWELS ADJUSTED 5 POSITIONS 12170996. Club tooth scape and straight line lever with oversprung and overcoil cut bimetallic balance and blued steel balance spring. Whiplash micro-regulator on balance cock.
Diameter 51mm

USA Patent No 760647 issued 24 May 1904 to Henri Sandoz of Tavannes, Switzerland, for a 'new winding and setting mechanism with an extremely limited number of parts. Its action is reliable and extremely simple'. See N.A.W.C.C. Bulletin June 1981, Whole Number 212, Vol XXIII, No 3.

PLATE 41

Plate 41
HELVETIA, General Watch Co, Bienne, Faucon 18, Switzerland

Chrome on brass open face case, snap-on bezel and back. White enamel dial, arabic numerals (luminous 3, 6, 9 and 12) with luminous five minute markers. Keyless wound, pull set. Top plate stamped GENERAL WATCH CO SWITZERLAND 1944533 32A HELVETIA SWISS MADE on barrel wheel. 15 jewel Swiss lever with club tooth scape wheel and straight line lever. Oversprung overcoil monometallic balance.
Diameter 52.1mm

Lever Pocket Watches – Swiss

PLATE 42A

PLATE 43

PLATE 42B

PLATE 42
PEERLESS, Stauffer & Co, Chaux de Fonds, Switzerland

Silver .925 half-hunter case No 766833, case mark CN (Charles Nicolet), imported hallmark London **1920**. White enamel dial, luminous hands and luminous 5 minute dot markings, black arabic numerals, inset sunk seconds dial. Keyless wound, hand set by lever pulled at edge of the dial at 23 minutes past position. Half plate lever movement. PEERLESS Swiss Patent 55231 and 636771 stamped on dial plate under edge of balance rim, Stauffer trade mark and SWISS MADE on top plate. Cut bimetallic balance with Breguet overcoil and spring. Index on balance cock.
Diameter 49.6mm

Ebauche by International Watch Co Savonette Calibre 84 83/4 ligne and made during the IWC manufacturing period 1921 (766301-766900). Swiss Patent 55231. Uhrenfabrik von J Rauschenbachs, Erben 636771, vormals International Watch Co, Schaffausen, Switzerland.

PLATE 43
Ward Bros (retailers), 249 High Road, Tottenham, H Williamson (manufacturers), Buren, Switzerland

Silver .925 case, imported hallmark London **1913**. Casemaker WHS No 101005. White enamel dial inscribed Ward Bros 249 High Rd Tottenham, roman numerals, gold spade hands, inset seconds. Keyless wound and set. 7 jewel lever escapement. Top plate stamped BUREN SWISS MADE 7 JEWELS. Monobimetallic balance with compensation screws. Index on balance cock.
Diameter 49mm

Buren was 'the only Swiss watch factory of English ownership' – H Williamson Ltd, 81 Farringdon Road, London, ref Horological Journal Vol LVIII, Jan 1916, pviii. Patents were taken out for a framework to facilitate removal of barrel, Patent 6577/Patent 12800 24 June 1901 by E S Edgar (H Williamson Ltd).

PLATE 44A

PLATE 44
PEERLESS, Stauffer & Co, Chaux de Fonds, Switzerland

Oxydised steel open face case No 553939 stamped II V. White enamel dial, black arabic numerals, blued steel spade hands, inset sunk seconds. Keyless wound and hand set by push piece at 4 minutes past hour position. Top plate stamped S & Co (under crown), the trade mark of Stauffer & Co. The dial plate stamped PEERLESS 513579. *Ebauche* made for Stauffer, probably by International Watch Co. IWC list 513401-514000 as within their 1911 period and with Lepine Calibre 63 12 ligne. 15 jewel lever with cut compensated balance with Breguet overcoil spring.
Diameter 32.3mm

PLATE 44B

Lever Pocket Watches – Swiss

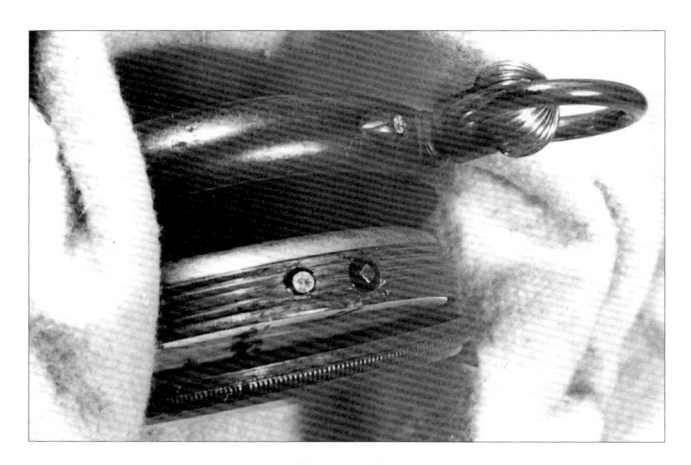

PLATE 45A

PLATE 45
Searle & Co (retailer), Lombard Street, London

Oxidised steel case of Borgel style. White enamel dial inscribed Searle & Co Lombard St London in upper centre, SWISS MADE below seconds dial, black roman numerals, blued steel spade hands, inset sunk seconds. Keyless winding, push piece to set hands at 3 minutes position. Movement removed by partially withdrawing winding button and unscrewing. Knurled bezel is pushed on and prevented from turning by locating pin at X o'clock position.
Diameter 46.8mm

François Borgel (late 19th century Geneva watchmaker) has UK Patent 20422 28 Oct 1891 (date claimed under section 103 of Patent Act AD 1883).

20,422. Borgel, F. Oct. 28, [*date claimed under Sec. 103 of Patents &c. Act, A.D. 1883*].

Watch cases. — The case A has an internally-screwed inturned rim which receives the externally-screwed ring B holding the movement. A bevel F to hold the glass G fits on the ring B. The stem D is cut flush with the ring and squared to engage a key D¹ extending through the pendant.

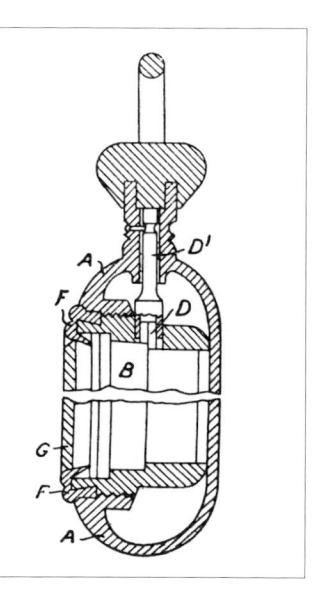

PLATE 45B

PLATE 46
International Watch Co, Switzerland

Nickel case hinged back glazed snap-on cover to movement, case stamped 22 M 100040. White enamel dial, narrow roman numerals with small arabic numerals at 5 minute position outside chapter ring. Keyless winding with push piece to set hands. High grade bar 17 jewel movement with wolf tooth. Club tooth scape wheel and oversprung overcoil bimetallic cut compensated balance with index on balance cock.
Diameter 53.3mm

Identified by referring to ébauche *books.*

PLATE 47
COLONIAL

Silver open face pocket watch, case hallmarked Birmingham **1926** and stamped ALD (Dennison Watch Case Co) No 515804 SPECIAL, hinged back and cuvette, snap-on bezel. White enamel dial inscribed JOHN ELKAN LTD 35 LIVERPOOL ST EC 69 CHEAPSIDE EC 70 LEADENHALL ST EC (he must have been proud of the diversity of his premises), Colonial SWISS MADE lower edge dial, black roman numerals. blued steel spade hands. Keyless wound and set. 15 jewel. Half plate movement with bimetallic cut compensation overcoil and oversprung balance. Club tooth lever scape wheel, straight line pallet. Index on balance cock. Top plate stamped SWISS MADE 15 JEWELS.
Diameter 49.5mm

Lever Pocket Watches – Swiss

PLATE 46A

PLATE 46B

PLATE 47A

PLATE 47B

Lever Pocket Watches – Swiss

PLATE 48A

PLATE 48B

PLATE 48
Paul Ditisheim Fabricant, La Chaux de Fonds, Switzerland

Silver .935 open face watch, hinged bezel back and cuvette, stamped PD, cuvette engraved Ankeruhr No 25568 PAUL DITISHEIM Hochste Austeichung in Genf Paris à London. White enamel dial, black roman numerals and blue steel spade hands, inset sunk seconds, PAUL DITISHEIM FABRICANT LA CHAUX DE FONDS inscribed upper centre dial. Keyless wound and push piece to set hands at 56 minutes position. 17 jewel gilded bar movement. Club tooth scape with straight line lever oversprung with overcoil cut bimetallic balance and blued steel balance spring with regulator index on balance cock engraved Avance Retard.

Diameter 53mm

Paul Ditisheim was born in 1868 and, having had an international training in the skills of watchmaking, he returned to Chaux de Fonds, Switzerland where he set up his own name manufactory. Later, with a partner, he founded the Vulcain factory, Vulcain & Volta Ditisheim & Co. He was particularly interested in fine adjusted watches and montres bijoux. Indeed his most important work was carried out in his quest for precision timekeeping. At the Universal Exhibition of Paris in 1900 he was awarded the Grand Prix for his general varied display. The most significant items were, however, the first public appearance of the first watches with Guillaume balances. Subsequent experimentation with balance springs, balances and the application of Elinvar and Invar led to his devising a simple ring balance of brass or other alloy adapted to take two small bimetal segments or 'affixes' attached at opposite points of the rim. Movements incorporating these achieved new records at many of the Observatory Trials. When his affix balance became a practical proposition he formed a company for its commercial development under the name of Solvil. He subsequently fell into disagreement as he felt that quality was being discarded for commercialism and resigned from the board in 1929 although the company retained his name. He also made a close study of watch oils and eventually collaborated with Dr Woog on the production of artificial oils. His contribution to horological literature was not insignificant.

'Paul Ditisheim - A Tribute to a Great Horologist' by Prof Torrens, Horological Journal *December 1945.*

Lever Pocket Watches — Swiss

PLATE 49

PLATE 50A

PLATE 49
SOLVIL, Paul Ditisheim Co, La Chaux de Fonds, Switzerland

Nickel open face with hinged bezel back and cuvette. White enamel double sunk dial with inset seconds, short black arabic numerals and blued steel spade hands, upper centre dial inscribed PAUL DITISHEIM SOLVIL. Keyless wound and pull to set hands with gold winding button. 17 jewel bar movement engraved 736404 with club tooth scape and straight line lever. Oversprung with overcoil monometallic balance (AFFIX – ELINVAR) and star and rack micro-regulator with white balance spring (Elinvar) and regulator index on cock engraved AF. Gold centre third and fourth wheels.
Diameter 49.5mm

London agents were (1992-1925) AL Fraissard, 34-35 High Holborn, London WC and subsequently (by 1936) Louis Braham, 25 Hatton Garden.

PLATE 50
Paul Ditisheim, La Chaux de Fonds, Switzerland

Silver open face, snap-on bezel, hinged back and cuvette, hallmarked imported London **1927**. Casemaker's mark ALF. .925 silver bow hallmarked imported London. White enamel dial, short black roman numerals and blued steel spade hands, inset sunk seconds dial with arabic numerals and 10 second intervals. Up and down subsidiary dial inset upper centre. Keyless wound and pull to set. Bar movement engraved 738644 (on barrel bridge) PAT COMPENS + No 98234 (on the centre bar). 17 jewel club tooth straight line lever movement with monometallic Elinvar balance with Ditisheim affix. Overcoil and oversprung Elinvar balance spring with star and rack index on balance cock engraved SR AF.
Diameter 50mm

PLATE 50B

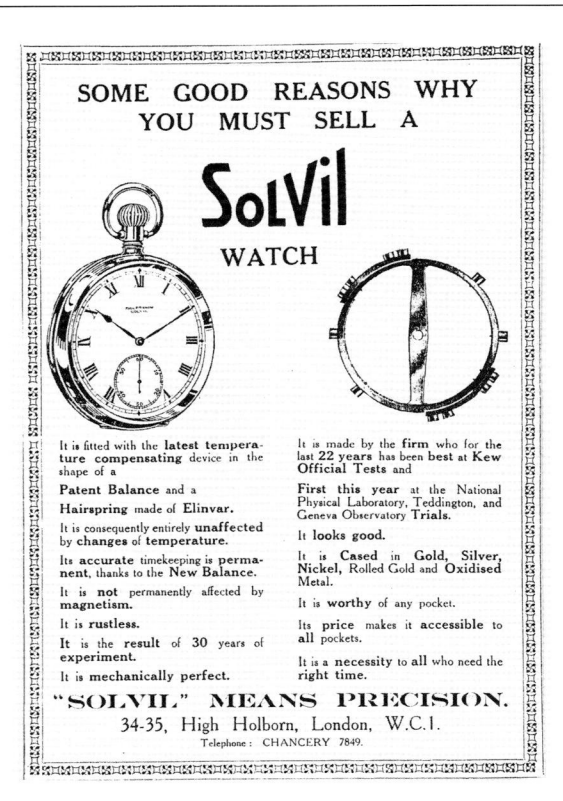

PLATE 50C

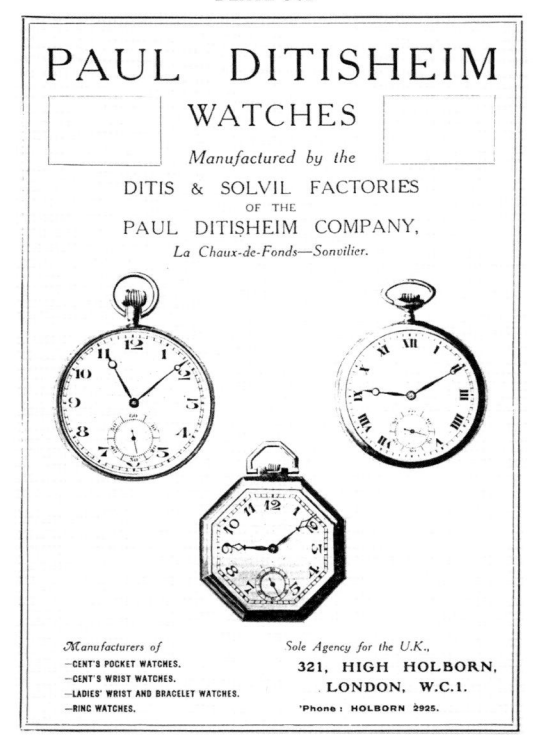

PLATE 50D

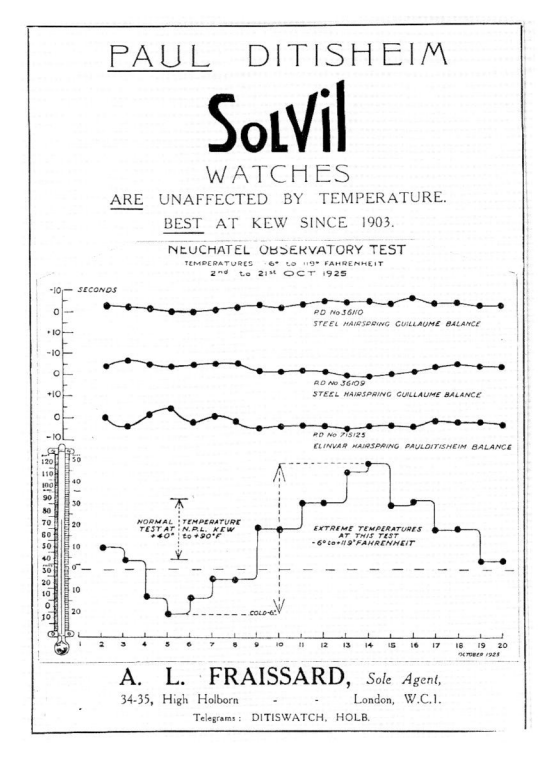

PLATE 50E

All from the *Horological Journal,* 1922

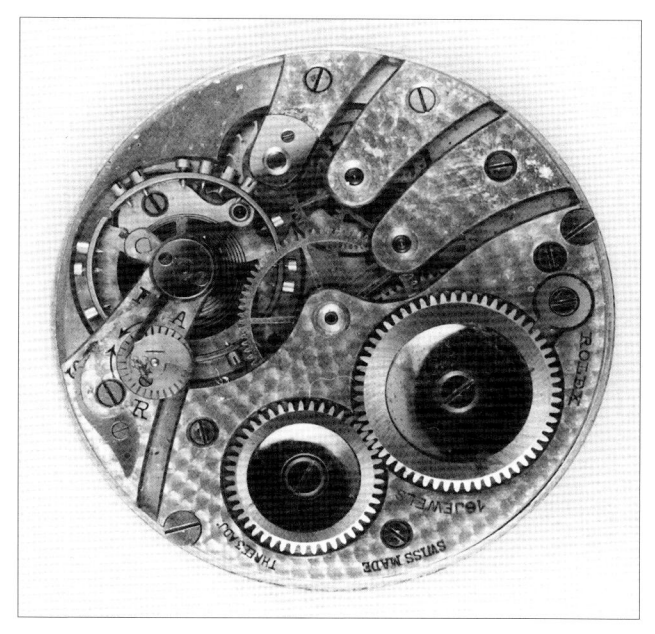

PLATE 51

PLATE 52

PLATE 51
THE LONDON

White enamel dial inscribed upper centre ROBINSON BROS The LONDON, short black roman numerals, blued steel spade hands, inset seconds dial. Dust ring to movement. Keyless wound. The top plate stamped ROLEX SWISS MADE 16 JEWELS THREE 3 ADJ. Bimetallic brass/steel cut balance with compensation screws. Blued steel Breguet overcoil balance spring. Snail micrometer index on balance cock. Endstones to balance and scape arbors.
Diameter 42.5mm

Founded by Hans Wildorf in 1878. Little seems to have been recorded regarding the pocket watches made by this firm; far more appears on their wristwatches.

PLATE 52
Brooklyn Watch Co

White enamel dial inscribed BROOKLYN WATCH CO, black roman numerals, blued steel spade hands, sunk seconds dial. Key wound going barrel. 7 jewel Swiss lever full plate movement with undersprung steel balance. Hand set by key to square on cannon pinion. Top plate stamped with registered trade mark three mast sailing ship B W Co and BROOKLYN WATCH CO No 1306873.

Marque de Fabrique No 1092 registered 24 Jan 1884, Petitpierre & Cie, Kaufleute und Fabrikanten, Chaux de Fonds. The Brooklyn Watch Co were wholesale retail distributors of jewellery, watches and watch supplies and never made watches. They purchased watches from Swiss and American watch companies and merely had their name added to the dial.

Chapter Four
American Lever Watches

America was the home of mass-production. This system of manufacture, characterised by the large scale production of machine-made interchangeable parts, admirably lent itself to the making of watches. Special mention has to be made of Aaron L Dennison who is regarded as the founder of the American watch factories. Inventing the Dennison standard gauge in 1840, he was by 1849 able to start manufacturing machinery for the production of watches on the interchangeable parts system. Further pioneering the system in 1850, he founded the firm of Dennison Howard & Davies which was later to become the giant Waltham Watch Company of Massachusetts. They were to produce nearly 35 million watches over the following century. Their obvious success dating from the mid-1800s was quickly followed by many other companies with varying results. The following chart compiled by LD Stallcup of Nashville, Tennessee, USA in 1943 demonstrates this point.

The unfolding story of the rise and subsequent dominance of the American watch industry is well documented. As these were factory-produced artefacts, records were kept of production figures of particular models, serial numbers, dates etc. Catalogues were produced to assist the repairer when ordering interchangeable parts from other material houses. With this wealth of source material it is not surprising to find a proliferation of detailed reference works for the present-day collector either providing a general overview or specifically on the output of one particular manufacturer.

It would be a grave mistake to assume that American mass-produced watches are of an inferior quality. There are admittedly those that are categorised as the 'dollar watch' which were intended to 'put a watch into virtually every American's pocket' and which were obviously made as economically as possible. The Waltham Riverside Maximus, the Premier Maximus and the Premier Vanguard displayed the highest grade finish to their movements. Similarly many models manufactured by the Elgin, Howard, Hamilton and Illinois factories were again to a high standard. By 1893 those made for the railroads had to comply with the strictest of standards.

They had to:
  be of 18 or 16 size
  have a minimum of 17 jewels
  be adjusted to 5 positions (a sixth was added later)
  keep time plus or minus 30 seconds per week
  be adjusted to temperature 40 to 95 degrees F
  have a double roller
  be lever set
  have winding stem at 12 o'clock
  have a plain arabic dial with heavy hands

It was not only the parts of the movements that were interchangeable. Cases were often made for several movement manufacturers by a common casemaking company or at best the cases were interchangeable with a number of movements. Quite often it is necessary to consult the reference lists of serial numbers, dates etc of a manufacturer in order to be certain that the movement in question is in the style of case originally intended for that calibre of movement. The quality of material used varies widely. Silver cases are marked 'coin' or 'silver'. Cases made of nickel, copper and manganese often have names alluding to silver – The Keystone Watch Case Company made the Silveroid, the Philadelphia Watch Case company the Silverode, the Fahys Watch Case Company the Silverore, while the Dueber Watch Case Company made the Silverine. Many of the 'gold' cases are actually gold filled or rolled, that is to say a brass sheet is sandwiched between two sheets of gold. The thickness of gold is indicated by the 'guaranteed' number of years the gold should wear before the brass shows through – ten years, fifteen, twenty or twenty-five being the commonly stated times. The solid gold cases can be as low as 8ct or as high as 18 ct.

The catalogue prepared by Donald Hoke on the comprehensive collection of American Pocket Watches housed in the Time Museum at Rockford, Illinois and *American Watchmaking; a Technical History of the American Watch Industry 1850-1930* by MC Harrold both provide an excellent background against which to acquire more detailed information on individual companies. Roy

Ehrhardt has written and published a series of books on such companies as Waltham, Elgin, Illinois, Hamilton etc as well as a further book on identification and price guides related to American watches. Further reading would include:

Almost Everything You Wanted to Know About American Watches and Didn't Know Who to Ask by GE Townsend
Encyclopedia of the Dollar Watch by GE Townsend
American Railroad Watch by GE Townsend
American Pocket Watch Identification and Price Guide 1830-1980 by R Ehrhardt and W. Meggers
Complete Price Guide to Watches series by Cooksey Shuggart
The Watch Repairers Manual 4th ed by Henry B Fried
History of the American Watch Case by Warren H Niebling
Bulletin of the National Association of Watch and Clock Collectors Inc, Columbia, PA, USA

History Chart of American Horology

THE history of watch manufacture in America is clearly shown in the chart below. It shows that out of the seventy or more enterprises launched since 1850 only four remain in operation. The chart was devised and drawn by L. D. Stallcup of Nashville, Tennessee, who is secretary-treasurer of his local Watchmakers' and Jewellers' Association, a regular reader of of the JOURNAL and a member of the B.H.I.

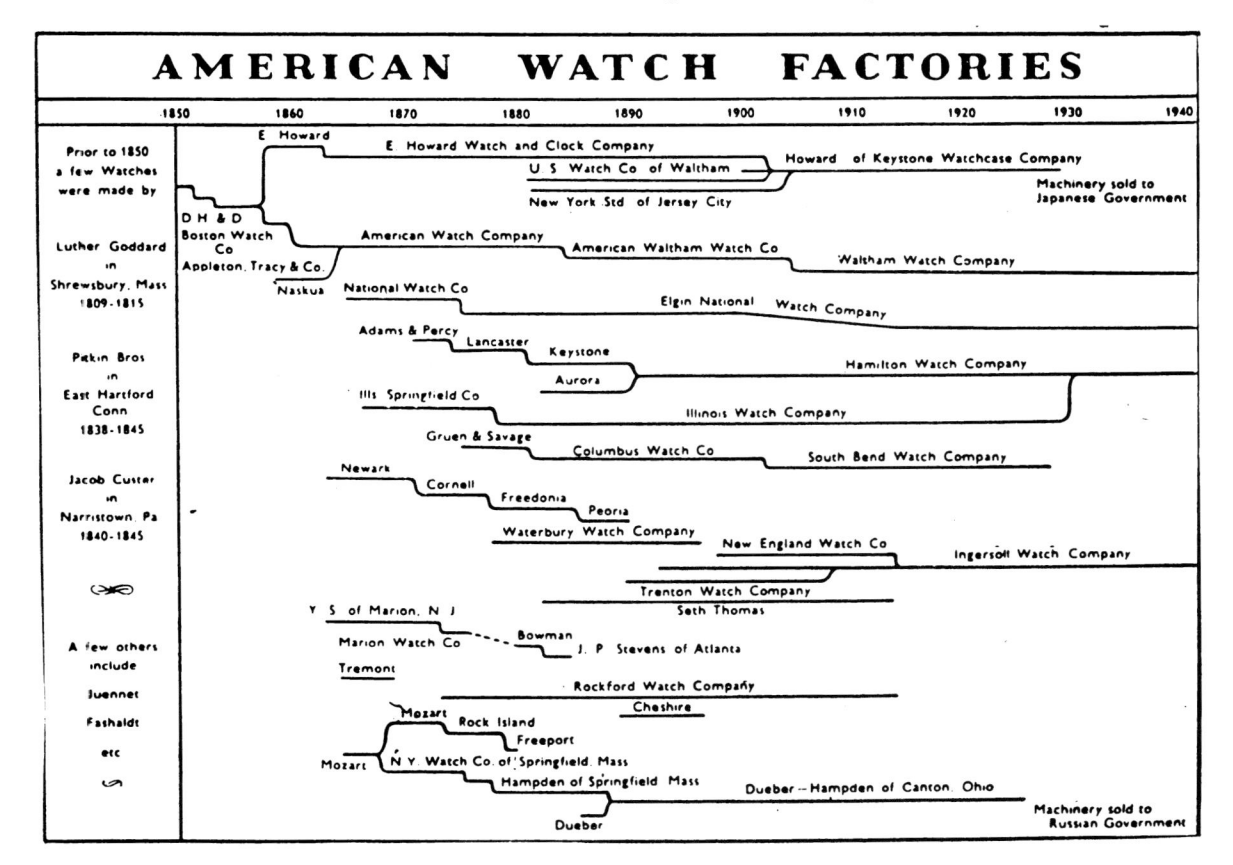

PLATE 1. *Horological Journal* July 1943, page 172

American Lever Watches

Condensed Table illustrating different sizes of
MAINSPRINGS used in WALTHAM WATCHES.

A. W. W. Co's Number.	Style.	Name.	Description of Movements, Barrels, &c.
2200A		For Series of 1862.	20 Size, ¾ Plate, 1st Series, Key Winding Movements, with Stratton's Patent Gilt Barrels. Hole on End. Hook in Barrel.
2200B		For Series of 1862.	20 Size, ¾ Plate, 2d Series, Key Winding Movements, with Regular Gilt Barrels. With T End.
2200		For Series of 1859.	18 Size, ¾ Plate, Key Winding Movements, with Regular Gilt Barrels. With Brace.
2201		For Series A. Model of 1857.	18 Size, Full Plate, Key and Hunting Stem Winding Movements. (Formerly known as Old Model) With Regular Gilt Barrels. Old Style with Brace.
2202		For Series A. Model of 1857.	18 Size, Full Plate, Key and Hunting Stem Winding Movements. (Formerly known as Old Model) With Regular Gilt Barrels. New Style with T End.
2203		For Series C. Model of 1877.	18 Size, Full Plate, Key and Hunting Stem Winding Movements, with Regular Gilt and Nickel Barrels (formerly known as New Model) Wide T End.
2204		For Series D. Model of 1879.	18 Size, Full Plate, Open Face, Stem Winding Movements, with Narrow, Regular Gilt and Nickel Barrels. (Formerly known as New Model Open Face) Narrow T End.
2205		For Series F. and G. Models of 1884.	18 Size, Full Plate, Open Face, Stem Winding Pendant and Lever Setting, also Key Winding; and Hunting Stem Winding Lever Setting Movements, with Regular Gilt and Nickel Barrels. Wide T End.
2206		For Series B. "Crescent Street." Model of 1870.	18 Size, Full Plate, "Crescent Street," Key Winding Movements with Narrow, Regular Gilt Barrels for Stop Works. Narrow T End.
2207		For Series B. "Crescent Street." Model of 1870.	18 Size, Full Plate, "Crescent Street" Key and Hunting Stem Winding Movements, with Wide, Regular Gilt Barrels. Wide T End.
2208A		For Model of 1860.	16 Size, ¾ Plate, 1st Series, Key Winding Movements, with Stratton's Patent Gilt Barrels. Hole on End. Hook in Barrel.
2208B		For Model of 1862.	16 Size, ¾ Plate, 2d Series, Key Winding Movements, with Regular Gilt Barrels. Narrow T End.
2208C		For Series H. Model of 1868.	16 Size, ¾ Plate, 1st Series, Hunting Stem Winding Movements, with Narrow, Regular Gilt Barrels. Narrow T End.
2208		For Series I and J. Models of 1872.	16 Size, ¾ Plate, 2d Series, Hunting and Open Face Stem Winding Movements, with Regular Gilt and Nickel Barrels. Narrow T End.
2209		For Adams Street and Crescent Garden: Model of 1870.	14 Size, ¾ Plate, Key Winding Movements, with Regular Gilt Barrels. With T End.
2210		For Series K, L, M, and N. Model of 1874.	14 Size, ¾ Plate, Hunting and Open Face Stem Winding Movements, with Narrow, Regular Gilt and Nickel Barrels. Narrow T End.
*2211		For Series O and P. Model of 1884.	14 Size, ¾ Plate, Hunting and Open Face Stem Winding Lever and Pendant Setting Movements, with Wide, Regular Gilt and Nickel Barrels. Wide T End.
2212		For Model of 1861.	10 Size, ¾ Plate, 1st Series, Key Winding Movements, with Narrow, Patent Gilt Barrels. Narrow with Brace.
2213		For Model of 1861.	10 Size, ¾ Plate, 2d Series, Key Winding Movements, with Wide, Patent Gilt Barrels. Wide with Brace.
2214		For Model of 1874.	10 Size, ¾ Plate, 3d Series, Key Winding Movements, with Regular Gilt Barrels. Wide T End.
2215		For Model of 1873.	8 and 6 Size, ¾ Plate, 1st Series, Key and Stem Winding Movements, with Regular Gilt and Nickel Barrels. With T End.
2216		For Models of 1882 and 1887.	1 and 0 Size, ¾ Plate, 1st Series, Stem Winding Lever and Pendant Setting Movements, with Regular Gilt and Nickel Barrels. With T End.
2217		For Model of 1889.	6 Size, ¾ Plate, 2nd Series, Stem Winding Lever and Pendant Setting Movements, with Regular Gilt and Nickel Barrels. With T End.
2218		For Model of 1888.	16 Size, ¾ Plate, 1st Series, Stem Winding Lever and Pendant Setting Movements, with Steel Barrels. Hole on End. Hook in Barrel.
2219		For Model of 1890.	6 Size, ¾ Plate, Stem Winding and Pendant Setting Movements, with Steel Barrels. Hole on End. Hook in Barrel.
2220		For Model of 1891.	0 Size, ¾ Plate, 2d Series, Stem Winding and Pendant Setting Movements, with Steel Barrels. Hole on End. Hook in Barrel.
2221		For Model of 1891.	Double Naught (00) Size, ¾ Plate, Stem Winding and Pendant Setting Movements, with Regular Nickel Barrels. With T End.
2222		For Model of 1892. Vanguard Series.	18 Size, Full Plate, Stem Winding Pendant and Lever Setting Movements, with Steel Barrels. Hole on End. Hook in Barrel.
2223		For Model of 1895.	14 Size, Stem Winding Lever and Pendant Setting Movements, with Steel Barrels. Hook on End. Hole in Barrel.
2224		For Model of 1894.	12 Size, ¾ Plate, Stem Winding and Pendant Setting Movements, with Steel Barrels. Hook on End. Hole in Barrel.
2224A		For Model of 1894.	12 Size, ¾ Plate, Stem Winding and Pendant Setting Movements, with Steel Barrels. Hole on End. Hook in Barrel.
2225		For Model of 1895.	14 Size, ¾ Plate, Stem Winding Lever and Pendant Setting Movements, with Steel Barrels. Hole on End. Hook in Barrel.
2226		For Model of 1898.	Jewel Series, ¾ Plate, Stem Winding and Pendant Setting Movements, with Steel Barrels. Hole on End. Hook in Barrel.
2227		For Model of 1899.	16 Size, ¾ Plate, 2d Series, Stem Winding and Pendant Setting Movements, with Steel Barrels. Hole on End. Hook in Barrel.
2228		For Model of 1900.	0 Size, ¾ Plate, 3d Series, Stem Winding and Pendant Setting Movements, with Steel Barrels. Hole on End. Hook in Barrel.
2230		For Model of 1907.	0 Size, ¾ Plate, 4th Series, Stem Winding and Pendant Setting Movements, with Steel Barrels. Hole on End. Hook in Barrel.

Regular Quality Mainsprings (all sizes) 4/- per dozen.
        Resilient   ,,   ,,   ,,   5/-   ,,
"Unit"   ,,   ,,   ,,   ,,   5/6   ,,

* 2211 also used in 14 size Full Plate, Key Wind.

PLATE 2. Waltham Watch Company 1910 Catalogue

WALTHAM WATCHES.

No. 625. 16 Size.
No. 625. Nickel. Seventeen Jewels, Settings, Exposed Pallets, Cut Expansion Balance, Meantime Screws, Patent Detachable Balance Staff, Patent Breguet Hairspring, Hardened and Tempered in Form, Patent Micrometric Regulator, Tempered Steel Safety Barrel, Exposed Winding Wheels, Red Gilded Centre Wheel.

No. 620. 16 Size.
No. 620. Nickel. Fifteen Jewels, Settings, Exposed Pallets, Cut Expansion Balance, Meantime Screws, Patent Breguet Hairspring, Hardened and Tempered in Form, Patent Detachable Balance Staff, Patent Micrometric Regulator, Tempered Steel Safety Barrel, Exposed Winding Wheels, Red Gilded Centre Wheel.

Marquis. 16 Size.
MARQUIS. Gilded. Fifteen Jewels, Settings, Exposed Pallets, Cut Expansion Balance, Patent Breguet Hairspring, Hardened and Tempered in Form, Patent Detachable Balance Staff, Tempered Steel Safety Barrel, Exposed Winding Wheels.

English Hall Marked. 9-carat Gold. "Scolloped."		"SUN." 14-carat Gold Filled. Guaranteed to wear 25 years.		"MOON." 10-carat Gold Filled. Guaranteed to wear 20 years.		"STAR." Guaranteed to wear 10 years.		English Hall-marked. Sterling Silver Cases.		Nickel Cases.	
Hunting.	Open Face.	Hunting.	Open Face. Jointed or Swing-Ring.	Hunting.	Open Face.	Hunting.	Open Face.	Hunting.	Open Face. Dome or Screw B.&B.	Open Face. Screw Bezel. Swing-Ring.	Open Face, Screw Back and Bezel.
£ s. d.	£ s. d.	£ s. d.	£ s. d.	£ s. d.	£ s. d.	£ s. d.	£ s. d.	£ s. d.	£ s. d.	£ s. d.	£ s. d.
14. 7. 3	11.17. 0	8. 5. 0	7.19. 6	7.16. 3	7. 8. 6	6.11. 9	6. 9. 6	6. 9. 6	6. 6. 6	5.19. 0	5.14. 6
13.14. 9	11. 4. 9	7.12. 9	7. 7. 3	7. 4. 0	6.16. 0	5.19. 3	5.17. 3	5.17. 3	5.14. 3	5. 6. 9	5. 2. 3
13. 3. 3	10.13. 3	7. 1. 3	6.15. 9	6.12. 6	6. 4. 6	5. 7. 9	5. 5. 9	5. 5. 9	5. 2. 9	4.15. 3	4.10. 9
12. 5. 0	9.15. 0	6. 3. 0	5.17. 6	5.14. 3	5. 6. 3	4. 9. 6	4. 7. 6	4. 7. 6	4. 4. 6	3.17. 0	3.12. 6
11.11. 3	9. 1. 3	5. 9. 3	5. 3. 9	5. 0. 6	4.12. 6	3.15. 9	3.13. 9	3.13. 9	3.10. 9	3. 3. 3	2.18. 9
11. 5. 6	8.15. 6	5. 3. 3	4.17. 9	4.14. 6	4. 6. 9	3.10. 0	3. 7. 9	3. 7. 9	3. 4. 9	2.17. 3	2.12. 9
10.18. 6	8. 8. 6	4.16. 3	4.10. 9	4. 7. 6	3.19. 9	3. 3. 0	3. 0. 9	3. 0. 9	2.17. 9	2.10. 3	2. 5. 9

Extra for Demi-Hunting 18-ct., 15-ct., 9-ct. and Gold Filled, 9/- ; Silver, 6/9.
Extra for 9-carat "Heavy" cases, Hunting, 20/3, and Open Face 18/6, on price of regular Hunting and Open Face.
Hunting 9-carat "Special" 11/-, and Open Face 15/6 less than price of regular Hunting and Open Face.

All orders accepted subject to prices prevailing at date of delivery.

PLATE 3. Waltham Watches price list catalogue December 1915

American Lever Watches

PLATE 4A

PLATE 4B

PLATE 4
American Watch Co, Waltham, Mass, USA

Silver open face case hallmarked Birmingham **1884**, hinged bezel back and cuvette. Casemaker Alfred Benson, marked AS. White enamel dial signed A W Co WALTHAM, black arabic numerals 1-12 and small red arabic numerals 5-60 at 5 minute intervals outside chapter ring, narrow spade blued steel hands, small inset sunk seconds dial. Keyless wound and lever to set under edge of dial at 1 o'clock position. Three-quarter plate. 15 jewels. Straight line lever movement and club tooth scape wheel. Oversprung cut bimetallic balance and blued steel balance spring. Heart piece attached to index moving between calliper micro-regulator. Top plate engraved Am Watch Co Waltham Mass 2298941. Safety pinion pillar plate stamped under balance rim WOERD'S PATENTS.
Diameter 48.7mm

Serial number 2298941 corresponds to manufacturing period 1883. Ehrhardt lists No 2298901-229850 Size 14, Model 74, Grade American W Co 3/4 plate, 15 jewels, Setting SE, Open face, Grade Material P.

Charles Vander Woerd, an immigrant from Leyden, Holland, was a distinguished inventor and machinist and worked in the machine shop at Walthams from 1857, leaving in 1859 to join the Nashua Company but returning to Waltham when Nashua was sold. He became the superintendent of the whole Waltham Watch factory in 1875 and finally retired from the company in 1887 at the age of sixty-six. A number of patents were taken out in his name.

USA Patents CV Woerd:

May 21	1866	65034	Jul 27	1880	230596
Aug 13	1867	67692	Feb 6	1883	271965
Oct 5	1869	95547	Jun 30	1885	320992
Mar 29	1870	101398	Apr 27	1886	340850
Dec 27	1870	110614	May 11	1886	341786
Aug 22	1871	118415	Dec 7	1886	354002
Apr 6	1875	161725			

PLATE 5A

PLATE 5B

PLATE 5
American Watch Co, Waltham, Mass, USA

Silver pair-case, hinged bezel and inner back, hallmarked London **1884**. Casemaker's mark JW. Antique style bow. White enamel dial, black roman numerals, gold thin spade hands, inset sunk seconds dial. Keywound and key set cannon pinion square. Top plate engraved WALTHAM MASS Pat Pinion 2703123. Full plate. 7 jewels. Going barrel lever movement with club tooth scape wheel and tangential lever. Oversprung three arm monometallic balance with rounded rim to balance and blued steel balance spring. The regulator index on balance cock.
Key size 5. Diameter 53.8mm

Serial No 2703123 corresponds to manufacturing period 1885. Ehrhardt lists 2703001-2703700 Size 14, Grade Home, Full plate, 7 jewels, Keywound, Grade Material U.

PLATE 6
American Watch Co, Waltham, Mass, USA

Silver open face case, snap-on bezel and hinged back and cuvette. Casemaker's mark AB (Alfred Benson – Dennison Watch Case Co Ltd). White enamel dial, black roman numerals, black spade hands. Key wound and key set. 7 jewels. Three-quarter club tooth scape wheel and lever movement with straight line lever. Oversprung with overcoil with blued steel balance spring and three arm gold balance. Regulator index on engraved balance cock.

PLATE 6

Top plate engraved A M WALTHAM CO WALTHAM MASS 3049213 SAFETY PINION.
Key size 4. Diameter 39mm

Serial number 3049213 corresponds to manufacturing period 1887. Ehrhardt lists 3047001-3051000 6 size, 73 model, Grade Am W Co, 3/4, 7-11 jewels, Keywound Breguet spring, Grade Material U.

American Lever Watches

PLATE 7A

PLATE 7B

PLATE 7
RIVERSIDE, American Watch Co, Waltham, Mass, USA

Silver open face case, hinged bezel back and cuvette. Cuvette engraved Henry W Bedford 69 Regent ST London, black roman numerals, blued steel spade hands, inset sunk seconds dial. Keyless winding, lever to set hand at 1 o'clock position. Three-quarter plate. 13 jewels. Engraved on the top plate RIVERSIDE AM WATCH CO WALTHAM MASS 3590934 SAFETY PINION ADJUSTED. Club tooth scape wheel and straight line lever. Oversprung and overcoil cut bimetallic balance with blued steel balance spring. Regulator index on balance engraved FS.

Diameter 48mm

Henry William Bedford was the agent for the Waltham Watch Co. Serial No 3590934 corresponds to manufacturing period 1886/1887. Ehrhardt lists 3590901-5591000 14 size, Model 84, Riverside, stem wound, Breguet spring to balance, 13 jewel lever, Open face. Grade material A. Patent regulator.

PLATE 8
RIVERSIDE, American Waltham Watch Co, Waltham, Mass, USA

Silver open face case hallmarked Birmingham **1899**. Hinged back and cuvette and snap-on bezel, engine-turned back with milled edge to body. Casemaker AB (Alfred Benson). White enamel dial inscribed AWW Co Waltham Mass in upper centre, black roman numerals, blued steel spade hands, inset sunk seconds. Keyless wound and pull to set. Top plate engraved 6614711 AWW Co RIVERSIDE WALTHAM MASS ADJUSTED 17 JEWELS. 17 jewels in gold settings. Club tooth scape wheel with straight line lever. Oversprung and overcoil cut bimetallic balance with blued steel balance spring with star and rack micro-regulator on an engraved cock.

Diameter 50.5mm

Serial number 6614711 corresponds to manufacturing period 1895. Ehrhardt lists 6614001-6615000 16 size, Riverside, 17 jewels, Setting PS, Breguet spring, Open face, Grade material P.

PLATE 9
American Waltham Watch Co, Waltham, Mass, USA

Silver open face case hallmarked Birmingham **1899**, hinged back bezel and cuvette. Casemaker AB (Alfred Benson). White enamel dial inscribed AWWCo WALTHAM MASS upper centre, black roman numerals, gold spade hands, inset seconds dial. Key wound and keyset cannon pinion square. 7 jewels. Full plate movement engraved on top plate SAFETY PINION 10607837 Am Watch Co WALTHAM MASS. Club tooth scape wheel and tangential lever with oversprung overcoil blued steel balance spring with regulator index and engraved balance cock.

Key size 7. Diameter 57.6mm

Ehrhardt lists 10607837, Size 18, Model 83, Export Grade, 7 jewels, Keywind, Grade material U.

PLATE 8

PLATE 9A

PLATE 9B

American Lever Watches

American Lever Watches

PLATE 10
American Waltham Watch Co, Waltham, Mass, USA

Silver engraved open face case, hallmarked Birmingham **1901**, hinged back and cuvette and snap-on bezel. Casemaker AB (Alfred Benson). Pendant maker AW. White enamel dial inscribed AWW Co WALTHAM MASS upper centre, narrow black roman numerals, narrow blued steel spade hands. Keyless wound and pull to set. 7 jewels. Three-quarter plate. Club tooth scape wheel and straight line lever movement. Oversprung cut bimetallic balance with gold quarter screws. Blued steel balance spring. Regulator index on engraved cock. Top plate inscribed AM WATCH Co WALTHAM MASS SAFETY BARREL 6798035.

Serial number 6798035 corresponds to manufacturing period 1898. Ehrhardt lists 6798001-6799000, Size 6, Model 90, Grade Export, 7 jewels, Breguet spring, Grade materials U.

PLATE 10A

PLATE 10B

200

PLATE 11

PLATE 12

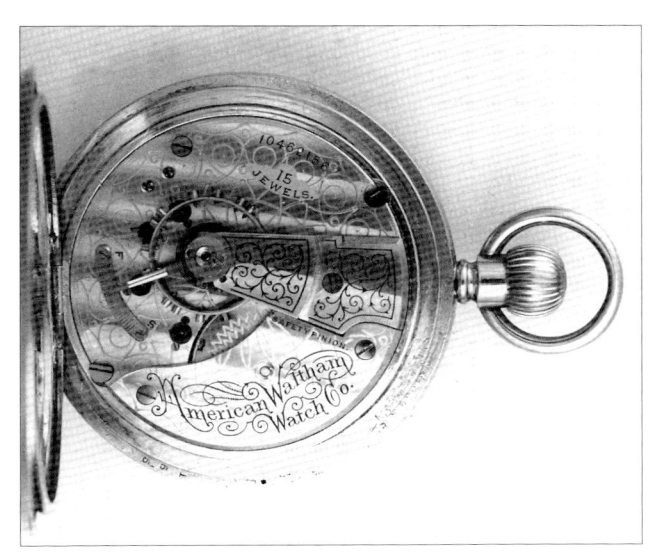

PLATE 11
American Waltham Watch Co, Waltham, Mass, USA

14ct gold open face filled to last 25 years Dennison case, snap-on bezel, hinged back and cuvette. White enamel dial, black roman numerals, blued steel spade hands, inset seconds dial. Keyles wound and pull to set hands. 15 jewels. Full plate club tooth scape wheel and tangential lever movement with oversprung cut bimetallic balance and blued steel balance spring. Top plate engraved American Waltham Watch Co SAFETY PINION 15 JEWELS 10462156. Regulator index over balance rim to silver quadrant.

Diameter 54.3mm

Serial Number 10462156 corresponds to manufacturing period 1901. Ehrhardt lists 10459501-10462500 18 size, Model 83, No 820, 15 Jewels, Grade material U.

PLATE 12
PS BARTLETT, American Waltham Watch Co, Waltham, Mass, USA

Gold filled open face, Fahys Montauk case, guaranteed 20 years, screwed bezel and back. White enamel dial inscribed WALTHAM upper centre with sunk centre and sunk inset seconds dial, narrow black roman numerals I-XII and arabic numerals 13-24, narrow blued steel spade hands. Keyless wound and pull to set. 17 jewel. Full plate club tooth lever movement with tangential pallets and oversprung with overcoil and cut bimetallic balance and blued steel balance spring. Top plate inscribed PS BARTLETT WALTHAM 10568363 17 Jewels ADJUSTED and 8 pointed star and rack micro-regulator on engraved balance cock.

Diameter 50.5mm

Serial number 10568363 corresponds to manufacture period 1901. Ehrhardt lists 18 size, 83 Gauge, No 87, PSB, 17 jewels, Patent regulator, Open face, Grade Material A. Joseph Fahys' watch case factory was at Sag Harbor, Long Island, New York. Montauk Trade Mark No 10372 registered 26 June 1883.

American Lever Watches

PLATE 13A

PLATE 14

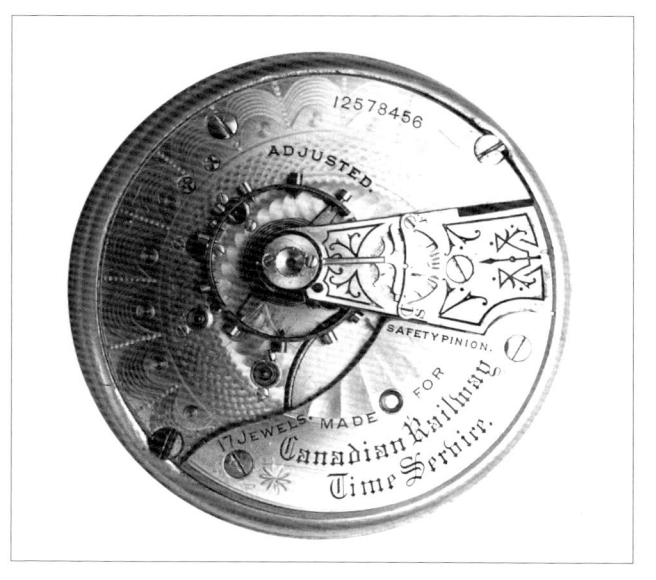

PLATE 13B

AW Co WALTHAM upper centre, narrow black roman numerals, narrow spade hands, inset seconds dial. Keyless wound and pull to set. Full plate. Club tooth wheel with tangential lever and oversprung and overcoil cut bimetallic balance with blued steel balance spring. Star and rack micro-regulator. 17 jewel in gold settings. Top plate inscribed SAFETY PINION 12578456 ADJUSTED MADE FOR CANADIAN RAILWAY TIME SERVICE 17 JEWELS.
Diameter 57.9mm

Serial number 12578456 corresponds to manufacturing period 1904. Ehrhardt lists 12578001-12578500 18 size, Model 83, Special, 17 jewels, Open face, Grade Materials A, Patent micro-regulator, Cut bimetallic balance.

PLATE 14
American Waltham Watch Co, Waltham, Mass, USA

Gold filled open face Moon grade Dennison case, hinged back and cuvette which is stamped AMERICAN WALTHAM WATCH COMPANY around Waltham fish trade mark. White enamel dial, WALTHAM upper centre, black roman numerals and gold spade hands, inset sunk seconds. Keyless wound and pull to set. Top plate engraved AWW Co WALTHAM MASS 17 JEWELS 13090040. Club tooth scape wheel, straight line lever. Oversprung and overcoil cut bimetallic balance and blued steel balance spring with patented whiplash regulator.
Diameter 50.90mm

Serial number 13090040 corresponds to manufacturing period 1904/1905. Ehrhardt lists 13088501-13090500 16 size, Model 99, No 625, 17 jewels, Open face, Patent regulator.

PLATE 13
American Watch Co, Waltham, Mass, USA

Silveroid open face case, screw bezel and swing-out movement (withdraw winder first!). White enamel dial,

American Lever Watches

PLATE 15A

PLATE 15
American Watch Co, Waltham, Mass, USA

Oxydised steel open face case, snap-on bezel and hinged back. White enamel dial inscribed SCHIERWATER & LLOYD LIVERPOOL, short black roman numerals, gold spade hands, inset seconds dial. Keyless winding and pull to set. 7 jewels. Three-quarter plate. Club tooth scape wheel with straight line lever movement. Oversprung cut bimetallic balance and blued steel balance spring with regulator index on engraved balance cock. Top plate engraved AM Watch Co WALTHAM MASS BOND ST 13931808.
Diameter 48.8mm

Serial number 13931808 corresponds to manufacturing period 1904/1905. Ehrhardt lists 13929001-13937000 Size 14, Model 1904, Bond Street, 7 jewels, Open face, Grade Material U.

PLATE 15B

American Lever Watches

PLATE 16

PLATE 17A

PLATE 17B

PLATE 16 and COLOUR PLATE 9 (page 7)
VANGUARD, American Waltham Watch Co, Waltham, Mass, USA

14 ct gold filled open face Dennison case, Sun grade, hinged back and cuvette. White enamel dial, black numerals, gold spade hands, inset seconds dial. Keyless wound and pull to set. Barrel plate stamped 14111687 and centre portion of top plate stamped 23 JEWELS ADJUSTED 5 POSITIONS VANGUARD WALTHAM MASS. Gold centre wheel and movement jewelled to the centre arbor with endstones to pallet and scape arbors with jewels in gold settings with gold screws. Club tooth scape wheel with straight line lever. Oversprung with overcoil cut bimetallic balance adjusted to 5 positions and blue steel balance spring with star and pinion micro-regulator on balance cock.
Diameter 50.7mm

Serial number 14111687 corresponds to manufacturing period 1905 Ehrhardt lists 14111501-14112300 16 size, Model 99, Vanguard Grade, 23 jewels, Patent regulator, Open face, Grade material P.

PLATE 17
P S BARTLETT, American Waltham Watch Co, Waltham, Mass, USA

Gold filled open face case stamped FORTUNE WARRANTED AWWC 20 years, screw back and bezel. White enamel dial inscribed WALTHAM upper centre, black roman numerals I-XII and black arabic 13-24, narrow blued steel spade hands, inset sunk seconds. Keyless wound and pull to set. Club tooth scape wheel and tangential lever. Oversprung and overcoil cut bimetallic balance and blued steel balance spring. 8 pointed star and rack micro-regulator. Top plate inscribed P S Bartlett WALTHAM MASS 17 Jewels ADJUSTED SAFETY PINION 16010796.
Diameter 55.2mm

Serial number 16010796 corresponds to manufacturing period 1907/1908. Ehrhardt lists 16009501-16012000 Size 18, Model 83, PSB, 17 jewels, Style OF, Grade material A.

PLATE 18

PLATE 18
American Waltham Watch Co, Waltham, Mass, USA

Silver open face Dennison case hallmarked Birmingham **1912** with engine-turned back and milled edge to body. White enamel dial inscribed upper centre Waltham USA, short black roman numerals, blued steel spade hands, inset sunk seconds dial. Keyless wound and pull to set hands. Top plate engraved AWWCo WALTHAM MASS 15 JEWELS 17562427. Club tooth scape wheel, straight line lever. Oversprung and overcoil cut bimetallic balance. Blued steel balance spring with star and rack micro-regulator on balance cock.

Diameter 50.7mm

Serial number 17562427 corresponds to manufacturing period 1909. Ehrhardt lists 17562001-17564000 16 size, Model 1908, Grade No 620, 15 jewels, Patent regulator, Open face Style, Grade U.

American Lever Watches

PLATE 19A

PLATE 19B

PLATE 20A

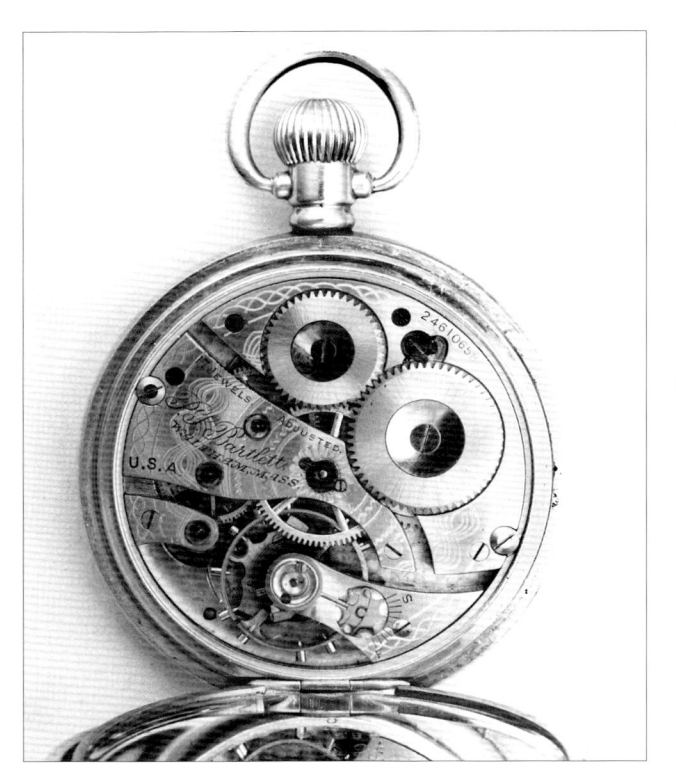

PLATE 20B

206

PLATE 21A

PLATE 21B

PLATE 19
RIVERSIDE, American Waltham Watch Co, Waltham, Mass USA

Gold filled open face Dennison case. White enamel dial inscribed Waltham upper centre, black roman numerals, blued steel spade hands, inset sunk seconds dial. Keyless wound and pull to set. Club tooth straight line lever escapement. Oversprung and overcoil cut bimetallic balance with blued steel balance spring with patent star and rack micro-regulator. Top plate engraved AWWCO WALTHAM MASS RIVERSIDE 19 jewels ADJUSTED 18108533.
Diameter 50.5mm

Serial number 18108533 corresponds to manufacturing period 1912. Ehrhardt lists 16 size, Model 1908, 19 jewels, Patent regulator, Open face, Grade Material P.

PLATE 20
P S BARTLETT, American Waltham Watch Co, Waltham, Mass, USA

Hunter Dennison case, Moon grade, Dennison bow. White enamel dial inscribed Waltham USA, black roman numerals, blued steel spade hands, inset sunk seconds. Keyless wound and pull to set. Club tooth scape wheel and straight line lever. Oversprung with overcoil bimetallic balance. Star and rack micro-regulator on balance cock. Top plate engraved P S Bartlett WALTHAM MASS USA 17 jewels ADJUSTED 24610658.
Diameter 51mm
Serial number 24610658 corresponds to manufacturing period 1924. Ehrhardt lists 16 size, Model 1908, PSB, 17 jewels, Whiplash regulator, Hunting case style, Grade Materials A.

PLATE 21
VANGUARD, American Waltham Watch Co, Waltham, Mass, USA

'Sturdy Nickel Silver' open face case with screw back and bezel. White enamel dial inscribed WALTHAM VANGUARD 23 JEWELS, black arabic numerals 1-12 and smaller black arabic 13-24 within the chapter ring, blued steel spade hands, inset seconds dial. Keyless wound with lever to set hands at 11 o'clock under edge of dial. Club tooth scape wheel and straight line lever. Oversprung and overcoil monometallic balance with poising screws, two of which are gold, and movement compensated for six positions. White coloured balance spring. Micro-regulator. Top plate engraved 30562747 ADJUSTED POSITION 23 JEWELS.
Diameter 41.9mm

Serial number 30562747 corresponds to manufacturing period 1941. Ehrhardt lists 30560001-30566000 Size 16, Model 1908, Grade No 1623, 23 jewels, Setting LS, Style OF, Grade Material P.

PLATE 22A

PLATE 22B

PLATE 23

PLATE 22
PREMIER, American Waltham Watch Co, Waltham, Mass, USA

Base metal open face Keystone case, screwed back and bezel. Black enamel dial inscribed upper centre WALTHAM in stylised white arabic numerals, steel spade hands, inset seconds dial. Keyless wound, pull to set. Club tooth scape wheel and straight line lever. Oversprung and overcoil monometallic balance with compensation screws and adjusted to temperature and six positions. Regulator index on balance cock. Top plate engraved WALTHAM PREMIER 16s USA ADJ TEMP 3 POs NINE JEWELS 30898891.

Diameter 52mm

Serial numbers 30890001-30900000 correspond to manufacturing period 1939-1941. 16 size, Model 08, Grade No 1609, 9 jewel, Setting PS (pendant set), Style OF, Grade Material U. See War Office Dept Technical Manual TM9 1575 published 6 April 1945.

PLATE 23
Elgin National Watch Co, Aurora, Illinois, USA

Coin silver full hunter case with gold hinges. White enamel dial, black roman numerals and blued steel narrow spade hands, inset seconds dial with small arabic numerals at 10 seconds intervals. Keyless wound and setting lever at 5 o'clock position under bezel. Top plate engraved Elgin Nat'l Watch Co 579154 Patent Pinion. Club tooth scape wheel and tangential lever. 11 jewels. Oversprung two arm bimetallic compensated balance with blued steel balance spring. Index regulator over the balance rim to silver quadrant on the top plate.

Diameter 56.4mm

Serial number 579154 corresponds to manufacturing period 1877/8. Ehrhardt lists 579001-580000 Grade 10. Total production 273,000. No of runs 81, first number 577001, last number 630900. 18 size, Hunting case, Gilded plate, Model 2, Lever, 11 jewels.

Plate 24
Elgin National Watch Co, Illinois, USA

Gold filled Keystone open face case with screw ring to hold movement from back. White enamel dial, Elgin Nat'l Watch Co upper centre, narrow black roman numerals, narrow blued steel spade hands, inset sunk seconds. Keyless wound and pull to set. Top plate engraved Elgin Nat'l Watch Co Elgin Ill Safety Pinion 5368476. 7 jewels. Full plate. Club tooth scape wheel with tangential lever and oversprung cut bimetallic balance and blued steel balance spring. Regulator index on balance cock engraved FS.

Diameter 55 mm

Serial number 5368476 corresponds to manufacturing period 1893/1894. Ehrhardt lists 5362001-5369000, Grade 73, 7 jewels. Total production 659,000 in 117 runs, first 224000 and last 7607000.

Plate 24a

Plate 24b

American Lever Watches

PLATE 25A

PLATE 25B

PLATE 26A

PLATE 26B

PLATE 27A

PLATE 27B

PLATE 25
Elgin National Watch Co, Elgin, Illinois, USA

Base metal open face Keystone case, screw back and bezel, back engraved 298451XX. Black dial inscribed ELGIN, luminous arabic numerals and luminous spade hands, inset seconds dial. Keyless wound and pull to set. Top plate engraved ELGIN NAT'L WATCH CO USA 41294476 NINE JEWELS. Half plate. Club tooth straight line lever movement with monometallic compensated balance. Oversprung and overcoil Elinvar spring balance. Regulator index on balance cock engraved FS.
Diameter 51.8mm

Serial number 41294476 corresponds to manufacturing period 1942. Ehrhardt lists 41294001-41299000 Grade 594, 16 size, 9 jewels. Total production 78,000. Number of runs 10 first starting 41233001 and last 42206000.

PLATE 26
Elgin National Watch Co, Elgin, Illinois, USA

Base metal open face Keystone case, screw back and bezel, outside back engraved HS 3 (Hydrographic survey), inside back stamped CASED AND TIMED BY ELGIN NATIONAL WATCH CO. White enamel dial, black arabic numerals 1-12 hours and small 5-60 minutes, blued steel spade hands, centre sweep seconds. Keyless wound and pull to set. Top plate engraved B W Raymond 21 JEWELS ADJUSTED 5 POSITIONS TEMPERATURE ELGIN NAT'L WATCH CO 41754056. Polished club tooth scape and polished straight line lever. Oversprung and overcoil. Blued steel balance spring with cut bimetallic balance and whiplash micro-regulator.
Diameter 51.7

Serial number 41754056 corresponds to manufacturing period c.1942. Ehrhardt lists 39261001-42079000 GDES ETC. Total production run 20,800 in 9 runs.

PLATE 27
Hamilton Navigation Master Watch, Hamilton Watch Co, Lancaster, PA, USA

Base metal open face Keystone case, screw back and bezel, the back engraved M FR's PART No 39928/SERIAL No 1038-1942/HAMILTON WATCH CO/AM 6 B/60. White plastic dial with Government Property Mark and 1038 upper centre and HAMILTON below centre, black arabic hour numerals and small black arabic 5-60 minutes at five minute intervals outside chapter ring, broad spade blued steel hands, centre sweep seconds. Keyless wound and pull to set. Top plate engraved HAMILTON WATCH CO USA 3992B US GOVT ADJ TEMP AND 6 POSITIONS 22 JEWELS. The pillar plate stamped 3C1041. Half plate. Club tooth and straight line lever with oversprung and overcoil monometallic balance with compensation screws. Elinivar balance spring and whiplash micro-regulator. Balance stud of pentagonal form in a pentagonal hole.
Diameter 52mm

Note Elinvar Patent was taken out in 1931. Hamilton 3992B introduced 15 October 1940.

American Lever Watches

PLATE 28A

PLATE 28B

PLATE 28
Hamilton 24 Hour Dial, Hamilton Watch Co, PA, USA

Base metal open face Keystone case, screw back and bezel, back of case engraved SERIAL No AF 42-10962/CONTRACT No W 535ac-28072/HAMILTON WATCH CO. Matt black dial inscribed G C T, white 24 hour arabic numerals and white broad spade hands, centre sweep seconds. Keyless wound and pull to set hands. Top plate engraved 22 JEWELS ADJ TEMP AND 6 POSITIONS HAMILTON WATCH CO USA 4992B US GOVT. Pillar plate stamped 4012505. Club tooth straight line lever escapement with oversprung and overcoil monometallic compensation balance and Elinivar balance spring with pentagonal balance stud and whiplash micro-regulator.

Diameter 52mm

War Dept Technical Manual TM9-1575 published 6 April 1945 p 84-103 for Model 992B Hamilton.

PLATE 29
Keystone Watch Co, Philadelphia, Pennsylvania, USA

14ct gold open face, hinged dome (cuvette) and back Cuvette engraved E HOWARD WATCH CO BOSTON PRESENTED TO HARRY LAUDER BY THE ROTARIANS OF TEXARKANA USA. Casemaker Keystone Watch Case Company. White enamel dial inscribed HOWARD, broad black roman numerals, inset seconds dial, distinctive hour and minute hands. Barrel inscribed E HOWARD WATCH CO BOSTON USA 1210025. 17 jewels. Pillar plate inscribed PAT 'D' 12 ADJUSTED TEMPERATURE 3 POSITIONS. Keyless wound and pull to set. Bar movement with club tooth straight line lever escapement with cut bimetallic compensated balance. Oversprung with overcoil balance spring. Whiplash style of regulator.

Diameter 46mm

The 1919 edition of Keystone Watch Case Co catalogue reprinted by Manfried Trauring states that in 1903 the E Howard Watch and Clock Company sold all rights to the manufacture of E Howard watches to the Keystone Watch Case Company of Philadelphia, Pennsylvania. They designed an entirely new and different style of movement which had no resemblance to the old E Howard & Co movements. These watches started with serial number 845001. This new company purchased the factory of the defunct United States Watch Company at Waltham, Mass and in 1905 began making watches which were engraved E Howard Watch Co, Boston, Mass.

Texarkana is situated near the Arkansas border of Texas. Harry Lauder, a famous Scottish comedian, was born 4 August 1870. He toured the British Dominions and the USA many times and during his career was an ardent Rotarian. He was knighted in 1919 and died 26 February 1950.

American Lever Watches

PLATE 29A

PLATE 29B

PLATE 29C

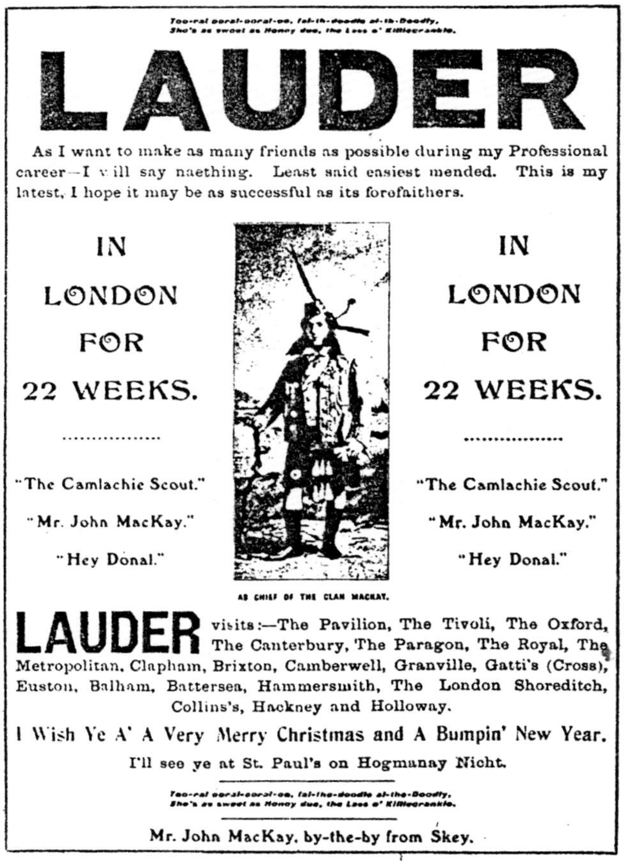

PLATE 29D

213

American Lever Watches

PLATE 30

PLATE 30
New York Standard Watch Co, New York, USA

Gilt brass hunter case, hinged front and back covers, cuvette, milled edge to body. White enamel dial, NEW YORK STANDARD WATCH Co USA upper centre, black roman numerals, blued steel tulip hands, inset seconds dial. Keyless wound, lever to set at V position. 7 jewels. Three-quarter plate. Club tooth straight line lever movement. Oversprung monometallic balance with compensation screws and blued steel balance spring. Regulator index on distinctive balance cock planted in mid line and engraved SAFETY PINION. Top plate stamped NEW YORK STANDARD WATCH Co 912295.
Diameter 54.5mm

A prefix number was added to the serial number after the first 10,000 watches were made. Total production 70,000,000 New York Standard Watch Co (1887-1925).

PLATE 31
Rockford Watch Co, Rockford, Illinois, USA

Gilt on brass open face case stamped ILLINOIS WATCH CASE Co ELGIN USA. White enamel dial, ROCKFORD WATCH Co USA upper centre, black arabic numerals, blued steel spade hands, inset seconds dial. Keyless wound and pull to set. 7 jewels. Half plate. Club tooth scape wheel and straight line lever movement. Overcoil cut bimetallic balance with blued steel balance spring and whiplash microregulator on balance cock engraved FS. Top plate inscribed 686173 ROCKFORD WATCH Co USA.
Diameter 51mm

Serial number 686173 corresponds to manufacturing period 1907.

PLATE 31A

PLATE 31B

Rockford Watch Co equipment was purchased from Cornell Watch Co. Two of their employees, CW Parkes and PH Wheeler, came to work from Rockford. Factory completed 1876 and first watch marketed 1 May 1876. In 1896 name changed to Rockford Watch Co Ltd. Closed in 1915 (Shuggart and Engle Complete Guide to American Clocks Book 5, p249).

Chapter Five
Duplex Watches

The English duplex escapement appears to be based on a patent No 1311 granted to Thomas Tyrer 1 Jan 1782, 'a horizontal escapement for a watch to act with two wheels, being a new and very great improvement on horizontal watches. No specification enrolled'. The Rev H L Nelthropp writing in his *Treatise on Watchwork Past and Present* outlines clearly the earlier history of the invention by Dr Robert Hooke of a new escapement called the two balance, one which attracted the attention of Jean Baptiste Dutertre (1715-42) of Paris who modified this and a form of duplex also assessed by Pierre Le Roi about 1750.

The duplex watch at the beginning of the nineteenth century was surpassed only by the chronometer escapement in performance, but had the advantage of not tripping. It was only superseded after 1850 by the more highly developed forms of the lever escapement. Whereas the Continental form of duplex had two coaxial wheels on the same arbor, the English had one wheel for both locking and impulse – the locking teeth extending radially from the rim and the impulse teeth vertically arranged. This complicated scape wheel required careful making and also precision jewelling of the scape arbor. Although an excellent performer in the hands of skilled makers, of whom the best known exponent was McCabe, the main problem was that wear easily occurred at the top of the teeth and the movement was likely to set when subjected to any jolting.

A special form of duplex is sometimes seen, usually from the Fleurier district of Switzerland. This escapement, designed by Charles Edward Jacot, became popular in the mid-nineteenth century for export to the Chinese market, assuming the name of Chinese duplex. The movement was heavily engraved and the locking teeth of the scape had a forked end, made to escape at alternate swings. The appearance therefore in a centre seconds watch was of the watch beating in one second steps.

The Waterbury duplex watch was not only of considerable interest from a technical and manufacturing aspect but also has a colourful history which has endeared the watches to the collector.

In 1875 Jason R Hopkins (1818-1902), Washington DC, designed a watch which he thought could be made for 50 cents. He was able to get the help of Benedict and Burnham Brass Manufacturing Co, Waterbury. Hopkins subsequently sold his ideas to M Fowle of Auburndale, Mass for $10,000. J Hopkins Patent 179019 20 June 1876 had been assigned to FB Fowle, but the venture failed, Mr Fowle losing $250,000. The original watch cost $10 and about 1,000 were made but few were sold as most of them were scrapped. This watch had a lever set for hands, 5 jewel lever escapement, rotated once in 2½ hours and winding was at the centre with a 4ft long spring.

In 1876 at the Philadelphia Centennial Exhibition, the one hundredth anniversary of the Declaration of American Independence, a Mr DAA Buck exhibited alongside a big Corliss Engine a miniature but complete steam engine that could be fitted under a thimble. In the spring of 1877 Mr Locke saw this miniature engine in the window of a watch repair shop of Mr Daniel Azro Ashley Buck and found that the engine weighed 15 grains, stood ⅝in. high and had 148 parts including 52 screws. The boiler was filled with three drops of water. Very impressed by this, Locke commissioned the making of a watch, long spring, rotary movement and chronometer escapement for which he paid $100. This was similar to the Hopkins watch and proved unsatisfactory, but a second design was conceived by Buck whilst recovering from an illness. This model had a duplex escapement and was a great success.

Buck, in designing the Waterbury duplex, was able to exploit its advantages and mitigate the disadvantages. Cheapness could be achieved as no lever, detent or pallets were required and the skill in designing a method of stamping out the duplex scape wheel in a press meant that, when wear did occur, a cheap replacement wheel was available to restore the satisfactory performance of the watch.

It was completed in the autumn of 1877. In January 1878 Edwards Locke again approached Benedict who recognised the potential of this model and agreed a loan of $8,000 plus use of some upper rooms at the factory of Benedict and Burnham. On 21 May 1878 Daniel Azro Ashley Buck, watchmaker of Worcester, Mass, applied for a series of patents.

The first watches were not ready till December 1878, after which they sold rapidly and were marked Benedict and Burnham Co. The first 1,000 were said to be poor runners due to a defect in the brass sheet used for stamping plates,

Duplex Watches

PLATE 1A

PLATE 1B

but subsequent batches were 90% satisfactory. This model, which came to be known as Series A, had a skeleton dial, nickel plated brass case, bezel and snap-on back. Some backs were said to be celluloid, for demonstration purposes. The balance was stamped out. The spring was nine ft long with 140 turns to wind up from the rim. The watch had a duration of 30 hours with the whole movement turning once per hour. The celluloid covered paper dial was glued to the plate (Hart's patent September 1883). Particular emphasis was placed on the good stamping out of all the functional parts in contrast to the expensive hand finishing of English and Swiss movements.

In 1885 Series B and C were introduced and it was stated that any factory repair on watches would not be more than 50 cents. By 1888 the F series had been introduced and production was running at 1,000 per day. When the Series I Trump was introduced in May 1889 production was claimed to be 1,500 per day.

It is interesting to note the comments of the opposition.

The despised Waterbury is steadily creeping into recognition by the watchmakers. The Company have just completed two new calibres – an 18 ligne and a 15 ligne movement. The first has a seconds hand and both are jewelled in the balance holes, have gilt plates and set hand at the pendant. Like the Waterbury short wind issued a few months ago the new watches discard the rotating movement and other features of the original conception except the duplex form of escapement which is retained.

The long wind series A to E were discontinued in 1891 after an estimated production of between 600,000 and 1000,000 units.

Comments from the German trade were no less kind – 'The Waterbury Watch is an ordinary worthless American watch named after the town in Connecticut where it was made'. During the years 1881-1886 the German market was literally flooded with this watch made out of stamped tin parts. The sale price was ten marks ($2.40) each.

Further alphabetical progression of the manufacturing series continued – a series W Addison model being presented in the 1896 catalogue. Series R to W could be bought in either rolled gold or silver cases in silk-lined presentation boxes.

By 1898 a further model had been introduced and marketed under the new company of the New England Watch Company. This company was to fail in 1912 when it was taken over and bought by Robert H Ingersoll Bros who had also been producing a watch for 5 shillings (a dollar).

Readers are referred for a full history to Charles S Crossmann's classic reference work *The Complete History of Watchmaking in America*, *American Pocket Watches* by Dr Hoke and *American Watchmaking* by MC Harrold.

Readers wishing to pursue further study of the history of American companies other than Waterbury to use the duplex are advised to consult 'American Watchmaking – A Technical History of the American Watch Industry 1850-1930' by Michael C Harrold, Supplement No 14, Spring 1984 to the *Bulletin* of the National Association of Watch and Clock Collectors.

PLATE 2A

PLATE 1
Charles Hall, 162 Fleet Street, London

Nickel open face double bottom case. White enamel dial, black roman numerals, broad spade gold hands, inset seconds dial. Top plate inscribed CHAs HALL 162 Fleet St 160 (movement number). Key wound and key set cannon pinion square. Full plate fusee movement with duplex scape wheel. Flat brass balance rim. Undersprung with diamond endstone balance staff. Bosley regulator. Foliate engraved balance cock with grotesque design engraved Slow Fast.
Key size 5. Diameter 55mm

Charles Hall 1810-20, Clockmakers Company, son of Joseph Hall, Newton Street, Holborn, a labourer, was apprenticed to Joseph Wyer, Norman Street, St Luke, a watch spring maker, on 3 September 1810. He was free of the Clockmakers Company by redemption in 1815. Britten 9th ed lists him as working as a watchmaker at 162 Fleet Street between 1817 and 1819, and 118 Chancery Lane in 1820.

PLATE 2
John Barwise, St Martins Lane, London

Silver full hunter double bottomed case with gold hinges, hallmarked London **1820**. Casemaker's mark WW (either William Watson, 67 Red Lion Street, Clerkenwell or William Webb, 54 Great Sutton Street, Clerkenwell). White enamel dial with black roman numerals and gold spade hands, inscribed upper centre BARWISE LONDON 8281 which is repeated on top plate, inset seconds dial. Key wound and set. Full plate fusee and chain. 8 jewelled movement with duplex escapement with diamond endstone balance staff and end jewels to the scape wheel arbor. Undersprung broad flat steel balance and blued steel balance spring and regulator from plain triangular balance cock over the balance rim to silver quadrant engraved alongside SLOW FAST.
Key size 5. Diameter 55.6mm

John Barwise was an important watch and chronometer maker active between 1790-1843. Britten 9th ed lists him as working with Weston Barwise at 29 St Martins Lane between 1820-1842, Barwise & Sons 24 St Martins Lane 1819-23. He was associated with the taking out of Alexander Bain's patent for electric clocks in 1841. He was chairman of the British Watch company between 1842 and 1843.

PLATE 2B

Duplex Watches

PLATE 3B

PLATE 3A

PLATE 4B

PLATE 4A

PLATE 5A

PLATE 3
Parkinson & Frodsham, 4 Change Alley, London

Silver gilt half hunter case hallmarked London **1823**. White enamel dial inscribed upper centre PARKINSON & FRODSHAM LONDON with black roman numerals, gold spade hands, large inset seconds dial. Top plate engraved Parkinson & Frodsham Change Alley London 1502. Key wound and set. Full plate duplex movement with fusee and chain maintaining power and jewellery to scape arbor with diamond endstone to balance staff. Oversprung cut bimetallic balance with blued steel balance spring. Freesprung.
Key size 4. Diameter 53mm

Mainspring setting up on top plate, see Liverpool Runner - The Fusee Lever Watch *by Dr Robert Kemp, 1st ed p60. William James Frodsham set up business with William Parkinson 1801 at 4 Change Alley, Cornhill. Listed in Pigot's Directory 1823/4. In 1828 Parkinson and Frodsham opened a branch in 54 Castle Street, Liverpool. See* The Frodshams *by Dr Vaudrey Mercer.*

PLATE 4
John Cross, Charterhouse Square, London

Silver open face double bottomed case hallmarked London **1838**, engine-turned back and body. Casemaker's mark JD (corresponds to case mark of John Dyer listed PO Directories 1836, 1838 at 55 Bath Street, St Lukes, but it should be noted that there is also a casemaker James Dow, 54 Percival Street, Goswell Street working at that period) repeated on pendant. Dustcap stamped RE. White

PLATE 5B

enamel dial, black roman numerals, gold spade hands, inset sunk seconds dial. Key wound and set. Full plate. 13 jewel movement engraved on top plate John Crofs Charterhouse Square London No 30739. Fusee and chain with maintaining power, steel duplex scape wheel with diamond endstones to balance staff and endstones to scape wheel arbor. Cut bimetallic oversprung balance with Pennington's double T balance and steel index over the balance rim to silver quadrant.
Key size 5. Diameter 51mm

1832 Post Office Directory lists John Cross at 44 Charterhouse Square and 1836 Directory at 41 Charterhouse Square.

PLATE 5
Edward John Dent, 61 Strand, 33 Cockspur Street, 34 Royal Exchange, London

18ct gold open face case, snap-on bezel and back, hallmarked London **1854**. Casemaker mark AN (Adolphe Nicole, 14 Soho Square, London). White enamel dial inscribed EJ DENT LONDON, black roman numerals, blued steel Breguet style hands, subsidiary seconds dial. Top plate engraved EJ DENT WATCHMAKER TO QUEEN LONDON Patent No 14452. Nicole keyless winding with push piece at 54 minutes position. Three-quarter plate. 10 jewels duplex movement with endstones to the duplex scape arbor. Oversprung bimetallic balances. Regulator index on small triangular cock scroll engraved and SF.
Diameter 34.3mm

This is an early example of Nicole winding. The movement should be compared with DENT 12907 and Robert Roskell 65110 (Plates 6 and 7). Reference Patent No 10348, Keyless Mechanism, 1844. Adolphe Nicole, Dean Street, Soho, Post Office Directory 1852 p1434. Dent *by Vaudrey Mercer pp275-279.*

Duplex Watches

PLATE 6A

PLATE 6B

PLATE 6
Edward John Dent, 61 Strand, 33 Cockspur, 34 Royal Exchange, London

18ct gold plain open face case with engraved monogram hallmarked London **1859**. Casemaker's mark AN (Adolphe Nicole, 14 Soho Square). Gold engine-turned dial, black roman numerals, blued steel moon hands. Top plate engraved EJ Dent Watchmaker to the Queen London No 12907. Early keyless winding with push piece at 51 minute position. Three-quarter plate duplex movement with jewelling to third wheel and ruby endstones. Oversprung cut bimetallic balance. Regulator index on scroll engraved cock. *This is a rare early keyless wind and duplex watch by a good English maker.*
Diameter 43mm

PLATE 7A

PLATE 7B

220

Duplex Watches

PLATE 8A

PLATE 8B

PLATE 7
Robt Roskell, Church Street, Liverpool

18ct gold open face case, snap-on bezel and back, engine-turned decoration back, hallmarked London **1860**. Casemaker AN (Adolphe Nicole, 14 Soho Square, London). White enamel dial inscribed ROBT ROSKELL LIVERPOOL, black roman numerals, blued steel narrow spade hands, inset sunk seconds dial. Top plate inscribed Robt Roskell Patent LIVERPOOL No 65110. Keyless Nicole winding and set push piece at 54 minute position. Three-quarter plate going barrel 12 jewel duplex movement with endstones to scape arbor. Oversprung and cut bimetallic balance with compensation screws. Blued steel balance spring. Polished steel regulator index on small scroll engraved cock and SF.
Diameter 48.3mm

PLATE 8
Waterbury Skeletonised Series A, Benedict and Burnham, Waterbury, Connecticut, USA

Nickel open face case, snap-on bezel and back. Annular paper dial with skeletonised centre to dial, narrow black roman numerals, narrow black spade hands. Keyless wound

PLATE 8C

with Series A long wind movement. Back cover to movement engraved with Waterbury registered trade mark. Duplex escapement.
Diameter 50.8mm

Waterbury Series A movements manufactured from December 1879 with Series B following in 1885. UK Patent 2283 Buck DAA 7 June 1878.

Duplex Watches

PLATE 9A

PLATE 9B

PLATE 9D. *The Graphic*, 26 December 1885

PLATE 9
Waterbury Duplex Series F, Waterbury Watch Co, Waterbury, Connecticut, USA

Nickel open face case, snap-on bezel and back, inside back stamped in ink Guaranteed 2 years P8. Varnished paper dial inscribed WATERBURY WATCH CO upper centre dial with Waterbury registered trade mark below, black roman numerals and narrow blued steel spade hands. Keyless wound and push button to set hands. Silvered movement also stamped with trade mark and PATENTED FEB. 5. 1884 SERIES F stamped above barrel wheel on bridge and THE WATERBURY WATCH CO WATERBURY CONN USA DUPLEX ESCAPEMENT stamped below balance rim on pillar plate. Uncut imitation compensated balance.

Diameter 52.8mm

THE LADIES' WATERBURY

SHORT WIND 17/6 — 17/6 STEM SET

AT LAST

ALSO

SERIES "E"	to retail	10s. 6d.
SERIES "F," SHORT WIND, STEM SET ...	,,	15s.
SERIES "J," SUNK SECONDS, SHORT WIND	,,	17s. 6d.

TRADE ONLY SUPPLIED.

FOR TERMS, &c.—WATERBURY WATCH (SALES) CO., Limited, 7, SNOW HILL, LONDON.

THE HOROLOGICAL JOURNAL.
TERMS FOR ADVERTISING.

Payment must in all cases be made in advance at the following rates:—

	PER MONTH.	PER YEAR.
Whole Page	£3 0 0	£24 0 0
Half do.	1 12 0	12 0 0
Quarter Page	0 17 0	6 0 0
One-eighth do.	0 9 0	3 0 0

A page advertisement measures 8 in. by 5¼ = 42 square in.

Column Advertisements.—For each in. in depth 4s. 6d. per month, £1 10s. per year. Special Rate for Situations Vacant and Wanted. Articles for Sale or Required, 16 Words, Sixpence.

Advertisements should reach the Office, 35, Northampton Square, E.C., not later than the 23rd of the month. Cheques and Post Office Orders to be made payable to F. J. BRITTEN.

D. BUCKNEY, 5, King Sq., London, E.C. } For Keyless Watches, in Open Face, Crystal, and Hunt Cases, made on Special Gauged Principle.

HOROLOGICAL JOURNAL VOL XXXI   xix

PLATE 9C. *Horological Journal* Volume XXXI, 1888

Duplex Watches

PLATE 10

PLATE 10
Waterbury Duplex Series I THE TRUMP, Waterbury Watch Co, Waterbury, Connecticut, USA

Nickel open face case, snap-on bezel and back. White enamel dial, black roman numerals and narrow spade hands, inset seconds dial. Keyless wound and set. Back of movement engraved Waterbury Trade Mark Waterbury USA THE TRUMP Series I. Duplex escapement. Plain brass flat rimmed balance. Regulator index to scale on top plate towards centre.
Diameter 53mm

PLATE 11
Waterbury Duplex Series K 'Addison', Waterbury Watch Co, Waterbury, Connecticut, USA

Gilt brass open face case, snap-on engraved bezel and back. White enamel dial, Waterbury trade mark top centre, black roman numerals, narrow blued steel spade hands, inset seconds dial. Keyless winding and set. Three-quarter plate movement engraved ADDISON PATENTED USA K and word TRADE MARK. Duplex escapement. Compensated oversprung balance. Scroll engraving to balance cock. Regulator index over balance rim to graduations on top plate.
Diameter 52mm

PLATE 11A

PLATE 11B

Duplex Watches

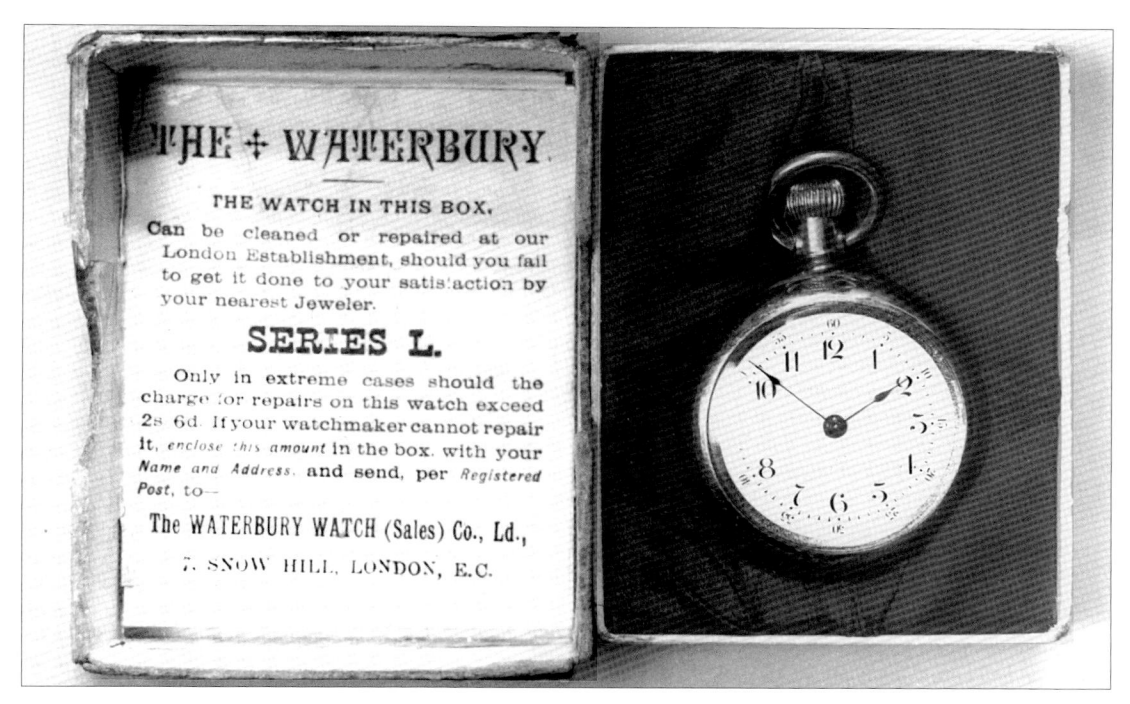

PLATE 12A

PLATE 12B

PLATE 12
Waterbury Duplex Series L, Waterbury, Connecticut, USA

Polished nickel case, snap-on bezel and back. Stencilled on back GUARANTEED FOR 2 YEARS. White enamel dial, black arabic numerals, blued steel spade hands, five minute marking, small arabic numerals outside chapter ring. Keyless wound and push to set hands. Top plate engraved Waterbury trade mark Patented Series I. The pillar plate stamped with list of patent dates. Duplex escapement with oversprung imitation compensation balance. Regulator index to graduations on top, plate engraved SLOW FAST. Original box and label with the comment that 'Only in extreme cases should the charge for repairs on this watch exceed 2s 6d'.

Diameter 41.8mm

London address The Waterbury Watch (Sales) Co Ltd, 7 Snow Hill, London EC.

Duplex Watches

PLATE 13
Waterbury Duplex Series R, Waterbury, Connecticut, USA

Sterling silver open face case, snap-on bezel and back, engine-turned back. White enamel dial with Waterbury trade mark upper centre and inset seconds dial, black roman numerals, narrow blue spade hands. Top plate engraved WATERBURY TRADE MARK USA and word PATENTED around centre arbor, R stamped near winding work. Duplex escapement. Oversprung imitation uncut bimetallic balance with blued steel balance spring. Scroll engraved cock. Index over balance rim to graduation near centre of movement on top plate.
Diameter 48mm

PLATE 13A

PLATE 13B

Duplex Watches

PLATE 14
Waterbury Duplex AMBASSADOR series, New England Watch Co, Waterbury, Connecticut, USA

Gilt open face case guaranteed 20 years, snap-on bezel, hinged back and cuvette. White enamel dial and inset seconds, black roman numerals, narrow spade hands. Keyless wind, push winder to set hands. Nickel movement with engine-turned pattern and inscribed N E W Co Ambassador. Duplex escapement. Imitation uncut compensated balance with blued steel balance spring. Small regulator index on triangular balance cock engraved FS.
Diameter 45.9mm

A new model was introduced in 1898 which was marketed by the New England Watch Company. The Company failed in 1912.

PLATE 14A

PLATE 14B

Duplex Watches

PLATE 15A

PLATE 15B

PLATE 15
Lady's Duplex Pocket Watch,
New England Watch Co, Waterbury,
Connecticut, USA

Light blue enamel case with fleur-de-lis design, snap-on bezel and back. Inside back cover stamped STERLING SILVER and NE within heart. Rotating oval gold plated bow and pendant. Opaline dial, blue arabic numerals and fancy hands. Duplex movement engraved on top plate with decorative engine-turned pattern and NEW Co. Imitation compensation oversprung balance. Small index on triangular balance cock.

Diameter 30mm

PLATE 15C

Chapter Six
Pin Pallet Lever Watches

The pin pallet escapement in its original form was devised by Louis Perron (1779-1836). It had the advantage of better locking with less recoil and was easier to make. It was displaced on the Continent by the widespread use of the cylinder escapement, but had a rebirth when adapted by Georges-Frédéric Roskopf with subsequent modification as the basis for a mass-produced cheap watch.

Plate 1. Pin pallets

Plate 2. Pin pallet scape wheels

Pin Pallet Lever Watches

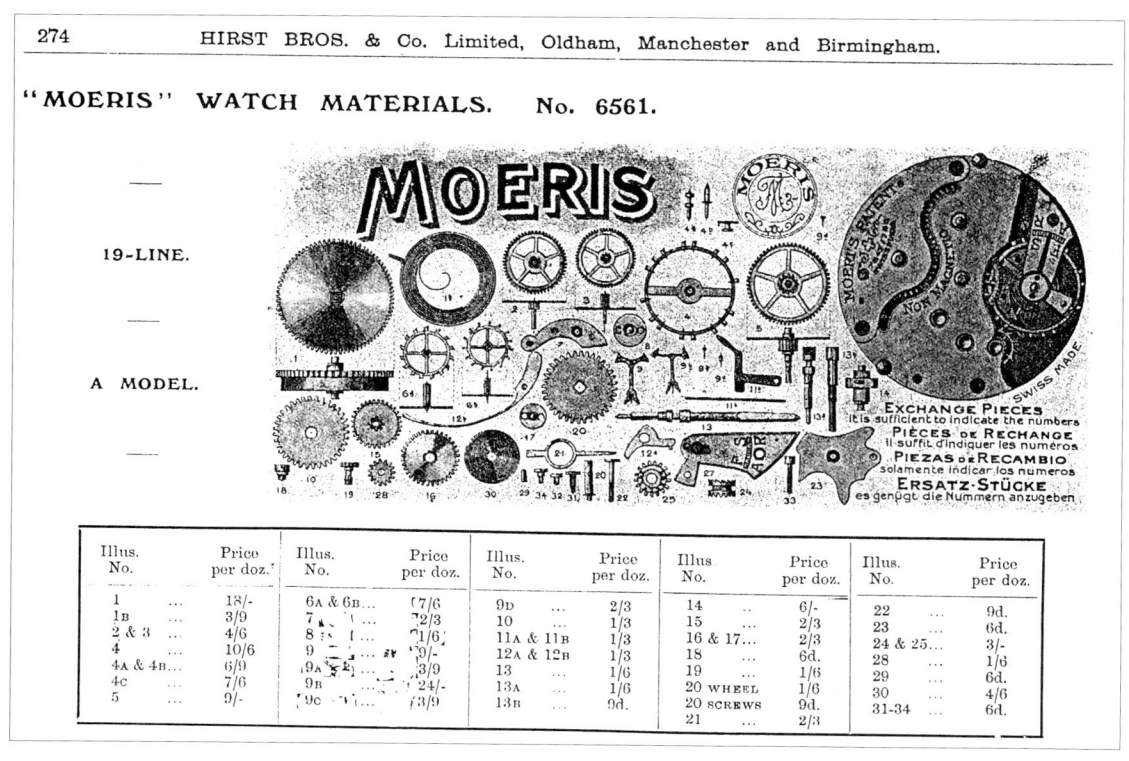

> 274   HIRST BROS. & Co. Limited, Oldham, Manchester and Birmingham.
>
> **"MOERIS" WATCH MATERIALS.   No. 6561.**
>
> 19-LINE.
>
> A MODEL.
>
> | Illus. No. | Price per doz. | Illus. No. | Price per doz. | Illus. No. | Price per doz. | Illus No. | Price per doz. | Illus. No. | Price per doz. |
> |---|---|---|---|---|---|---|---|---|---|
> | 1 | 18/- | 6A & 6B | 7/6 | 9D | 2/3 | 14 | 6/- | 22 | 9d. |
> | 1B | 3/9 | 7 | 2/3 | 10 | 1/3 | 15 | 2/3 | 23 | 6d. |
> | 2 & 3 | 4/6 | 8 | 1/6 | 11A & 11B | 1/3 | 16 & 17 | 2/3 | 24 & 25 | 3/- |
> | 4 | 10/6 | 9 | 9/- | 12A & 12B | 1/3 | 18 | 6d. | 28 | 1/6 |
> | 4A & 4B | 6/9 | 9A | 3/9 | 13 | 1/6 | 19 | 1/6 | 29 | 6d. |
> | 4C | 7/6 | 9B | 24/- | 13A | 1/6 | 20 WHEEL | 1/6 | 30 | 4/6 |
> | 5 | 9/- | 9C | 3/9 | 13B | 9d. | 20 SCREWS | 9d. | 31-34 | 6d. |
> | | | | | | | 21 | 2/3 | | |

PLATE 3A. *Fourth Wide Awake Catalogue*, 1915

PLATE 3
E & M Goldston, Edinburgh

Silver 0.935 open face case, engine-turned back with plain centre, hinged back and cuvette, snap-on bezel. White enamel dial, E & M Goldston Edinburgh upper centre, black roman numerals, gold spade hands, inset seconds. Keyless wound and push piece to set at 4 minutes to the hour position. Top plate engraved MOERIS PATENT + 7547/780 NON MAGNETIC. 4 jewelled pin pallet escapement of Moeris design with adjustable pallets and eccentric banking pins. Oversprung monometallic balance with imitation compensation screws. Brown balance spring and regulator index on balance cock engraved FS AR.

Diameter 51mm

MOERIS mark Deposé No 17816 1940 18 October Fritz Moeris, St Imier, Brevet No 7547, Montre Perfectionée Exposé d'Invention dated 31 July 1894 issued 14 Nov 1893, Moeris et Jeanneret à St Imier, Berne, Switzerland.

PLATE 3B

Pin Pallet Lever Watches

PLATE 4

PLATE 4
H Samuel, Manchester

Silver face case, hinged bezel, back and cuvette, hallmarked 0.935 silver and Swiss Bear mark inside, outer cover engraved HS within shield. White enamel dial with sunk centre, black roman numerals, luminous 5 minute dots, gilt spade hands, inset seconds. Key wound and set at back. Three-quarter plate MOERIS movement stamped top plate H SAMUEL MARKET STREET MANCHESTER SWISS MADE NON MAGNETIC AND 7547 + on the pillar plate below balance rim. 4 jewel pin pallet escapement with adjustable pallets and adjustable eccentric balance pins. Oversprung monometallic balance with imitation screws Brown balance spring and regulator index on engraved balance cock and FS AR.
Key size 5. Diameter 50.7mm

PLATE 5
Westclox, Canada

Chrome on brass open face case, snap-on bezel and back. Silvered dial inscribed upper centre Westclox POCKET BEN, black arabic numerals, black pointer hands, inset sunk seconds. Keyless wound and pull to set. Top plate stamped 2 56. Barrel cover stamped MADE IN CANADA WESTERN CLOCK CO LTD. Unjewelled straight line pin pallet lever movement with two arm flat brass oversprung balance and regulator index riveted to balance bridge which forms part of the top plate on which the regulator scale is stamped FS.
Diameter 49mm

Westclox introduced 1932 (see The Watch that made the Dollar Famous by G Townsend page 31 plate 80). The Western Clock Company became Westclox, a division of General Time Corporation, in 1936, a Talley Industries Company. In 1962 #7 was redesigned as Model 9001.

PLATE 5A

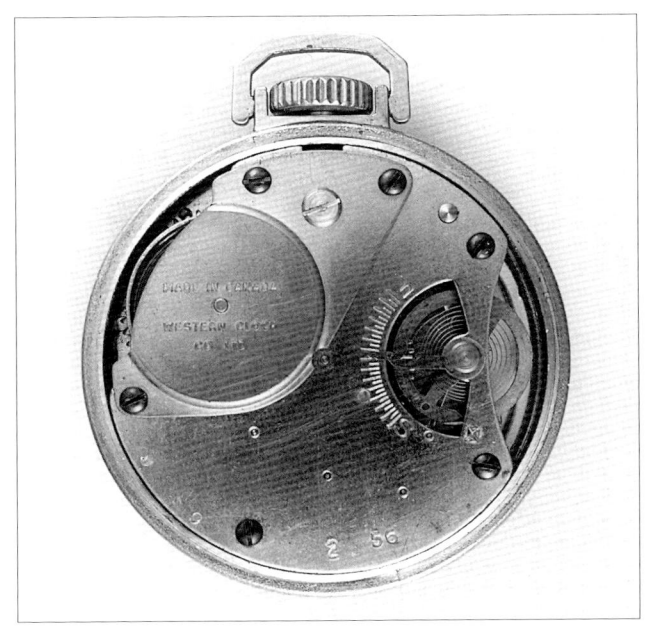

PLATE 5B

231

Pin Pallet Lever Watches

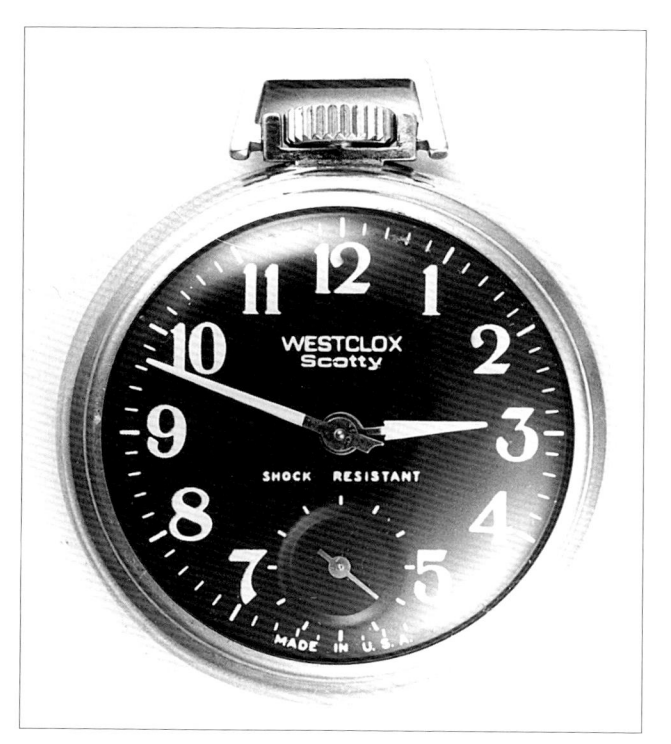

PLATE 6A

PLATE 6B

PLATE 7A

PLATE 7B

PLATE 8A

PLATE 8B

PLATE 6
Westclox, USA

Nickel chrome open face case. Black painted dial inscribed WESTCLOX Scotty SHOCK RESISTANT MADE IN USA, luminous arabic numerals and pointer hand, inset sunk seconds. Keyless wound and set. Unjewelled pin pallet movement stamped 10 on dial plate under balance rim. MADE IN USA stamped on barrel bridge. *Note method of retaining movement by spring clip across diameter of case.*
Diameter 50.4mm

Westclox Model 9001 (Scotty) introduced 1962.

PLATE 7
Ansonia Clock Company, USA

Nickel plated open face case, snap-on bezel and back. White varnished card dial inscribed with Ansonia trade mark upper centre and ANSONIA CLOCK COMPANY MANUFACTURERS USA lower edge, black roman numerals, blued steel spade hands, inset seconds dial. Keyless wound and push winding button to set. Full plate unjewelled pin pallet lever movement. Oversprung two arm stamped out brass imitation compensation balance. Blued steel balance spring with blued steel regulator index on to scale on top plate. There is no separate balance cock forming part of the top plate.
Diameter 52.5mm

See Ansonia Clocks Catalogue 1914 for model Chauffer Roman. See The Watch that made the Dollar Famous by G Townsend for movement patented 17 April 1888.

PLATE 8
Thomas Haller, Schwennigen, Germany

Nickel plated brass open face case, snap-on bezel and back. Cream lacquered card dial with Thomas Haller trade mark lower centre, black roman numerals, blued steel spade hands, inset seconds dial. Keyless wound and push piece to set at 5 minutes past position. Unjewelled straight line pin pallet lever. Oversprung three arm brass balance and blued steel balance spring. Regulator index on to scale on skeletonised top plate. There is no separate balance cock.
Diameter 53.5mm, 17mm thick

Pin Pallet Lever Watches

PLATE 9
Thomas Haller, Schwennigen, Germany

Nickel plated brass open face case, snap-on bezel and back. Cream painted dial with Thomas Haller registered trade mark as issued from 1850 upper centre, black roman numerals, gilt fancy Louis hands, small inset seconds dial. Skeletonised top plate stamped with trade mark. Keyless wound with push piece at 5 minute position to set hands. Pin pallet lever with oversprung monometallic balance. Regulator index pointing to scale on top plate. No separate balance cock. *A very unusual and early pin pallet movement number 4264.*
Diameter 58.2mm, 21.4mm thick

PLATE 9A

PLATE 9B

PLATE 9C

PLATE 10A

PLATE 10
Kienzle Uhrenfabriken AG, Schwennigen/Neckar, Germany

Chromium plated brass open face case, snap-on bezel and back, engine-turned design back. Cream painted dial inscribed FOREIGN lower edge, black arabic numerals, gilt skeletonised spade hands, inset seconds dial. Keyless wound and push piece to set at 5 minute position. Three-quarter plate unjewelled straight line pin pallet movement. Oversprung two arm flat rimmed brass balance and red brown balance spring. Regulator index riveted to balance bridge with index to scale on top plate.

Diameter 51mm

See Rudolf Flume Catalogue 1888-1937 and 1958 G2/3, p92, Nr 144998.

PLATE 10B

Pin Pallet Lever Watches

PLATE 11A

PLATE 11B

PLATE 11C

PLATE 11
Kienzle Uhrenfabriken, Schwennigen, Germany

Chrome plated open face case, snap-on bezel and back. Swing ring attached to watch to enable use as desk watch. Silvered dial inscribed KIENZLE upper centre, ANTIMAGNETIC AND MADE IN GERMANY lower edge, black arabic numerals, blued steel spade hands, inset seconds dial. Top plate stamped 146/00e KIENZLE GERMANY NO(0) JEWELS UNADJUSTED. Keyless wound with pull to set hands. Three-quarter plate straight line pin pallet lever escapement with oversprung two arm flat steel balance. White balance spring. The regulator index on balance cock engraved + -. Compare with Novoris Twin by Oris Watch Company (Plate 23).

See illustration of a very similar watch in Robert Pringle and Sons Wilderness Catalogue November 1934 entitled 'Utility' Pocket or Desk watch. Chromed strut case No 4495 quoted as 10/6d (ten shillings and sixpence trade price – old style currency).

PLATE 12
Thiel, Ruhla, Germany

Nickel on brass open face case, snap-on bezel and back. Varnished card dial with sunk seconds inscribed RAILWAY TIME and MADE IN GERMANY below inset seconds dial, black arabic numerals, blued steel spade hands. Keyless winding and push piece to set hands at 5 minute past position. Top plate with engine-turned straight line design. Unjewelled straight line pin pallet escapement. Oversprung two arm brass balance stamped out in imitation of compensation balance. Red brown balance spring. Blued steel regulator index on to scale on top plate.

In 1862 Siegmund Thiel started a small tobacco pipe factory in Ruhla in the midst of the Thuringian Forest area of Germany. Later with his sons Christian and Georg Thiel they expanded to a small manufactory making small metal articles. Following the Franco-Prussian war in 1870, Ernst (brother of Christian) introduced the manufacture of children's toy watches. These had mainly come from France. The highly successful export of this

236

Pin Pallet Lever Watches

PLATE 12A

PLATE 12B

product, especially to America, was reflected in the increase of the work-force from 70 to 200. In spite of the death of Christian, Ernst, now joined by his third brother Reinhold, was able by extensive travelling abroad to build up the demand even more. Responding to the request to produce a toy watch for children that could go for an hour, he found that by only a small modification it would be possible to produce a watch that would go for twelve hours. Unfortunately this appeared to price the watch out of the market for toy watches. Largely through the inspiration and efforts of Emil Durer, a director of the firm, further modifications resulted in a suitable cheap and serviceable watch being produced for adults. This was marketed in 1891.

Mass-production methods in Germany were still largely in their infancy and it was necessary initially to design and create the requisite machinery. The workshop was reorganised. Dr Reinhold Thiel, the eldest son of Ernst, joined the company in 1897. Like his father before him, he fully explored the overseas market and eventually in 1910 established a retail outlet in London. He returned to Germany to become a director of the factory in 1912. In spite of the disasters of the 1914-1918 First World War and the liquidation of their assets in this country, Thiel were able by energetic marketing to survive the depression. They increased watch production to 6,000 watches per day and in 1927 were able to re-establish their London branch as Thiel Bros Ltd. The production of toy watches ceased in 1929 — parents could now afford real watches for their children. Their registered trade mark was The Champion.

PLATE 12C

237

Pin Pallet Lever Watches

PLATE 13A

PLATE 13B

PLATE 14A

PLATE 14B

Pin Pallet Lever Watches

PLATE 15A

PLATE 15B

Plate 13
Thiel, Ruhla, Germany

Chromium plated open face case, snap-on bezel and back. Cream lacquered card dial with sunk seconds inscribed Federal and FOREIGN below inset seconds dial, black arabic numerals, black steel spade hands. Keyless wound and pull to set hands. Three-quarter plate unjewelled straight line pin pallet lever. Oversprung two arm balance stamped out in imitation of a compensation balance. White balance spring. Regulator index pointing to scale on top plate stamped RS AR.
Diameter 51.3mm

Plate 14
Thiel, Ruhla, Germany

Chrome on brass open face case, snap-on bezel and back. Cream varnished card dial inscribed 'Services' SCOUT and GERMAN MADE lower rim, black Roman 1-12 hour numerals with 13-14 in black arabic numerals around outer chapter ring, steel spade hands and small inset seconds dial. Keyless wound and push piece to set at 5 minute position. Top plate engine-turned in straight line. Three-quarter plate unjewelled pin pallet tangential lever movement. Oversprung two arm brass balance. Yellow brown balance spring. Regulator index on to scale on top plate stamped SF RA.
Diameter 49.1mm

Plate 15
Anon, Germany

Chrome on brass open face case, snap-on bezel and back. Silvered engine-turned pattern centre of dial inscribed Alert and MADE IN GERMANY around inset seconds dial, black arabic numerals, blued steel spade hands. Keyless winding push piece at 4 minute position. Three-quarter plate unjewelled straight line pin pallet lever movement. Stamped out brass imitation compensated balance. Oversprung blued steel balance spring. Regulator index pointing towards centre pinion but scale is on the balance cock which is inscribed Alert Germany.
Diameter 51.3mm

Trade mark Alert registered 2 Nov 1911 by Hoefen a Enz, Germany.

239

PLATE 16A

PLATE 16B

PLATE 16
Anon, Switzerland

Nickel on brass full hunter case, engine-turned back and front covers with milled edge to body. Pendant with split push for front and back case springs. Antique pattern bow. White enamel dial with garland inscribed BAGDAD, black Turkish numerals, blued steel spade hands. Key wound and set from back. Three-quarter plate Roskopf type pin pallet lever movement with 4 ruby red 'jewels' as decorative nonfunctional endstones. Tangential pallet lever. Oversprung three arm brass balance. Red brown balance spring. Small regulator index on balance cock engraved SR FA.
Key size 7. Diameter 54.4mm

Turkish numerals were outlawed by Kemal Ataturk in 1928.

PLATE 17
Anon

Silvered brass full hunter case with heraldic design inside back and front covers incorporating imitation English hallmark. Pendant with split push for front and back case springs. White enamel dial, again with heraldic design, black Turkish numerals, copper Louis style hands. Key wound and set from back. Three-quarter unjewelled Roskopf movement with right-angle adjustable pin pallet lever and depthing. Oversprung three arm brass balance with blued steel balance spring. Index on balance cock stamped RA SF.
Key size 9. Diameter 55mm

PLATE 18
Mohertus Trading Co, Austria

Chrome plated open face case, snap-on bezel and back, sunburst pattern to back. Cream painted luminous dial with picture of 4.8.4 steam locomotive inscribed RAILWAY TIMEKEEPER SHOCK-PROTECTED SPECIALLY-EXAMINED MADE IN AUSTRIA on lower edge, black arabic numerals, black pointer hands. Keyless wound, push winder to set hands. Top plate stamped MOHERTUS TRADING CO MADE IN AUSTRIA NO JEWELS. Pillar plate stamped 6 near pallet tail. Three-quarter plate unjewelled pin pallet lever. Oversprung three arm flat rimmed brass balance. Red brown balance spring. Plain index screwed on balance cock which is engraved SF RA.
Diameter 50.5mm

Pin Pallet Lever Watches

PLATE 17A

PLATE 17B

PLATE 18A

PLATE 18B

241

Pin Pallet Lever Watches

PLATE 19A

PLATE 19B

PLATE 19
Anon, Switzerland

Chrome plated open face case, snap-on bezel and back, the back engraved 2.8.6 locomotive steaming to the left. Silver dial with 2.6.6 locomotive steaming to the right inscribed RAILWAY REGULATOR ZANCO SWISS MADE, black arabic numerals, blued steel modern style hands. Keyless wound and push to set. Three-quarter plate jewelled Roskopf type tangential pallet lever movement. Oversprung three arm flat brass balance. Yellow balance spring. Regulator index on balance cock engraved SF RA.
Diameter 49.3mm

PLATE 19C

Pin Pallet Lever Watches

PLATE 20A

PLATE 20B

PLATE 20
Amida SA, Grenchen, Switzerland

Nickel chrome plated on steel open face case, back cover engraved with steam 2.8.6 locomotive steaming to the left, snap-on bezel and back cover. Cream painted dial inscribed RAILWAY REGULATOR AMIDA SPECIALLY EXAMINED, SWISS MADE lower edge and locomotive steaming to the right, large arabic numerals, blued steel moon hands. Keyless wound, push button winder to set hands. Top plate stamped CAL 143 and trade mark of Oris Watch Co (OWC in oval) SWISS MADE. Three-quarter plate late Roskopf type unjewelled pin pallet lever with tangential pallet. Oversprung three arm flat rimmed brass balance. Red brown balance spring. Small regulator index on balance cock stamped RA.
Diameter 49.8mm

PLATE 20C

Pin Pallet Lever Watches

PLATE 21A

PLATE 21B

PLATE 21C

**PLATE 21
Anon, Switzerland**

Nickel chrome on steel open face case, engraved back cover with monoplane with undercarriage flying above clouds, snap-on bezel and back. Silvered aluminium dial inscribed SHOCKPROOF LEVER and SWISS MADE lower edge, HFCB on fusilage of monoplane flying to left, H on tail, black arabic numerals 1-12 and small numerals 13-24, blued steel taper baton hands. Keyless wound and push button to set. Top plate stamped SWISS MADE. Pillar plate stamped A below balance rim. Three-quarter plate later Roskopf type movement. Oversprung three arm flat rimmed brass balance. Red brown balance spring and small index regulator stamped Br + and screwed to the balance cock engraved SF RA.
Diameter 49.4mm

**PLATE 22
Anon, Switzerland**

Chrome plated open face case, snap-on bezel and back, back engraved racing car and driver facing left. Cream painted aluminium dial inscribed SUPERIOR MOTOR TIMEKEEPER LEVER SWISS MADE with SWISS MADE lower edge and racing car facing left, black arabic numerals, luminous taper baton hands. Keyless wound and push button to set. Three-quarter plate unjewelled pin pallet late Roskopf type lever movement with tangential lever. Oversprung flat rimmed three arm brass balance. Red brown balance spring. Small index screwed to balance cock engraved SR FA. Pin pallet movement clips into case without case screws.
Diameter 49.6mm

244

Pin Pallet Lever Watches

PLATE 22A

PLATE 22B

PLATE 23A

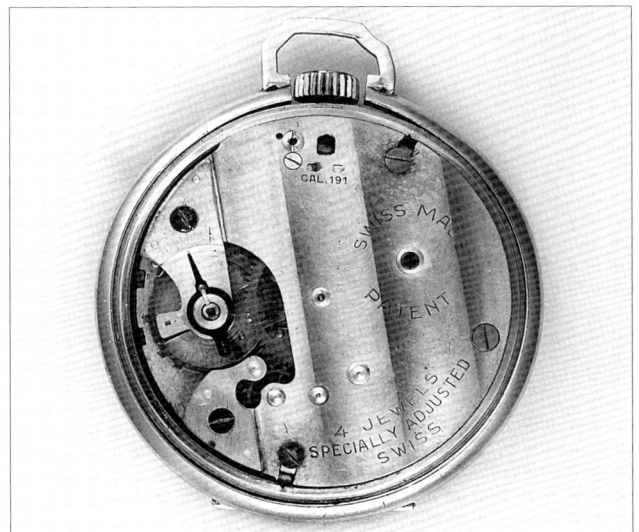

PLATE 23B

PLATE 23
NOVORIS TWIN, Oris Watch Co, Holstein, Baselland, Switzerland

Pure nickel chrome open face case, snap-on bezel and back, engine-turned pattern to back. Swing ring to act as stand. Silvered dial inscribed NOVORIS TWIN and SWISS MADE lower edge, luminous arabic numerals and luminous arabic spade hands, inset seconds dial. Keyless wound and pull to set hands. Top plate inscribed CAL 191 PATENT SWISS MADE 4 JEWELS SPECIALLY ADJUSTED SWISS. Three-quarter plate straight line pin pallet lever movement with oversprung monometallic balance and white balance spring. Regulator index on balance cock with a shockproof spring clip on balance assembly (compare with Kienzle swing ring pin pallet, Plate 11).

Diameter 53mm with ring, 49.5 without ring

Novoris was registered mark of Compagne des Montres Oris SA. Registered Trade mark 136648 31 January 1951. International Trade mark 152468. London Agents for Oris – The Betima Co Ltd., 18 Bury Street, London EC3.

245

Pin Pallet Lever Watches

PLATE 24A

PLATE 24B

PLATE 24
SERVICES ARMY, Müller-Schlenker, Schwennigen, Germany

Nickel chrome open face case, snap-on bezel and back. Cream varnished card dial inscribed 'Services' ARMY with FOREIGN on lower edge, black roman numerals, blued steel thin spade hands, inset seconds dial. Keyless wound and set with push piece 1 o'clock position. Top plate with engine-turned zigzag pattern and inscribed 'Services' WATCH CO LTD FOREIGN. Unjewelled pin pallet lever movement with tangential lever. Oversprung three arm brass balance with blued steel balance spring. Regulator index on balance cock. Complete with original Services presentation box – original price 5/- (five shillings).

PLATE 25
SERVICES, Smith's Industries, England

Chrome open face case, snap-on back and bezel. Painted cream dial inscribed SERVICES trade mark (S within hexagon), MADE IN GREAT BRITAIN on lower edge, black arabic 2, 4, 8, 10, 12, luminous skeletonised hands, inset seconds dial. Keyless wound with push to set hands. Top plate stamped 72. Unjewelled full plate straight line pin pallet lever movement. Oversprung two arm brass balance and brown balance spring. Regulator index towards centre of movement and stamped SF.
Diameter 51.1mm

PLATE 26
Anon, Switzerland

Nickel open face case, hinged back and snap cuvette. Push piece in pendant for case spring. Dustcap pierced to show regulator index with S-shaped slide pivoted in centre. White enamel dial with sunburst superimposed on SOLATIME, black roman numerals, blued steel spade hands, inset seconds dial. Key wound and set at back. Top plate inscribed SWISS MADE. 14 jewel three-quarter plate tangential pin pallet lever movement. Oversprung half cut imitation bimetallic monometallic balance with red brown balance spring. Small blued index on balance cock.
Key size 7. Diameter 55.5mm

PLATE 25A

PLATE 25B

PLATE 26A

PLATE 26B

PLATE 27A

PLATE 27B

PLATE 27
Anon, Switzerland

Nickel plated brass open face case with engine-turned barleycorn design on back with central cartouche, snap-on bezel hinged back and cuvette. White enamel dial with RAILWAY TIMEKEEPER SPECIAL LEVER and SWISS MADE on lower edge, heavy black roman numerals, bronzed spade hands, inset seconds dial. Key wound and set from back. Three-quarter plate Roskopf type pin pallet lever movement with tangential lever. Three arm brass balance with blued steel balance spring. Regulator index on balance cock engraved RA SA.
Key size 8. Diameter 53.8mm

PLATE 28
Anon, Switzerland

Brass half hunter case stamped LWM REMONTOIR ACRE DE PRECISION on cuvette, GUARANTEED TO BE MADE WITH BEST MATERIALS AND TO LAST FIVE YEARS stamped inside front cover. White enamel dial inscribed SUPERIOR TIMEKEEPER SPECIALLY EXAMINED and SWISS MADE lower edge, black roman numerals, blued steel spade hands. Keyless wound, push piece to set hands at 2 o'clock position. Three-quarter plate unjewelled tangential pin pallet lever movement. Top plate stamped SWISS MADE. Oversprung three brass balance with blued steel balance spring. Regulator index on balance cock engraved RA SF.
Diameter 54.2mm

PLATE 29
Anon, Switzerland

Gilt brass open face case with medallion pattern on back and milled edge to body. Cream painted dial inscribed JOHN FORREST LONDON and SWISS MADE on lower edge, short roman numerals, gilt spade hands. Keyless wound with push piece to set hands at XI position. Top plate skeletonised. Pillar plate stamped Z. Tangential pin pallet lever escapement with oversprung brass imitation bimetallic balance. Blued steel balance spring. Small index on balance cock engraved RA SF.
Diameter 50.2mm

John Forrest had died in 1871 and in 1894 the Horological Journal *reported 'The plantiff who carried on business at Coventry and London was entitled by purchase to the name of John Forrest etc'. John Forrest, 29 Myddleton Street, London EC is listed in the* Official Illustrated Catalogue of the International Exhibition *1862 as exhibiting every description of pocket watches, various escapements and springs – London work.*

Pin Pallet Lever Watches

PLATE 28A

PLATE 28B

PLATE 29A

PLATE 29B

Pin Pallet Lever Watches

PLATE 30
INGERSOLL CROWN, Robert Ingersoll

Nickel chrome case. Black roman numerals, blued steel narrow spade hands. White card dial, INGERSOLL CROWN upper centre with inset seconds dial. Made in U.S.A. printed bottom of dial. Keyless winding.
Diameter 51.3mm

Movement number 21,472,538 corresponds to production period 1909. Movement stamped Patent.

Patented			No.
29 Jan 1901	*Ingersoll Watch Co.*	*A Bannatyne*	*666,997*
11 Apr 1905	*Ingersoll Watch Co.*	*E H Horn*	*787,041*
4 June 1907	*Ingersoll Watch Co.*	*E H Horn*	*855,950*
29 June 1909	*Ingersoll Watch Co.*	*E H Horn*	*926,329*

PLATE 30A

PLATE 30B

PLATE 30C

Pin Pallet Lever Watches

PLATE 31A

PLATE 31B

PLATE 31C

PLATE 31
Connecticut Watch Co., USA

Heavy nickel silver open face case by Illinois Watch Case Co numbered 932611, inside screwed back to case which carries a picture of a steaming cow catcher locomotive, screwed bezel to movement. White card dial inscribed WINNER with trade mark CTWCo, black arabic numerals, narrow blued steel hands, inset sunk seconds. Spotted engine-turning top plate. Unjewelled straight line pin pallet lever movement. Blued index pointing across balance to scale on top plate stamped FS. Blued balance spring two arm stamped out flat rimmed balance.

Diameter 56mm

In 1910 the Ingersoll firm in the USA was in difficulties. The expense of maintaining a policy of guaranteeing the watches and repair costs was a major cause of the subsequent failure of the firm. Many skilled workers had been lost to military duties in the 1914-1918 war. Management by a New York bank resulted in a loss of $800,000 in the first six months. Receivership followed in 1922 and America Ingersoll was purchased by the Waterbury Clock Company and renamed Ingersoll Waterbury Watch Co Ltd. The Waterbury Clock Company acquired all the American assets, trade name and goodwill and neither Robert nor Charles Ingersoll was allowed to engage in the watch trade again. Ingersolls in England continued, becoming Ingersoll Ltd in 1930.

Jewellers refused to sell Ingersoll watches but instead sold them under the name of Defiance and Winner.

The Ingersoll movement No 28383624 was made c.1910/11. For discussion of serial numbers, models and dates see NAWCC Bulletin 233, December 1984.

Patented				
29 Jan	1901	A Bannatyne USA Patent	666,997
11 Apr	1905	E H Horn	787,041
4 June	1907	E H Horn	855,950
29 June	1909	E H Horn	926,329
24 May	1910	F Wehinger	958,987

Pin Pallet Lever Watches

PLATE 32A

COLOUR PLATE 10 (page 8) and PLATE 32
SOUVENIR EUROPEAN WAR,
Robert Ingersoll Bros

Nickel chrome on brass open face case, snap-on bezel and dome. White paper dial displaying flags including Union Jacks and Tricolour, SOUVENIR EUROPEAN WAR INGERSOLL LONDON 1914 within inset seconds dial and WATCH PARTS MADE IN USA ASSEMBLED IN ENGLAND lower edge, black arabic numerals, blued steel spade hands, inset seconds dial with narrow blued steel spade hands. Keyless wound and set. Unjewelled pin pallet straight line lever movement. Imitation screws moulded in stamped-out monometallic balance. Regulator index on to scale on top plate numbered 37135186. Barrel bridge stamped ROBT H INGERSOLL & BRO. NEW YORK MADE IN USA PATENTED JAN 29 1901; APR 11 1905; JUN 4 1907: JUN 29 1909; MAY 24 1910; NOV 12 1912.

Diameter 51.5mm

PLATE 32B

PLATE 33A

PLATE 33B

PLATE 33
YANKEE RADIOLITE, Robert Ingersoll Bros, Ely Place, London EC

Nickel on brass open face case, snap-on back and bezel. Black paper dial inscribed INGERSOLL YANKEE RADIOLITE and PARTS MADE IN USA ASSEMBLED IN ENGd, luminous arabic numerals, luminous spade hands, inset seconds. Keyless wound and push to set. Barrel cover stamped ROBT H INGERSOLL & BRO NEW YORK MADE IN USA APRIL 11 '05; JUNE 4 '07; JUNE 29 '09; MAY 24 '10; NOV 12 '12; 2 '18. Unjewelled straight line pin pallet lever movement with two arm stamped-out oversprung balance. Blued index pointing to calibrations on top plate SF. *Number 53691834 relates to 1919 to 1920 manufacturing period.*
Diameter 50mm

PLATE 33C. *Tit-Bits*, 24 November 1923

Made for YOU!
INGERSOLL'S SUIT EVERY NEED & POCKET.

The Ingersoll Agent in your district has just the watch you need—accurate, reliable, thoroughly dependable—with the name Ingersoll plainly stamped on the dial—at a price that will conveniently suit your pocket. 10/6 to 45/-! Every model a guaranteed timekeeper made of first class materials and timed and tested to a high degree of accuracy. See the Ingersoll Agent to-day and get the watch that was made for *you*.

XMAS COMING!
No gift like an Ingersoll, nothing used so much, consulted so often, lasts so long. Give Ingersolls this year.

Look for the dealer who displays this sign:—
ACCREDITED AGENT FOR
INGERSOLL WATCHES & CLOCKS.

REFUSE SUBSTITUTES
Dealers who offer you a substitute have an ulterior motive for so doing. Insist on an Ingersoll and buy it at the sign of The Ingersoll Accredited Agent.
INGERSOLL WATCH CO., LTD.
386, Ingersoll House, Kingsway, London, W.C.2.

Yankee Improved 12/6
„ „ Radiolite 17/6
Other models 10/6 to 45/-
Ingersoll Alarm Clocks, 7/6 to 21/-

THE GUARANTEED *Ingersoll* 10/6 to 45/-

Pin Pallet Lever Watches

PLATE 34A

PLATE 34
MIDGET, Ingersoll Watch Co

Nickel open face case, snap-on back and bezel. White card dial inscribed INGERSOLL MIDGET and MADE IN USA lower edge, black arabic numerals, blued steel spade hands, inset sunk seconds dial. Keyless wound and push to set. Full plate unjewelled straight line lever escapement stamped on top plate 67200685 and on the barrel plate INGERSOLL WATCH CO MADE IN USA PATENTS JULY 21 1914; NOV 14 1922; NOV 12 1912; MAY 24 1910; SEP 10 1907; JUN 4 1907; MAR 5 1907. Two arm monometallic oversprung balance. Blued regulator index pointing to graduations on top plate towards centre arbor. Serial number corresponds to manufacturing period 1925/26.

Diameter 39.4mm

Ingersoll Midget was available in gold plate, gunmetal or solid nickel – the last mentioned was available in 1906 for 8/6d (8 shillings and sixpence). The lady's slim line Midget was introduced in 1910. By 1913 a model appeared with lugs on the case to facilitate conversion to a wristwatch by the addition of a strap. During the First World War leather straps with a cup to hold these small pocket watches were not uncommon.

PLATE 34B

254

Pin Pallet Lever Watches

PLATE 35A

PLATE 35B

PLATE 35
WATERBURY, Ingersoll Watch Co

Chrome open face case, snap-on bezel and back. Black painted dial inscribed Ingersoll WATERBURY and MADE IN USA around inset seconds dial, luminous arabic numerals, luminous spade hands. Keyless wound and push to set. Three-quarter plate stamped ROBT H INGERSOLL & BRO 4 JEWELS 891305. Two arm pin pallet straight line lever movement. Oversprung bimetallic balance. Regulator index on balance cock engraved FS.
Diameter 46.2mm

Note that the serial number 891305 does not follow the previous Ingersoll production that had reached 70,000,000.

PLATE 36
RELIANCE, Ingersoll Watch Co

Gold filled open face case, screw back inscribed INGERSOLL RELIANCE ILLINOIS WATCH CASE Co ELGIN USA GUARANTEED TEN YEARS 2765402. White enamel dial inscribed Ingersoll RELIANCE and MADE IN USA around inset seconds dial, black arabic numerals, blued steel moon hands. Keyless wound and pull to set. Three-quarter 7 jewel club tooth straight line lever movement. Top plate stamped ROBT H INGERSOLL & BROS USA 754657.
Diameter 49.5mm

Serial numbers for Ingersoll Reliance movements marked Robt H Ingersoll & Bros range from below 100,000 to near 1,000,000.

PLATE 36

The Ingersoll-Trenton watch factory was apparently limited to using this six figure stamping machine, the seven digit machine used up until 1917 no longer being serviceable. They were reluctant to invest in new expensive machinery.

Pin Pallet Lever Watches

PLATE 37

PLATE 38

PLATE 37
CORONATION ELIZABETH II,
Ingersoll Watch Co

Chrome open face case, snap-on bezel and back. Dial portrait of Queen Elizabeth II flanked by Union Jack and Commonwealth flags GREAT BRITAIN INGERSOLL LTD LONDON. CORONATION around the dial omitting the 6 and 12 positions in substitution of numerals, gilt hands, inset seconds dial with 1953 within, gilt hands. Pin pallet lever movement stamped MADE IN GREAT BRITAIN and numbered 531M. Contained in original presentation box with medallion inside and outside box.
Diameter 51.3mm

PLATE 38
MICKEY MOUSE, Ingersoll Watch Co

Nickel plated brass open face case, snap-on bezel and back. Varnished paper dial inscribed MICKEY MOUSE INGERSOLL and FOREIGN lower edge with characteristically coloured Mickey Mouse running beside rotating seconds dial that depicts three miniature Mickey Mouses. Large black arabic hour numerals 1-12, small black arabic hour numerals 13-24. Keyless wound and push to set Mickey's hands. Thiel (Ruhla Germany) unjewelled pin pallet straight line lever movement. Nickel top plate stamped INGERSOLL FOREIGN. Oversprung two arm balance with blued steel balance spring. Regulator index blued and pointing to graduation in top plate near centre arbor.
Diameter 50.3mm

Ingersoll also manufactured several models of Mickey Mouse (Ingersoll Waterbury Company). These show Mickey standing rather than running and can be differentiated by the length of the watch stem which becomes progressively shorter until Model 4 finishes up as a lapel button. Sold for a low profit margin at $1.33, it was an inferior product on old machinery.

Pin Pallet Lever Watches

PLATE 39

PLATE 40A

PLATE 39
INGERSOLL

Gilt open face case, snap-on bezel and back. Skeletonised dial with white annular dial with black arabic numerals, black pointer hands and concentric centre seconds. INGERSOLL SWISS MADE lower edge of dial. Keyless wound and pull to set. Pillar plate stamped 425S. Barrel bridge stamped AMIDA LIMITED. Top plate stamped DRAW SWISS 17 JEWELS SEVENTEEN. Club tooth straight line lever. Oversprung monometallic two arm balance and shock suspension with white balance spring. No index to regulator.
Diameter 37mm

PLATE 40
Anon

Chrome on brass open face case, snap-on bezel and dome stamped with medallion portrait of King George V and Queen Mary 1910 Jubilee 1935. Cream painted dial with sunk centre, black arabic numerals, moon blued steel hands, inset seconds dial. Keyless wound and pull to set hands. Believed to be Thiel made movement for Robert Ingersoll Ltd. Pin pallet straight line lever.
Diameter 50.2mm

PLATE 40B

257

Pin Pallet Lever Watches

PLATE 41A

PLATE 41
INGERSOLL

Chrome open face case, snap-on back and bezel, back of case with picture of eagle in flight and EAGLE. Painted dial depicting Jeff Arnold, the central character in the 'Riders of the Range' series in the *Eagle* boys' magazine and known to millions of radio listeners to Radio Luxembourg. White arabic numerals on yellow background, minute and hour hands black in centre third and painted white on the outer two thirds. Animated movement of the left-hand gun mounted on pallet arbor. INGERSOLL in large black letters below 12 o'clock position. Immediately below in red lettering JEFF ARNOLD. Keyless wound and push to set hands. Backplate stamped MADE IN Gt BRITAIN 5210M. Unjewelled pin pallet lever movement.
Diameter 51.3mm

Famous "Eagle" and Radio Characters on New Ingersoll Pocket Watches

TWO new action watches, the dials of which show figures popular with millions of young people, are now being made by Ingersoll Ltd., Ruislip, Middlesex. On one watch is shown Dan Dare, hero of the famous Hulton "Eagle" paper for boys, whose adventures are daily broadcast from Radio Luxembourg. The other shows Jeff Arnold, central character of "The Riders of the Range", which is heard by millions of radio listeners and also featured in the "Eagle". First supplies of the Dan Dare watch was released to dealers at the end of May, and the Jeff Arnold "Riders of the Range" watch will be released at the end of June.

Both are pocket watches, and retailing at 30s. The use of the magazine and radio characters is exclusive to Ingersoll. In the Dan Dare watch, the space adventurer is shown with a moving ray-pistol arm ; in the "Riders of the Range" model, the cowboy's six-shooter moves. Both dials are in full colour, and on the backs of the watches is the distinctive "Eagle" badge.

PLATE 41B

Company Results

Elkington & Co., Profit, £198,906 (£47,543), plus E.P.T. refund £33,976 (£431) ; taxation £125,555 (£26,781) ; 8 per cent (8).

The Ingersoll Group of Companies profits have again substantially risen. Mr. P. J. Morren, Chairman of Ingersoll Ltd., revealed in his Annual Statement that the Company earned a net profit of £114,126 in 1951, which is nearly 60 per cent more than the figure for the year 1950, when the Group Net Profit, at £71,446 was also a record.

PLATE 41C

PLATE 1

CHAPTER SEVEN

Roskopf Watches

Georges-Frédéric Roskopf was born 15 March 1813 at Niederweeler in the Grand Duchy of Baden, Switzerland, son of Johann Georg Roskopf (boucher and hotelier) and Maria Elizabeth Roskopf (née Gmelin). In 1829, at the age of sixteen, he travelled to Chaux de Fonds where he undertook a three year commercial apprenticeship with Mairet & Sandoz, a firm selling iron metal goods and horological sundries. In 1834, at twenty-one, he became apprenticed to J Biber, a watchmaker also in Chaux de Fonds. He remained with him for a year when he married Françoise Lorrimer (the daughter of Françoise Robert), a thirty-seven year old widow with two children from her previous marriage. Financed by his wife, he was able to set up his own watchmaking business in rue Léopold Robert. A son, Fritz Edouard Roskopf, was born the following year – 1836.

In 1849 he sold his business in Chaux de Fonds to become joint manager of Guttmann Frères, a watch manufactory in Warburg, at a salary of 5000F per annum (around £200). Here he was able to produce watches of his own design, namely *genre anglais* for Hamburg buyers, before leaving in 1855 to establish a joint venture with his son Fritz and Henri Edouard Gindraux – Roskopf, Gindraux & Cie. In 1858 this firm broke up with Gindraux leaving to become Director of the Neuchâtel School of Horology and Fritz to establish his own business in Geneva.

Georges-Frédéric Roskopf now conceived the idea of producing a pocket watch that could be sold for 20 francs but at the same time was strong, not easily damaged and a reliable timekeeper. He envisaged a watch with the minimum number of working parts, a simplified escapement for ease of production, as great a power as possible, the simplest of winding systems and a good quality unembellished case. To obtain this result, Roskopf made the following innovations:

The suppression of the centre wheel and direct gearing of the barrel with the third pinion; thus a very large barrel could be used passing beyond the centre of the plate and therefore giving a greater motive force.

The use of an independent escapement unit which facilitated the work and accelerated production.

The employment of a pin-pallet escapement.

This had come about following consultation with Jules Grossman of Le Locle School of Horology who had advised the use of this escapement rather than the cylinder. This was of a type conceived in 1798 by L Perron of Besançon, a watchmaker, but in the interests of economy using steel pins instead of pallet jewels and a specially shaped escape wheel of 18 teeth. The success of the final escapement is ably expressed by Paul Chamberlain in *Its About Time:*

> The steel pin pallets are made tapered, and staked into the pallet frame and apparently bent as might be required to correct any imperfection in the original staking. Also, to overcome any defect of planting or depthing of the scape wheel and pallet, the pillar plate is split and arranged for screw adjustment. The tendency of the oil to run off the pins is corrected by having the pin pallets bank on to the rim of the escape wheel in notches in which the oil pushed over the impulse face collects and in which the pin gets an oil bath each banking. Vast numbers of this type have fulfilled their destiny.
>   The application of the keyless winding, acting forwards only and eliminating the usual to-and-fro action in winding.
>   The fitting of the minute wheel and its pinion friction tight directly on to the barrel.
>   The suppression of stopwork and the application of the 'free' mainspring or Philippe mainspring.
>   Setting by turning the hands with finger, as in a clock.
>   The case in very thick white metal with no hinged joint on the back.

Nevertheless the aim was to use materials of excellent quality and pay the work-force a proper wage in order to maintain high quality workmanship. The difficulties that faced him have been well expressed by Eugène Jacquet and Alfred Chapuis in *The Technique and History of the Swiss Watch*. Initially he intended the dials to be made of strong white paper or cardboard, but in order to find a manufacturer who could supply paper that was free of active chemicals, ie Montgolfier Frères of Annonay, he was required to place large orders. This he was unable to do. M L Bovy of Geneva experimented on his behalf with cases of English brass but finally (1866) these were replaced with those of white metal – *maillechort* – German silver, an alloy of copper, zinc and nickel from Jurst in Berlin. The case was made with a joint to open in order to hand set the hands. The glass was convex. It was proposed that the back cover should not be hinged in order to avoid the egress of dust. The first cases were delivered from Malleray in 1876 with later cases being produced by Constant Hamel of Le Noirmont.

The first watch with these characteristics appeared in 1867 and was awarded a bronze medal at the Universal Exhibition of Paris in 1868 with further recognition at the Amsterdam Exhibition in 1867. He had achieved his goal – La Montre du Proletaire.

Roskopf was an *établisseur,* ie a watch manufacturer who bought and assembled all the component parts of a watch. He therefore acted as a co-ordinator in the production process as well as being a merchant for his own finished product. In March 1866 his request for *ébauches* from Emile Roulet of Sagne Eglise was turned down and he then turned to the Fontainmelon factory for help. They were not able, however, to agree a price. An agreement was made instead with Société d'Horlogerie de Malleray in August/November 1866 for 2,000 *ébauches* and *porte échappements*. Of the twenty-four dozen *porte échappements* received in January 1867 three-quarters had to be sent back for correction. Delivery of the *ébauches* began in February 1867 but the order for 2,000 was not completed until the December. By March 1870 the total number ordered had reached 20,000. Outside help was required in assembling – the motion work and planting the two train wheels was done by M Châtelain of Damprichard Doubs, France, while the escapements were assembled by Adolphe Jacot of Dombresson – all from parts supplied by Roskopf. From the available records it is clear that the six jewel escapements were supplied by JA Voiblet & Co of Chaux de Fonds and M. Erbeau of Lucens. The escapements were planted by M Gustave Rosselet à Verrières (ref A Taylerson). The first watches were available early in 1867. By the end of the year Roskopf considered adding a single but strong set hands mechanism which was made by Raiguel, Juillard & Co of Cortébert. Watches with this modification were introduced in 1870 for a slightly higher price (25 francs, around £1, which was at that date one week's wages).

Unfortunately a short time after 1873 G-F Roskopf was unable to continue his work due to deteriorating health following the death of his wife. He transferred the ownership of the entire company to Willie Frères and a highly trusted employee, Charles Léon Schmid. Georges-Frédéric died on 14 April 1889. His son continued his own business in Geneva. Since 1939 the manufacture of Roskopf watches has been carried out by the Roskopf Association.

Roskopf was unable to take out patent protection for his innovations in Switzerland as Swiss Federal protection was not to become available until 1888. He therefore had his imitators. Although produced from differing sources and dressed up in cases and with dials displaying superficially similar monograms, trade marks and patent numbers, it is very interesting to note the uniformity of the basic design of Roskopf type watches.

It has, however, been possible to trace patent protection in several other countries for watches made by Georges-Frédéric Roskopf.

French Brevet   No 54528   18 June 1862   Georges-Frédéric Roskopf – système de remontoirs aux pendants
Belgium Patent No 21988   3 Aug 1867
USA Patent   No 75463   10 Mar 1868
French Brevet   No 80611   25 Mars 1868   Georges-Frédéric Roskopf – un genre de montre avec porte échappements dite: Montre de proletaire and eventually in Switzerland:
18632 3 Jan 1899   Fritz Edouard Roskopf
20371 27 July 1900   Greder Frères à Longeau près Benne
26514 26 Mai 1903   (K)Roskopf & Cie Chaux de Fonds
28545 9 Feb 1904   (K)Roskopf & Cie Chaux de Fonds
29831 20 July 1904   H Rosskopf & Cie Holstein – Patent Cattin & Christian Holstein Manufacture Horlogerie (on watchpaper)
30353 23 Feb 1904   Vve Ch Léon Schmid & Cie Chaux de Fonds
90517 21 Mar 1921   Alex Dubois fab de Grenier Chaux de Fonds

Suggested Reading
'Historique et Technique de la a Montre Roskopf' by E Buffat published *Administration du Journal Suisse D'Horlogerie,* 2 rue Necker, 2, 1914
The Swiss Watch – Jaquet and Chapuis 1953 and reprints
L'Horloger A l'Etabli – Shultz and Kames p355-375
It's About Time – Chamberlain 1941 and reprints
Pin Lever Watch – Cutmore 1991
'The Roskopf Watch' – Taylerson *(Antique Collecting)*
The Story of the Roskopf Watch – Player
Practical Watch Repairing – De Carle 1946 and reprints
Repertoire des Marques de Fabriques Annuaire de l'Horlogerie Swiss 1945
Brevets d'Invention and Patent Office Specifications

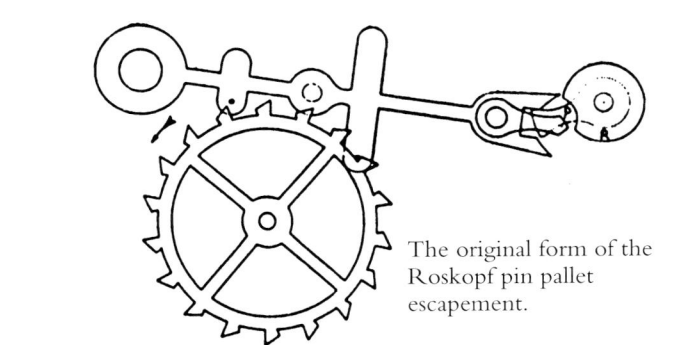

The original form of the Roskopf pin pallet escapement.

A diagram showing the Roskopf device for adjustment of wheel and pallet depth.

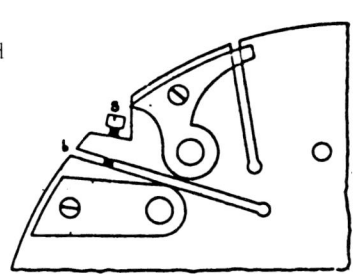

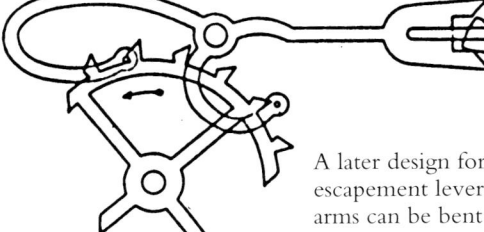

A later design for the escapement lever. The curved arms can be bent to correct positions.

PLATE 2. *Horological Journal,* April 1942

Roskopf Watches

PLATE 3A

PLATE 3B

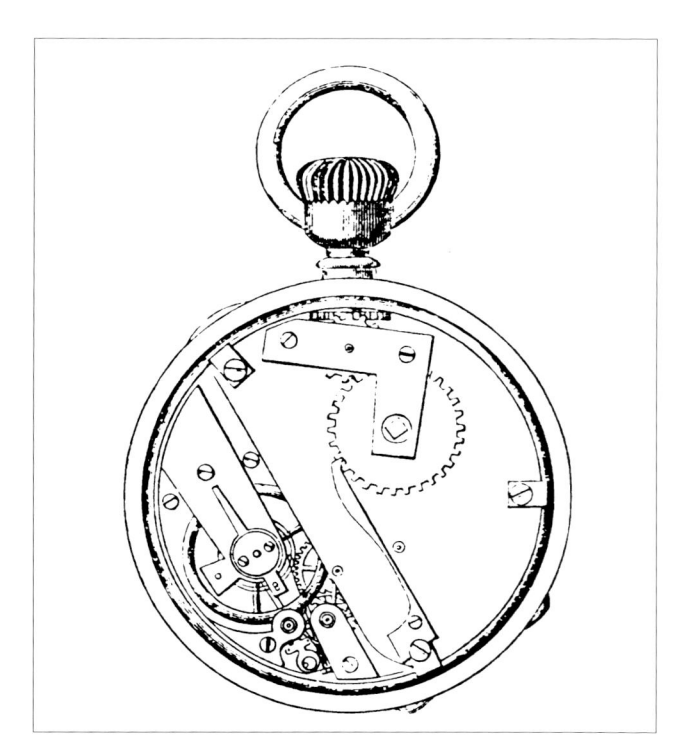

PLATE 3C. Illustration Eugène Buffat (1914) corresponds to French patent 25 March, 1868

PLATE 3
Georges-Frédéric Roskopf, Chaux de Fonds, Switzerland

Nickel open face case, hinged back and bezel c.1868. Casemakers possibly Hamel of Le Noirmont from Cortébert or Malleray. White enamel dial inscribed FESTONJEE FRANJEC BOMBAY, black roman numerals, heavy gilt brass spade hour hand with tail to minute hand. Keyless wound, hand set by pushing hands round. Half plate. 8 jewelled tangential pin pallet lever with adjustable depthing. Oversprung brass three arm balance.
Diameter 53.8mm

It is interesting to compare this watch with the illustration of two examples in Technique and History of the Swiss Watch *by E Jacquet and A Chapuis 1st ed, Plate 123. The first is an early* ébauche *supplied by the Cortébert factory to Roskopf and in the Cortébert collection – characterised by the curved edge to the top plate adjacent to the balance with a canted L-shaped barrel bridge. The second is an early Roskopf from the Chaux de Fonds Museum, probably a Malleray* ébauche *with straight edge to the top plate and squared positioning of the L-shaped barrel bridge. These examples again differ from the illustration in* La Montre Roskopf *by Eugène Buffat which corresponds to the French Patent 25 March 1868. It would seem that Roskopf was dependent on several different sources and finishers as demand for his watch rapidly grew. It would be good to hear of other examples in collections.*

PLATE 4A

PLATE 4
Roskopf Patent, Willie Frères/Ch Léon Schmid & Cie, Chaux de Fonds, Switzerland

Nickel open face case, snap-on bezel, hinged back and glazed movement, back of case engraved 18 models and inscribed MEDAILLE D'OR GENEVE 1896 in circle around Trade Mark ROSKOPF PATENT (L Schmid & Co). White enamel dial ★ROSKOPF PATENT★ in upper centre, roman numerals I-XII and small arabic 5-60 outside chapter ring, Louis style hands. Keyless wound and push piece to set at 57 minute position. Top plate with trade mark and barrel bridge engraved 30353 Swiss patent cross. Jewelled Roskopf pin pallet lever, WF stamped on tail of lever (Willie Frères). Adjustable depthing on pallets. Three arm oversprung brass balance. Regulator index on balance cock engraved ADVANCE RETARD.

Diameter 53mm

Swiss patent cross 30353 23 Fevrier 1904, Vve Ch Léon Schmid & Cie, Chaux de Fonds. Dispositif d'adjustement à frottement gras d'une roue dentée sur une parte de montre. GF Roskopf was succeeded by Willie Frères and former employee CL Schmid.

PLATE 4B

PLATE 5A

PLATE 5
G^(RE) Roskopf Patent 1a, Switzerland

Chrome on brass open face case, snap-on bezel and hinged back and cuvette, cuvette stamped Remontoire PERFECTIONNE FABRICATION SUISSE with five medals. Cream painted dial with black arabic numerals within individual yellow circles, Louis copper hands. G^(RE) ROSKOPF PATENT 1^a upper centre dial, SWISS MADE below subsidiary seconds dial. Keyless wound and pull to set hands. Barrel bridge engraved REPASSE EN SECOND. 14 jewel pin pallet lever with adjustable depthing tangential pallets. Three arm oversprung brass balance. Regulator index on balance cock engraved RA SF.

Diameter 47mm

Eberhard & Cie, Chaux de Fonds, Marque Deposé No 14141 Dec 18 1901 ROSKOPF SYSTEME la sure etiquette. G ROSKOPF & CIE Bale – transmission Meyer Grabes, Chaux de Fonds, Marque Deposé No 23356 19 Feb 1908 1a dans un circle.

PLATE 5B

PLATE 6A

**COLOUR PLATE 11 (page 8) and PLATE 6
G Rosskopf & Co Patent (Vittori & Cie),
La Chaux de Fonds, Switzerland**

Nickel open face case, snap-on bezel and back with glazed cover to movement. White enamel dial, G ROSSKOPF & Co trade mark in centre, G ROSSKOPF & Co PATENT below centre, grey arabic numerals each within individual small mauve circles, Louis style hands. Keyless wound movement with push piece to set hands at 55 minutes position. Top plate stamped SWISS MADE 10 RUBIS Rosskopf and barrel bridge stamped REPASSE EN SECOND REPASADO DE SECUNDO MANO. 10 jewel pin pallet. Roskopf type lever movement with tangential lever and three arm brass oversprung balance with blued steel balance spring. Regulator index on balance cock stamped RA SF.
Diameter 43.2mm

PLATE 6B

Roskopf Watches

PLATE 7A

PLATE 7B

PLATE 7C

PLATE 7
H Rosskopf & Co, Holstein, Switzerland

Nickel open face case, snap-on bezel, hinged back and cuvette. Back stamped H ROSSKOPF & CO PATENT, cuvette stamped the same plus ANCRE DU PRECISION. White enamel dial with trade mark upper centre, black arabic numerals, Louis copper hands. Keyless wound, push piece to set at 55 minute position. Barrel bridge engraved BREVET Swiss Patent Cross 29831 GARANTI VERITABLE. Roskopf type pin pallet lever escapement. Adjustable depthing to tangential lever, three arms to ?gilt balance. Regulator index on balance cock stamped RA SF.

Diameter 51.4mm

Brevet Patent Cross 29831 20 Juillet 1904 (stamped 12 Dec 1904) Cattin & Christian à Holstein, Bale, Compagne Suisse. See watchpaper.

PLATE 8A

PLATE 8B

PLATE 8C

PLATE 8
Louis Roskopf SA, Reconvillier Watch Co SA, Reconvillier, Switzerland

Nickel open face, snap-on bezel, hinged back and cuvette, the cuvette inscribed PATENT Louis ROSKOPF SA PATENT 1906 QUALITE SUPERIEURE ANTI-MAGNETIC VERITABLE Louis ROSKOPF, Roskopf trade mark and sailing scene with hunter and child carrying crossbow. White enamel dial, Louis ROSKOPF SA PATENT upper centre, black arabic numerals 1-12 and small red arabic numerals 13-24, Louis style hands. Centre seconds stop operated by slide in side of case to side of balance staff. Keyless wound and set hands with push piece at 11 o'clock position. Top of winding button stamped in relief LOUIS ROSKOPF SA. Top plate engraved ANTIMAGNETIC and Louis Roskopf trade mark. 8 jewel pin pallet escapement with adjustable depthing to tangential lever, three arm brass oversprung balance. Regulator index on balance cock engraved AVANCE RETARD. Barrel bridge engraved REPASSE EN SECOND REPASSADO DE SECUNDA MANO.
Diameter 52.6mm

PLATE 9B

PLATE 9A

PLATE 9
Louis Roskopf SA, Reconvillier Watch Co SA, Reconvillier, Switzerland

Nickel open face case, snap-on bezel, hinged back stamped Louis ROSKOPF SA PATENT and cuvette depicting a sailing scene with hunter and child carrying crossbow and inscribed Louis ROSKOPF SA PATENT 1906 QUALITE SUPERIEURE VERITABLE Louis ROSKOPF. White enamel dial inscribed Louis ROSKOPF SA PATENT in upper centre and ANTI MAGNETIC in red above subsidiary seconds dial, black numerals and small red arabic numerals 13-24 outside chapter ring, black roman numerals 1-12, black steel spade hands. Keyless winding, top of winding button stamped Louis Roskopf SA, hand set push piece at 11 o'clock position. Barrel bridge engraved REPASSE EN SECOND NON MAGNETIC LOUIS ROSKOPF SA PATENT. Pin pallet tangential lever escapement with adjustable depthing to the pallet, three arm oversprung brass balance with regulator index on balance cock engraved ADVANCE RETARD.
Diameter 54mm

PLATE 9C

PLATE 10A

PLATE 10B

PLATE 10
W Rosskopf & Co, La Chaux de Fonds, Switzerland

Nickel open face case with hinged bezel and back, the outside of back cover stamped W ROSSKOPF & Co CMG PATENT. Watch paper W ROSSKOPF & CO Uhrenfabrik La Chaux-de-Fonds. Glazed back of movement. White enamel dial inscribed ★W ROSSKOPF & Co PATENT★ in a semicircle with black roman numerals, gilt brass spade hands with decorative tail to minute hand. Keyless wound, push piece at 55 minute position to set hands. 8 jewel pin pallet lever movement with tangential lever and adjustable depthing. Pillar plate stamped 4. Three arm oversprung balance with blued steel balance spring and regulator index on balance cock engraved RS AD. Barrel bridge engraved Swiss Patent Cross DEPOSE 22061 Trade mark (W ROSSKOPF Co CMG).

Diameter 53.7mm

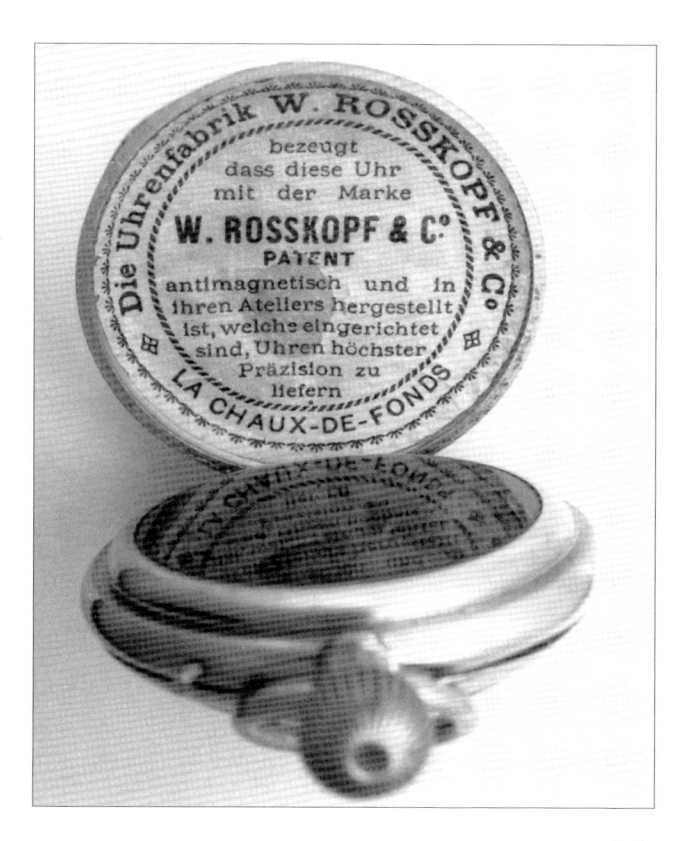

PLATE 10C

Roskopf Watches

PLATE 11A

PLATE 11B

PLATE 11
W Rosskopf & CO (Vittori et Cie), La Chaux de Fonds, Switzerland

Nickel open face case, snap-on bezel and hinged back and cuvette with ORNAMENTATION in small letters above five 'medals', W ROSSKOPF & CO PATENT trade mark repeated in larger capitals, W ROSSKOPF & CO PATENT below and GARANTI ANTI-MAGNETIQUE. Enamel white dial with luminous arabic numerals and skeletonised hands inscribed W ROSSKOPF & CIE PATENT ANTI-MAGNETIQUE. Keyless wind with push piece to set hands at 55 minute position. Special winding button with W ROSSKOPF & CO patent and trade mark on top of winding button. Barrel bridge engraved DEPOSE Swiss Patent Cross 22061. Circular trade mark. 14 jewel pin pallet tangential lever with adjustable depthing. Three arm brass oversprung balance. Regulator index on balance cock engraved RA SF.
Diameter 52.6mm

Interesting variation on these watches of the spelling of Roskopf ie Rosskopf on this particular example! Also appears to be some indecision regarding whether to have & CO or & CIE. On the cuvette the original & Cie has been changed by over-engraving 'ie' with 'o'.

PLATE 11C

Roskopf Watches

PLATE 12A

**COLOUR PLATE 12 (page 8) and PLATE 12
Roskopf Système Patent**

Nickel open face case, snap-on bezel, hinged back and glazed back to movement. White enamel dial inscribed ROSKOPF SYSTEME PATENT SWISS MADE in red lettering, black broad roman numerals and gilt spade hands – the minute hand with decorative tail. Keyless wound at pendant with push piece to set hands at 56 minutes position. Unjewelled pin pallet tangential movement with adjustable depthing. Oversprung three arm brass balance with blued steel balance spring. Regulator index on balance cock engraved AVANCE RETARD.
Diameter 46.8mm

PLATE 12B

PLATE 13

COLOUR PLATE 13 (page 8) and PLATE 13
Messaggero, Echappement Roskopf

Nickel open face case, snap-on bezel, hinged back with MESSAGGERO within belted circle stamped in centre of back. White enamel dial repeating MESSAGGERO trade mark, ECHAPPEMENT ROSKOPF, grey roman numerals 1-XII within mauve circles with red arabic numerals 13-24 outside chapter ring, copper hands. Keyless wound with push piece to set hands at 57 minute position. 8 jewel Roskopf pin pallet tangential lever movement with adjustable depthing. Barrel bridge engraved REPASSE EN SECOND REPASADO DE SEGUNDO MANO. Three arm oversprung brass balance with blued steel balance spring. Regulator index on balance cock engraved AVANCE RETARD.
Diameter 57mm

PLATE 14
Anonymous

Fine silver in the style of a half hunter case, snap-on back case stamped 906/4. Note XII is at the position of the button wind. In the true half hunter the winding button is at 3 o'clock position. Early form of keyless winding with push piece to hand set at 57 minute position. White enamel dial, black roman numerals and broad blued steel spade hands. 8 jewelled early Roskopf pin pallet tangential lever escapement with adjustable depthing. Three arm oversprung brass balance and blued steel balance spring. Regulator index on plain balance cock.
Diameter 52.7mm

PLATE 15
Système Roskopf Nautical

Nickel open face case, snap-on bezel, hinged back and glazed back of movement. Centre of back, the white enamel dial and top plate of movement inscribed with logo SYSTEME ROSKOPF NAUTICAL. Black arabic numerals 1-12 within chapter ring with red numerals 13-24 outside, Louis style copper hands. Keyless wound movement with push piece to set hands. 8 jewel pin pallet tangential lever escapement with adjustable depthing. Brass three arm oversprung balance with blued steel balance spring. Regulator index on balance cock engraved AVANCE RETARD.
Diameter 58.2mm

PLATE 14A

PLATE 14B

PLATE 15A

PLATE 15B

Roskopf Watches

PLATE 16A

PLATE 16B

PLATE 17

PLATE 16
Système Roskopf

Nickel open face case, snap-on bezel, hinged back and glazed cover to movement. Bezel engraved and back engraved depicting woman with a punt pole and man attending to lobster crate on boat. White enamel dial inscribed SYSTEME ROSKOPF, black arabic numerals 1-12 and smaller 13-24, copper Louis hands. Keyless winding, push piece to set hands at 56 minutes. Jewelled pin pallet tangential lever movement with adjustable depthing. Oversprung three arm brass balance. Blued steel balance spring, 4 stamped on pillar plate.

Diameter 54.2mm

PLATE 17
The People's Watch

Nickel open face case, snap-on bezel and back. Inside dome inscribed SOLID NICKEL. White enamel dial, black arabic numerals, blued steel spade hands, inset seconds dial. Keyless wound with push piece at 55 minute position. Top plate stamped SWISS MADE. Half plate Roskopf type movement with adjustable tangential pin pallet escapement. Three arm monometallic oversprung balance with blued steel balance spring and index on balance cock engraved RETARD AVANCE.

Diameter 51mm

Roskopf Watches

PLATE 18A

PLATE 18
Anonymous

Nickel open face case, screw-on back and bezel, bezel edge knurled. White enamel dial, luminous arabic numerals, luminous skeletonised spade hands. Keyless wound and push piece to set at 55 minutes position. Inscribed DEPOSE on barrel bridge and 4 JEWELS LEVER SWISS MADE on top plate. Roskopf type pin pallet escapement with adjustable depthing, slot in pillar plate. Imitation oversprung bimetallic balance with blued steel balance spring. Regulator index on balance cock engraved RA SF.
Diameter 58.6mm

PLATE 18B

PLATE 19A

PLATE 19B

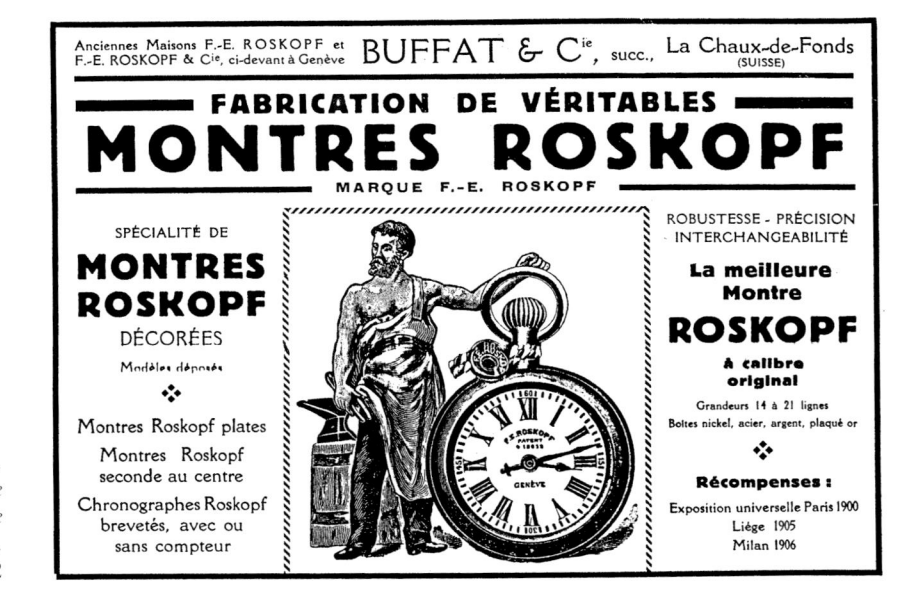

PLATE 19C.
Journal Suisse d'Horlogerie et de Bijouterie. Sept. 1922

PLATE 19
Best Patent Lever, Jequier Frères, Fleurier, Switzerland

Gilt brass open face case, snap-on bezel, hinged back, inside back cover stamped above five pointed star (Jequier Frères). Gilt brass engine-turned dial inscribed BEST PATENT LEVER SWISS MADE, black roman numerals, blued steel spade hands. Keyless wound with push piece to set at 11 o'clock. Top plate stamped SYSTEME PATENT ROSKOPF and barrel bridge inscribed REPASSE EN SECOND REPASADO DE SEGUNDA MANO. Half plate Roskopf type tangential pin pallet lever with adjustable depthing. Oversprung three arm brass balance with blued steel balance spring. Regulator index on balance cock engraved AVANCE RETARD.

Diameter 51.1mm

Chapter Eight
Chronographs

A chronograph is basically a timer that measures elapsed time. Today they are complicated with many sophistications but the early examples were little more than time counters. The invention of the chronograph has been attributed to H Rieussac of St Mande, a town near Paris. This instrument was used to time horse races over specific distances. A chronographic *compteur* of ten minute duration, dated 1822 and signed Rieussac is in the collection of the Musée de la Chaux de Fonds in Switzerland. Subsequently Rieussac made pocket chronographs. These had two movements, one for the timepiece and one for the chronograph.

Frederick Louis Fatton, who worked for Breguet, registered a UK patent (No 4645) on 9 February 1822 for an instrument with an inking pen dial. In this instance the dial was fixed with the inking pen the moving indicator. A mark was made on the dial when the hand stopped rotating. A second patent by Fatton (UK Patent No 4707) was taken out seven months later. This was similar in principle to the earlier instrument but applicable to watches. This is fully described in Newton's *London Journal* Vol 5, p281. An excellent description with extremely clear line drawings and explanation of the mechanism of Fallon's inking chronograph can be found in *The Art of Breguet* by George Daniels. A watch of this type was sold by Christie's New York on 28 April 1990 (Lot 522).

Further advances in the development of the chronograph were made by Winnerl, who produced his chronograph in 1831. This was still, however, a simple counter and could only record elapsed time from the beginning to the end of an event, although his modification allowed the second hand to be stopped and returned to the position it would have occupied if it had not been checked. There was no provision for re-setting in order for the time to be counted from zero – neither could an elapsed period greater than sixty seconds be recorded.

In 1844 Adolphe Nicole registered UK Patent No 10348 describing a heart cam mechanism for a seconds dial. This is also described in the *Repertory of Arts* Vol 7, p151. However, on 14 May 1862 UK Patent No 1461 was issued for the first chronograph with hands capable of returning at will to zero position (see over).

This was achieved by means of a three push button system and incorporating a castle ratchet. The timing mechanism can be started, stopped and reset without stopping the time train. Watches were exhibited at the London Exhibition of 1862, the first mechanisms being made by an employee, M Henri Piquet, a native of Geneva. Adolphe Nicole, a native of the Vallée de Joux in Switzerland, was working in 1840 at 80 Dean Street, London in partnership with Henri Capt. The partnership changed to Adolphe Nicole and Jules Capt in 1843 before their move to 14 Soho Square in 1868. In 1876 Adolphe Nicole established a partnership with his son-in-law (Sophus Emile Nielson) before forming a company in May 1888.

Initially the chronograph mechanism was under the dial but after 1868 this appeared on the top plate (back of the movement). At the same time the minute recording dial was added. In 1871 Kulberg, when writing in the *Horological Journal*, describes this form of keyless split seconds watch, but it was not until 1880 that the split second watch in its present form was to appear.

The watch timer is a type of stop-watch used to record elapsed periods of time. It is designed with balance oscillation of $\frac{1}{5}$, $\frac{1}{10}$, $\frac{1}{20}$, $\frac{1}{50}$ or $\frac{1}{100}$ second in order to record time intervals requiring this degree of accuracy. It is usually provided with an elapsed time minute recorder indicating duration of 30 or 60 minutes, but for special purposes an elapsed time of 12 to 24 hours may require to be displayed. Some mass-produced and cheap stopwatches, however, may be nothing more than an ordinary watch movement with a stop/start braking acting on the rim of the balance and operated by a slide in the band of the case or a push button at pendant. On the other hand, this method of operation was sufficiently refined by some makers to give satisfactory service for most purposes. Quite often resetting and zeroing mechanisms are also included. A good performance is obtained using non-magnetic and temperature corrected balance assemblies.

The split second chronograph is a special type of chronograph designed to enable it to act as a double chronograph. This is achieved by providing a second centre

1461. Nicole, A. May 14. 1862

Chronographs; stop-mechanism. — A central seconds hand on the arbor of a wheel f, Fig. 1, is started, stopped, and returned to zero by successive depressions of a push-piece i, which is pivoted at i^1 and carries a spring-pressed pawl i^2 to turn a ratchet-wheel h one tooth at each depression. A disc g attached to the ratchet-wheel is notched at each third tooth of this wheel, to receive projections on a lever d and springs k, l. When the push-piece i is first depressed, a notch in the disc g allows the lever d to be turned by a spring d^2, so that a fine toothed wheel e carried by the lever gears with the central wheel f; the wheel e is always in gear by another wheel c with a wheel b on the arbor of the ordinary seconds hand, and the central seconds hand is thus driven. On the second depression of the push-piece i, the wheel e is disengaged from the wheel f, and the notched disc g drops the spring k on the wheel f to stop it. A third depression of the push-piece i raises the spring k from the wheel f, and drops the end of the spring l on a notched heart cam m, on the arbor of the central seconds hand, to return this to zero. A minute hand may also be provided, loose on the arbor of the wheel f, and provided with a wheel to engage a pinion carried by the the wheels c, e, when these drive the wheel f; the minute hand is returned to zero by a heart cam engaged by the spring l. The ordinary minute hand is on an arbor passed through the arbor of the wheel f, and the hour hand is on a separate axis. The construction may be modified so that after the first depression, the rise of the push-piece i moves the ratchet-wheel a tooth, without further effect; but after a second depression, the rise of the push-piece causes the hand to return to zero.

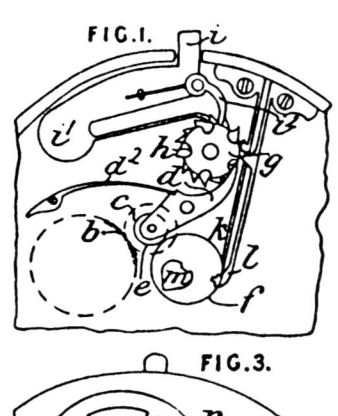

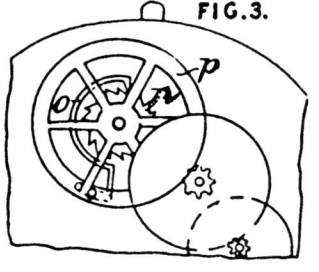

Short-time movements; keyless mechanism; mainsprings. — A time-keeper with stop-mechanism arranged as described above may be driven by a spring o, Fig. 3, acting on a wheel p geared to the arbor of the wheel b, Fig. 1, and to an escapement. The spring o is "wound" by its engagement with a ratchet-wheel n, Fig. 3, on the axle of the ratchet h, Fig. 1, when these wheels are turned by the action of the push-piece i. If the wheels are turned too much, the spring o slips over the teeth.

---

seconds hand concentric with and superimposed on or under the first but able to act independently. This enables the measuring of successive time intervals and is applied to both stop-watches, timers and full chronographs. The usual sequence of operation is as follows. At the commencement of timing the main pendant button is pressed, both hands moving forwards together At the end of the first time interval (eg as a racer completes a lap) the supplementary button on the case band controlling the split seconds hand is pressed, the split second hand stops, whilst the sweep seconds hand continues. Having read the time and recorded the first time interval the supplementary button on the side is again pressed which causes the split seconds hand to rejoin the sweep seconds hand. The sequence may now be repeated until the end of the race when timing stops for both hands on depressing the pendant button which, when pressed again, resets both hands to the zero position.

There are a number of interesting variations in the layout and design of the movements. Technical details of many of these can be found in the books listed in the bibliography on page 279. It is possible by inspecting the hands to have some indication as to the age. The older type of chronograph with the mechanism under the dial has the split second hand lying under the sweep seconds hand whilst in the later watches with split mechanism visible on the top plate (ie back of the movement) the split hand is placed on top of the sweep seconds hand.

A useful quick check to the satisfactory action without friction of the chronograph wheel may be made by comparing the amplitude of the balance when the chronograph is working both with and without the split seconds being in use.

In Great Britain the National Physical Laboratory at Teddington had been for many years the official body issuing certificates of performance of watches under certain controlled conditions. A watch was not only required to be of a constant rate but this had also to be constant in all positions. Likewise, performance over a range of temperature had to be established. Chronograph watches entered for a Class A or Class B certificate were not tested for rate until they had been examined and found to have a satisfactory action of chronograph mechanism. Tests were carried out over several periods of twenty-four hours each at 67° in the dial up position, 1. with chronograph disengaged, 2. with chronograph in action

PLATE 1. 1862

and 3. in split seconds chronograph watches with the hands split. The daily rate of 2. must not exceed more than five seconds the rate obtained under conditions 1. and 3.

It is essential that there should be precise action when the push piece of the chronograph is operated. The hand must start precisely and stop at once with no run on, jumping or lagging. Furthermore, the chronograph hand starting from precisely zero must be capable of returning exactly to zero and in the case of split seconds chronographs the hands must travel together exactly until split. The minute recorder dial reading must match exactly the elapsed time of 60 seconds.

Consideration of daily rate wards
maximum 40 marks
Position adjustment
maximum 40 marks
Temperature compensation
maximum 20 marks

A watch obtaining more than 80 marks out of the maximum 100 would be awarded a certificate endorsed 'Especially good'. Further details concerning the tests that were carried out at the National Physical Laboratory and the basis of the calculations are given in *Watch Adjustment and Repair* by FJ Camm, 1940.

Stop-watches and chronographs are generally associated in people's minds for use in the sporting world, from horse racing, yachting, running, water ski-ing through to heats in the Le Mans, but their application is manifold. Sir David Salomon (1851-1925) was seen on many occasions sitting in the back of his Peugeot motor car (capable of 15mph!) timing his chauffeur over the drive from Broomhill to Tunbridge Wells. The arrival of the steam train necessitated the chronometer with a dial to time trains between the measured quarter to one mile posts, a pursuit still repeated by today's traveller with the help of the complicated quartz wrist watch!

Scientists found them essential in research laboratories; they were used for manufacturing processes and production timing in the world of industry and for the taking of pulses by medical practitioners. For these many uses special dials are required. The smallest increments of time are measured – fifths to twenty-fifths or even one-hundredths of a second are all to be found. Obviously it is not just the markings on the dial that are required; it is the period of oscillation of one complete swing of the balance wheel that make the various calibrations of the dial possible. It is meaningless to have a dial marked in one-hundredths of a second if a single complete oscillation of the balance takes one-fifth of a second. The particular fascination of the chronograph is the ingenuity and varying methods devised in order to overcome problems. This invariably finishes up with a delicate and intricate mechanism that can only be admired and respected. The following illustrations of a selection of chronographs and stop-watches designed over the past 150 years will, it is hoped, demonstrate the gradual progression as well as illustrate the immense variety. It is not possible to include a detailed technical aspect of each watch shown, but if this should be felt of interest recourse to any of the following titles should prove rewarding:

A Guide to Complicated Watches by François Lecoultre
Britten's Watch and Clockmaker's Handbook, Dictionary and Guide 16th ed., Richard Good
Watch and Clock Making and Repairing, WJ Gazeley, Chapter 14
The Chronograph, Its Mechanism and Repair, B Humbert, 1950
American Watchmaking by M C Harrold, p99

Chronographs

PLATE 2A

PLATE 2
Payne & Co, 163 New Bond Street, London

Silver open face case, hinged bezel, back and cuvette. Milled edge to body, engine-turned barley decoration back. Casemaker's mark JS. White enamel dial inscribed Payne & Co 163 New Bond Street London Chronographe, small black arabic seconds and subsidiary sunk hour and minute dial with roman numerals, fine blued steel spade hands. Split seconds. Key wound and key set from back. 15 jewel Swiss movement. Three-quarter plate crab tooth duplex escapement with tangential lever with ringed tail. Oversprung compensated cut bimetallic balance and blued steel balance spring. *This is a rare example of a split seconds chronograph with an unusual escapement in so far as it beats seconds. See Scolnik Catalogue No 26 item 54 for a similar watch by Robin & Cie.*
Key size 4. Diameter 49.5mm

The premises of Payne Watch & Clock Maker are illustrated in the Tallis Street View for 1838 (Plate 2D opposite). Wm Payne & Co are listed in the Directories at this address between 1852-1915 as clock and watchmakers. They were undoubtedly also retailers and were one of the main stockists of Thomas Cole strut clocks.

PLATE 2B

Chronographs

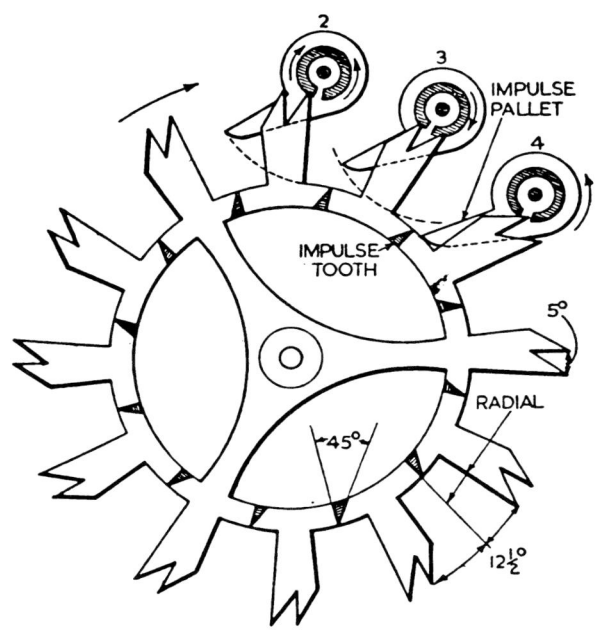

Figure 93. Crab-tooth duplex escapement. Beats seconds with a 14,000 train.

PLATE 2C. From *Clock and Watch Escapements* by W.J. Gazeley, 1st edition 1956

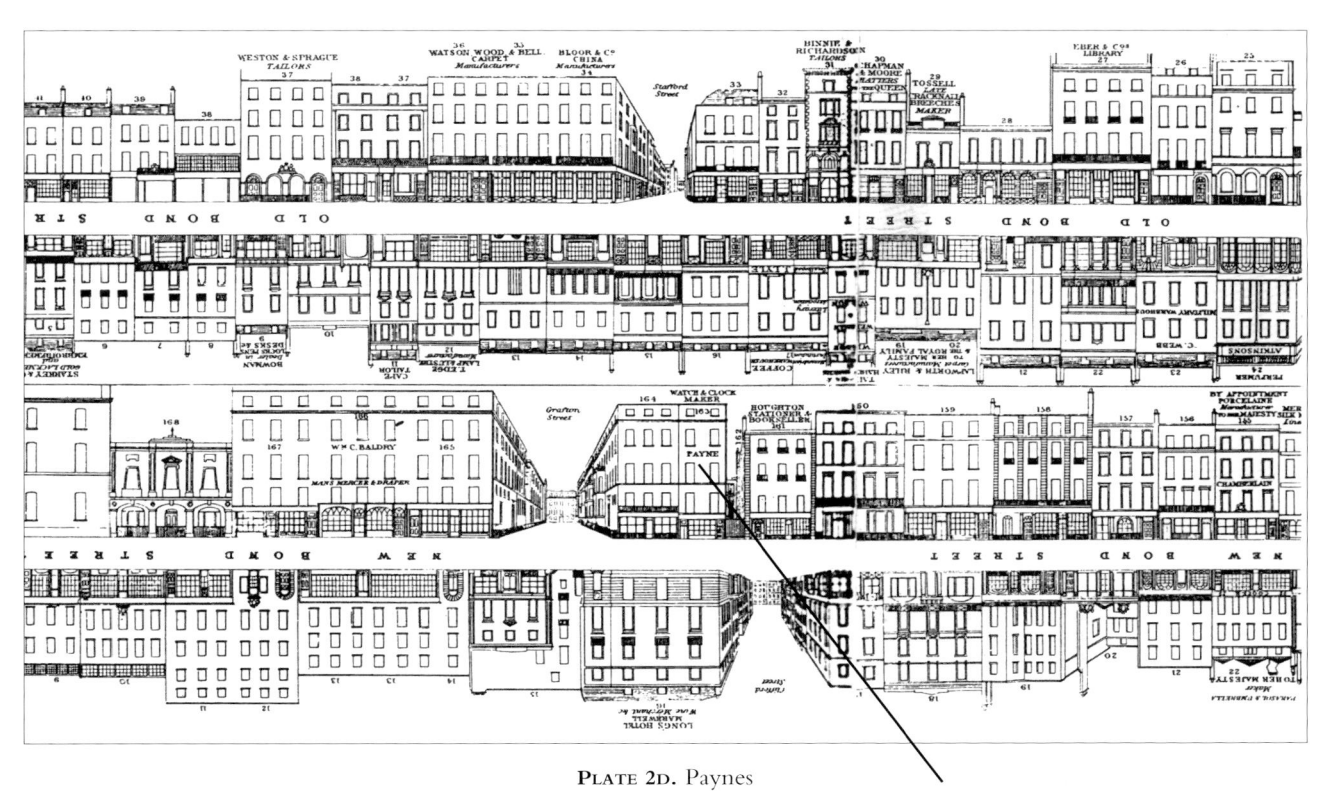

PLATE 2D. Paynes

Chronographs

PLATE 3A

PLATE 3B

8508. Baume, J. A., and **Baume, L. C. A.,** [*trading as* Baume & Co.], and **Chatelain, C. V.** May 22. 1889

Chronographs, split seconds. The stopped hand is set free to rejoin the advanced hand, without using the splitting side push-piece a second time, by the pendant pressure, which restores the latter hand to zero. When the lever *a* from the pendant gives the ordinary ratchet and castle wheel *b*, *b¹* the zero movement, an additional lever *o*, with spring *o¹* and pawl *o²*, turns the notched wheel, which applies and removes the stop arms of the restrained hand, and which can also be turned by the side push-piece.

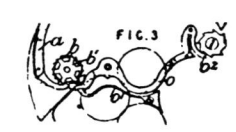

PLATE 3C

PLATE 3
Baume & Co, 21 Hatton Garden, London

18ct gold open face case numbered 3153, Swiss hallmark, hinged bezel back and cuvette. White enamel dial 0-60 centre seconds and inset 0-60 minute recorder with inset 0-60 running seconds dial, blued steel spade hands. Keyless winding, push piece to set at 4 minutes position. Top plate engraved SWISS MADE BAUME PATENT NO 8508 No 3153. 21 jewel club tooth straight line lever movement. Cut bimetallic balance. Blued oversprung and overcoil balance spring. Index on balance cock FS and foliate engraving.

Diameter 53.6mm

Only eighty of this type of watch were made in 1894. Kew Certificate issued at 78 marks. Subsequently a further twelve certificates were issued on this watch – the last being in 1950. The watch is the subject of UK Patent No 8508 applied for 22 May 1889 and accepted 29 March 1890 to Joseph Arthur Baume, Louis Celestin Alexander Baume and Charles Victor Châtelain, 21 Hatton Garden, London. See also Horological Journal *Vol XXXIII, May 1891, p129, p149.* Handbook Dictionary and Guide, *Britten 14th ed.*

Chronographs

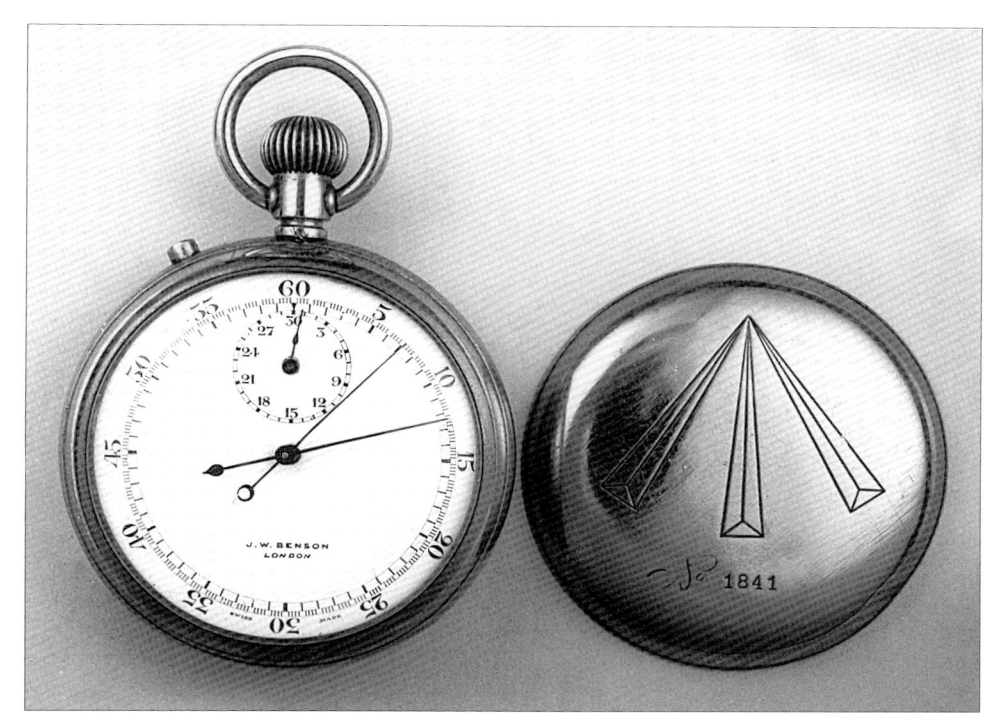

PLATE 4A

PLATE 4
J W Benson, London

Nickel open face case, snap bezel, back and hinged cuvette. Back engraved with Government Property broad arrow No 1841. White enamel dial inscribed J W Benson London Swiss Made, black arabic numerals, blued steel spade hands. Seconds dial 0-60 seconds. Minute recorder dial 0-30 minutes inset upper dial. Keyless wound at the pendant, push piece for split second mechanism at 55 minutes position. 13 jewel three-quarter club tooth lever movement. Oversprung and overcoiled compensated monometallic balance with screws and white balance spring. Polished regulator index on balance engraved FS.
Diameter 52.5mm

PLATE 4B

283

Chronographs

PLATE 5A

PLATE 5B

PLATE 5
Nero Lemania, Lugrin SA, St Orient, Switzerland

⅕ second split second chronograph with 30 minute recorder. Chrome open face case, snap-on bezel and back. White painted dial inscribed NERO LEMANIA (note stylised A in centre of Lemania), black arabic 0-60 centre seconds. Inset minute recorder 0-30 seconds and inset running seconds 0-60 seconds. Keyless wound at pendant and push pendant button to stop, start and reset chronograph with plastic push button at 55 minutes position to set split seconds hand. Top plate engraved 1900 Lemania Swiss. Pillar plate stamped Lemania Watch Co trade mark 2200081. 19 jewels. Half plate club tooth straight line lever movement with oversprung and overcoil monometallic compensated balance and blued steel balance spring. Regulator index on balance cock engraved RA SF.
Diameter 51mm

PLATE 6
Lemania, Lugrin SA, St Orient, Switzerland

⅕ second 0-60 split second chronograph with 30 minute recorder stop-watch. Nickel open face case, hinged back and cuvette, back engraved PATT 4 Government Property broad arrow 41694. White dial, black arabic numerals, centre seconds sweep, second spade hand, top black, bottom red. Keyless wound. Pillar plate numbered 27429. 9 jewels. Three-quarter plate movement with club tooth scape wheel and straight line lever. Oversprung compensated monometallic balance with blued steel overcoil balance spring. Regulator index on balance cock stamped RA SF.
Diameter 51.2mm

A high grade stop-watch. Reference George Jacob Catalogue 1949, No 50074, p298.

PLATE 7
SLAVA, Russia

Chrome open face screw back case. Silvered dial inscribed with cyrillic inscription transcribed as SLAVA. with vertical grain, black arabic numerals. 0-30 seconds outer scale and 0-30 minute totaliser on inner scale in ½ minute intervals.The centre split seconds dial 0-30 seconds. Top plate numbered 308359 and inscribed 20 KAMHEN. Half plate. 20 jewels. Rapid beating club tooth straight line lever. Monometallic balance with compensation screws. White oversprung and overcoil balance spring. Regulator index on balance cover and stamped - +.
Diameter 65.1

Watches of this type used to time experiments, photography and astronomical navigation and return to earth. See Russian Wristwatches *by Juri Levenburg, published 1995.*

PLATE 6A

PLATE 6B

PLATE 7A

PLATE 7B

Chronographs

FOR USE OF "TIMER" OR "ADJUSTER" ONLY.

THE NATIONAL PHYSICAL LABORATORY.

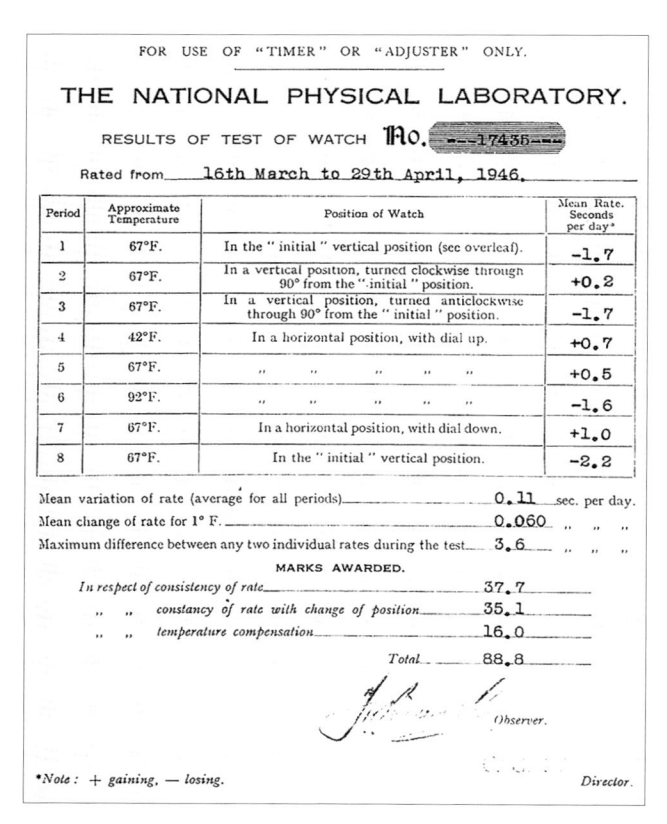

RESULTS OF TEST OF WATCH №. 17435

Rated from 16th March to 29th April, 1946.

Period	Approximate Temperature	Position of Watch	Mean Rate. Seconds per day*
1	67°F.	In the "initial" vertical position (see overleaf).	−1.7
2	67°F.	In a vertical position, turned clockwise through 90° from the "initial" position.	+0.2
3	67°F.	In a vertical position, turned anticlockwise through 90° from the "initial" position.	−1.7
4	42°F.	In a horizontal position, with dial up.	+0.7
5	67°F.	,, ,, ,, ,,	+0.5
6	92°F.	,, ,, ,, ,,	−1.6
7	67°F.	In a horizontal position, with dial down.	+1.0
8	67°F.	In the "initial" vertical position.	−2.2

Mean variation of rate (average for all periods) _____ 0.11 sec. per day.
Mean change of rate for 1° F. _____ 0.060 ,, ,, ,,
Maximum difference between any two individual rates during the test _____ 3.6 ,, ,, ,,

MARKS AWARDED.

In respect of consistency of rate _____ 37.7
,, ,, constancy of rate with change of position _____ 35.1
,, ,, temperature compensation _____ 16.0
Total _____ 88.8

_____ Observer.
_____ Director.

*Note: + gaining, − losing.

PLATE 8
Ulysse Nardin, Le Locle, Switzerland

Silver (.925) open face case, hallmark imported Glasgow **1936**, hinged back and cuvette. Case numbered 376441. White enamel dial inscribed CHRONOMETRE ULYSSE NARDIN LOCLE & GENEVE centre dial, No 16056 within inset running second dial and SWISS MADE lower edge, blue arabic numerals, blued steel spade hands. Movement stamped ULYSSE NARDIN LOCLE & GENEVE with 16056 on pillar plate adjacent to balance. Fully jewelled Guillaume balance gold compensation screws. Blued steel spring. Whiplash regulator. Complete with rating certificate, original card box with leather and blue silk and velvet-lined carrying case.
Diameter 54.5mm

Swiss Patent No 54714 was issued to Paul D Nardin, successeur de Ulysse Nardin, Le Locle, Switzerland, 21 January 1911.

PLATE 8c. National Physical Laboratory rating certificate for an identical watch numbered 17435

PLATE 8A

PLATE 8B

PL. 3

ULYSSE NARDIN, Locle et Genève
Chronométrie de Marine et de Poche
Maison fondée en 1846

CALIBRES DÉPOSÉS DE MONTRES NARDIN, SIMPLES ET COMPLIQUÉES

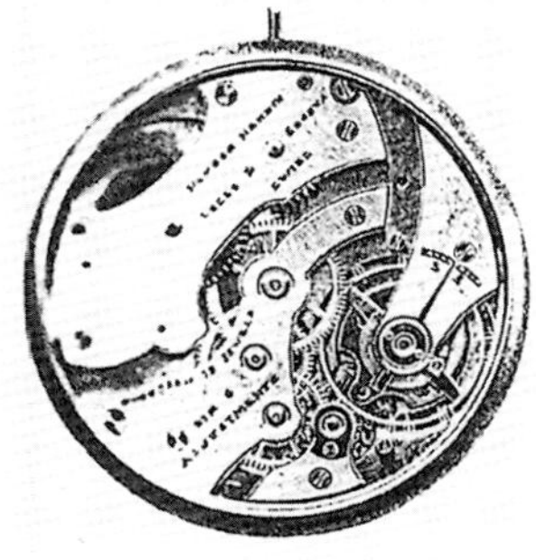

No 212. Calibre extra-plat Nardin déposé, grandeur: 17''', hauteur: 3,2 mm. (17/12).

No 213. Calibre Nardin déposé, grandeur: 17 à 20''', 12 et 16 size, hauteur: 4 - 5,3 - 6,2 mm.

No 212   *No 213*

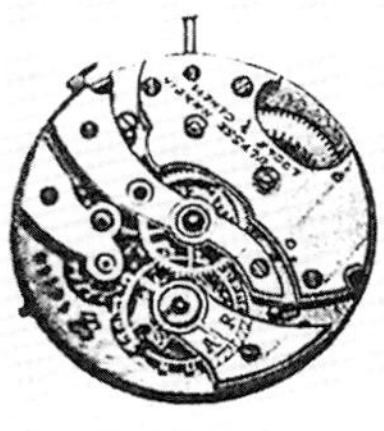

No 214. Calibre Nardin déposé, grandeur: 10 1/2, 11, 12''', hauteur: 4 mm.

No 215. Calibre Nardin déposé, grandeur: 19''', hauteur normale.

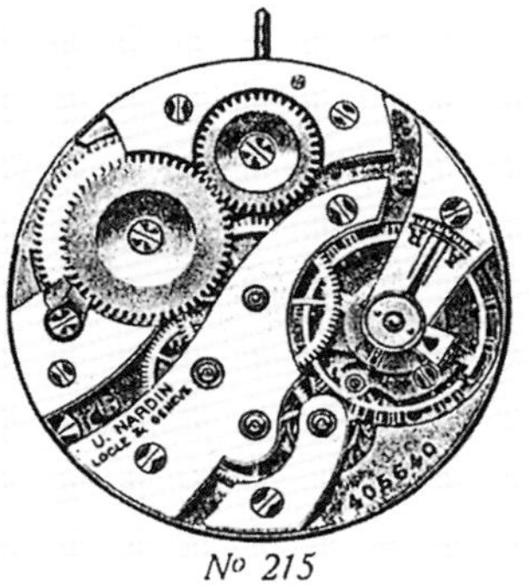

No 214   *No 215*

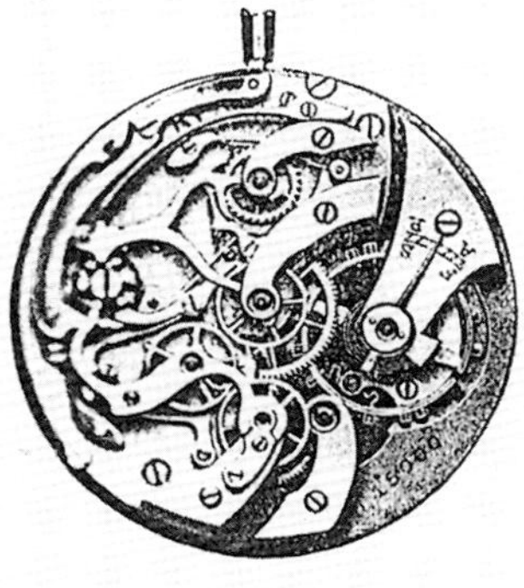

No 216. Calibre Nardin déposé, chronographe-compteur instantané breveté, grandeur: 19, 20''', hauteur: 5,3 - 6,2 mm.

No 217. Calibre Nardin déposé, chronographe-compteur et rattrapante, grandeur: 19, 20''', hauteur: 5,3 - 6,2 mm.

No 216   *No 217*

ULYSSE NARDIN, Locle et Genève, 7 Grands Prix

PLATE 8D. Catalogue Chronométrie Ulysse Nardin. Maison fondée 1846. Five generations of Le Locle Switzerland. Léonard, Ulysse, Paul D., Alfred, Ernest et Gaston Nardin and sons Claude, François & Raymond

Chronographs

PLATE 9A

PLATE 9B

PLATE 9
Omega, Switzerland

⅒ second split seconds with 30 minute recorder time of the day chronograph. Chromed outer case and nickel cuvette, snap-on bezel, hinged back and cuvette stamped OMEGA FAB SUISSE SWISS-MADE. White enamel dial inscribed OMEGA and trade mark, arabic numerals, blued steel spade hands. Inset recording minute dial 0-30 minutes upper centre. Running seconds dial 0-30 seconds lower centre. Keyless wound with push piece at 4 minutes past position to set hands. Half plate. 21 jewels. Club tooth straight line lever movement number 10829153. Pillar plate stamped 53.7T 1/10 under balance. Oversprung cut bimetallic balance. Blued steel overcoil balance spring. Regulator index on balance cock engraved RA SF. Complete with rating certificate 6 June 1950 to 5 June 1970, green leather and velvet-lined presentation case.

Diameter 65mm

These last two watches are fine watches in fine condition with supporting documentation – thus highly desirable items.

PLATE 10
Stauffer & Co, Switzerland

⅕ second split second chronograph with 30 minute recorder. Nickel open face case, snap-on bezel, hinged back and cuvette. White enamel dial, black arabic numerals, blued steel spade hands. Sweep centre split seconds dial 0-60 seconds with 30 minute recorder inset upper centre. Keyless wound. Top plate stamped SWISS MADE SS & CO with trade mark. 11 jewels. Three-quarter plate club tooth tangential lever movement with oversprung monometallic balance and white balance spring. Regulator index on engraved balance cock with RA SF. Movement number 315456.

Diameter 51.9mm

Stauffer, Son & Co was founded in 1830 in Chaux de Fonds, Switzerland. By 1860 they were trading as a partnership (Jules Stauffer and Francis Claude) at 12 Old Jewry Chambers, London EC. Stauffer subsequently retired with his place being taken by Charles Nicolet. They were listed in Directories at this period as wholesale watch manufacturers and importers. Claude retired in 1874 leaving Nicolet to take control of the firm. The Watchmaker Jeweller and Silversmith *reported on the watches, chronometers and repeating watches exhibited by them at the Inventions Exhibition of 1885. They moved to 13 Charterhouse Street, Holborn, London EC in 1887. A highly reputable firm.*

PLATE 11
Valjoux SA, Les Bioux, Switzerland

Split second chronograph watch used by the Air Ministry. Nickel open face case, snap-on bezel, hinged back and cuvette, back engraved AM 6B/107 205. White enamel dial inscribed INOK 205. White enamel dial calibrated 0-60 seconds within chapter ring and 0-360 outside. 30 minute recorder upper centre and 0-60 running lower centre. Keyless wound. 19 jewels. Half plate club tooth straight line lever movement with oversprung cut bimetallic balance and red brown overcoil balance spring. Regulator index on balance cock engraved FS AR. Calibre 24.

Diameter 56.8mm

Chronographs

PLATE 10A

PLATE 10B

PLATE 11A

PLATE 11B

289

Chronographs

PLATE 12A

PLATE 12B

PLATE 13A

PLATE 13B

PLATE 14A

PLATE 14B

PLATE 12
Minerva, Robert Frères, Vallon de St Imier, Switzerland

1/10 second timer recording ½ and 1 minutes up to 15 minutes. Open face case, snap-on bezel, hinged back and cuvette, back engraved ₣ 162 TP 1/10 GW 4579. White enamel dial, black arabic numerals 0-30, 31-60, red 15 minute recording dial upper centre, blued steel hands. Keyless wound, push pendant button for start, stop and reset. Top plate stamped 647216 and pillar plate with Minerva trade mark. 9 jewels. Half plate club tooth straight line lever movement with oversprung two arm monometallic balance and white balance spring. Regulator index on balance cock engraved FS AR.
Diameter 51.3mm

Compare movement with George Jacob Catalogue 1949, Nr 60023, p318.

PLATE 13
Leonidas, St Imier, Switzerland

1/10 second timer 15 minute recorder stop-watch. Chrome open face case, snap-on bezel and back. White enamel dial, SWISS MADE within chapter ring, black arabic numerals 1-30, 30-60, red in 1/10 seconds markings, blued steel hands. Keyless wound at pendant and push start, stop and reset. TALIS with three five-pointed stars engraved on winding wheel. 7 jewels. Club tooth straight line lever movement with flat rimmed monometallic oversprung balance. Regulator index on balance cock stamped FS AR. Spare parts compartment under glass plate set into main plate should contain balance staff, two return springs and two different sized screws. Leonidas using Excelsior Park patent but under transparent cover to the compartment.
Diameter 49.8mm

Swiss No 100230, 24 May 1922, publié 16 Juillet 1923. Class 71f, Mouvement d'horlogerie, Les Fils de Jeanneret-Brehm, German DRP 388992. USA George Jacob Catalogue 1949, p313, No 60,005, Leonidas 18/121. Tallis watch marketed by REICHENBERG & Co Ltd, 44 Hatton Garden, London EC1.

PLATE 14
Heuer & Co, Switzerland

1/5 second time 30 minute recorder. Chrome open face case, snap-on bezel, hinged back and cuvette. Case stamped HEUER Ed HEUER & CO SWISS, back cover engraved WS/B/6 ICI Billingham Division E24. White enamel dial with Heuer trade mark, black numerals 5-60, inset minute recording dial 0-30 minutes, blued steel pointer hands. Keyless wound, slide at 55 minute position to stop and start, press pendant button to reset. Top plate engraved ED. HEUER & CO SWISS SEVEN 7 JEWELS UNADJUSTED. Half plate club tooth straight line lever movement with oversprung two arm flat rimmed monometallic balance and blued steel balance spring. Regulator index on balance cock engraved AR FS.
Diameter 51.3mm

291

Chronographs

PLATE 15A

PLATE 15B

PLATE 15
Lemania, Lugrin SA, Switzerland

¹⁄₁₀ second timer 15 minute recorder stop-watch. Chrome open face case, snap-on bezel, back and cuvette. White enamel dial inscribed NERO LEMANIA lower centre, SWISS MADE, black arabic numerals 0-30 and 30-60 red, blued steel pointer hands, 0-15 minute recorder dial inset upper centre. Keyless wound, push stop/start and reset at pendant. 7 jewel. Half plate club tooth straight line lever escapement with two arm flat rimmed oversprung monometallic balance. Blued steel balance spring. Regulator index on balance cock engraved RA SF. Top plate engraved 1258955 LEMANIA SWISS and LWC within cross 5020.
Diameter 50.2mm

PLATE 16
Cabot Watch Co

¹⁄₁₀₀ second timer 3 minute recorder stop-watch. Chrome open face case, snap-on bezel. Movement lifts out of front of case which is stamped CWC BASE METAL SWISS MADE. White painted dial, Government property black arrow CWC with 2 and 1 either side, Made in Switzerland, black arabic numerals, centre sweep seconds 0-100 repeated thrice, 3 minute recorder dial upper centre. Keyless wound and push pendant stop/start and reset. Top plate stamped CWC SWISS MADE SEVEN 7 JEWELS UNADJUSTED. Pillar plate stamped Baumgartner Frères trade mark calibre 411. Club tooth straight line lever movement with oversprung three arm monometallic balance with serrated edge to balance rim to assist in a positive stop/start action. Regulator on balance cock without any index. A triangular shockproof system fitted.
Diameter 53.5mm

This is a Baumgartner Frères ébauche (Calibre 411). It is understood that CWC refers to the Cabot Watch Company, a holding company buying government watches rather than to the Cortébert Watch Company.

PLATE 17
Breitling Watch Co Ltd, G Léon Breitling, Montbrillant, Chaux de Fonds, Switzerland

¹⁄₁₀ second timer 15 minute recorder stop-watch. Chrome open face case, snap-on back, inside back engraved SPRINT MADE BY BREITLING SWISS MADE. Cream painted dial inscribed SPRINT 1/10 Swiss, black arabic numerals 1-30, 30-60 red, 15 minute recording dial upper centre marked in ½ minutes. Keyless wound, push pendant button to stop/start and reset. 1 jewel. Unadjusted pin pallet movement with straight line lever. Oversprung three arm monometallic balance and white balance spring. Regulator on balance cock but no index provided. Pillar plate stamped Baumgartner trade mark Calibre 411 SWISS MADE. Top plate stamped BREITLING WATCH LTD 1 JEWEL UNADJUSTED.
Diameter 51.6mm

Baumgartner Frères ébauche (Calibre 411).

Chronographs

PLATE 16A

PLATE 16B

PLATE 17A

PLATE 17B

293

Chronographs

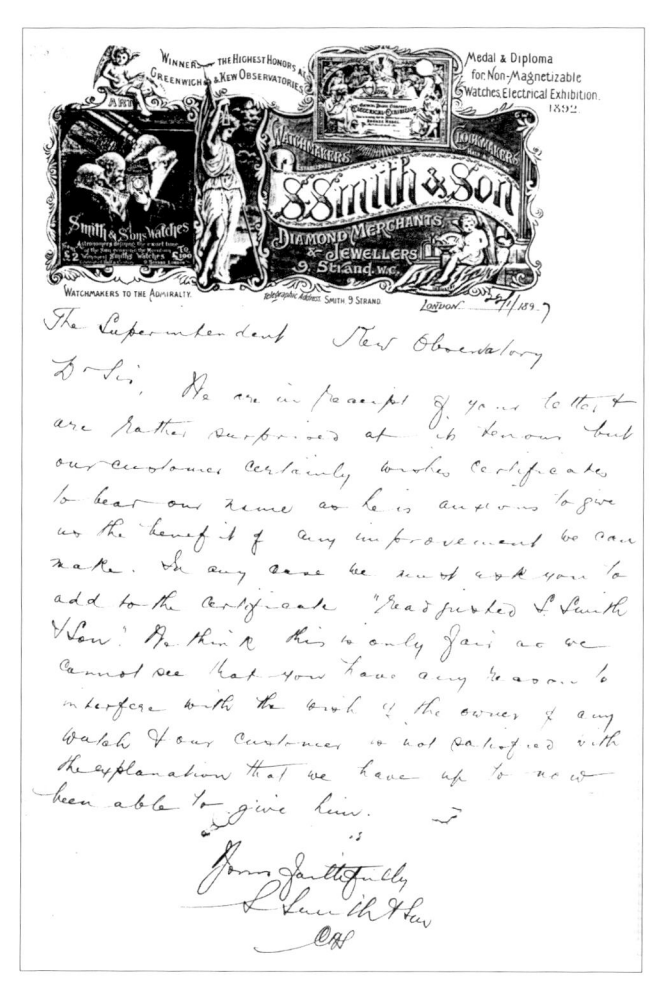

PLATE 18A

PLATE 18B

PLATE 18
S Smith & Son Ltd (retailer), **9 Strand, London WC**

Nickel open face case, snap-on bezel, hinged back and cuvette. White enamel dial overwritten with S SMITH & SON LTD 9 STRAND WC LONDON SWISS MADE, small arabic numerals 5-60 seconds with inset minute recording dial 0-30 minutes upper centre. Keyless wound, pendant push stop/start and reset. 17 jewels. Half plate club tooth and straight line lever movement engraved on top plate 12405 +. Oversprung monometallic balance with compensation screws and white balance spring. Regulator index on balance cock engraved SF.
Diameter 48.5mm

Swiss Patent No 12405, Class 64, 4 July 1896 6hp Mechanisme de chronograph Henchoz Frères Locle Manditair – Imer-Schneider E Genève.

PLATE 18C

Chronographs

PLATE 19A

PLATE 19
Minerva, Robert Frères, Vallon de St Imier, Switzerland

⅕ second timer 30 minute recorder stop-watch. Chrome on base metal open face case, snap-on bezel and cuvette. White enamel dial, black arabic numerals 0-60 seconds and 30 minute recorder upper centre inscribed PATENT MINERVA FA. JOH. E. POST AMSTERDAM below centre. Keyless wound, stop/start by slide in band and reset at pendant. Engraved MINERVA on barrel winding wheel and top plate engraved SWISS SEVEN 7 JEWELS UNADJUSTED MINERVA WATCH CO. Pillar plate stamped Minerva trade mark alongside balance. 7 jewel. Half plate club tooth straight line lever movement. Oversprung monometallic balance and blued steel balance spring. Regulator index engraved FS AR.
Diameter 49.1mm

PLATE 19B

Chronographs

PLATE 20A

PLATE 20B

PLATE 20
Moise Drefuss Ltd (Wholesale Importers), 88 Hatton Garden, London EC1

⅕ second timer 30 minute recorder stop-watch. Nickel open face case, snap-on bezel, hinged back and cuvette. White enamel dial inscribed SWISS MADE, black arabic numerals 5-55 60, red 30 minute recording dial upper centre, sweep centre seconds dial 0-60 seconds, blued steel hands. Slide at 55 minute position to stop/start. Press pendant button to reset. Keyless wound. 7 jewels. Half plate club tooth straight line lever movement. Oversprung two arm flat rimmed oversprung monometallic balance with white balance spring. Regulator index on balance cock engraved FS AR.
Diameter 55.6mm

George Jacob Catalogue 1949, p313, Nr 60.008, illustration of Montbrillant 18 Calibre 518 Montbrillant Watch Co (Geo Léon Breitling), Chaux de Fonds, Switzerland (manufacturer). As this watch was issued in 1940 for military use it must have been imported before the declaration of the Second World War (3 September 1939).

PLATE 21
Minerva, Robert Frères, Vallon de St Imier, Switzerland

German Airforce ¹⁄₁₀₀ second timer 78 second recorder stop-watch. Nickel open face case, snap-on bezel, back and cuvette. Engraved on back insignia comprising swastika with circle with bird above, M below. White enamel dial, black arabic numerals, 20, 40, 60, 80, 100, 200, 300 red and thrice repeated, 78 second recorder inset upper centre, blued steel spade hands, Minerva trade mark lower centre. Keyless wound. Top plate numbered 1519145. Pillar plate stamped .9145. 7 jewel. Half plate club tooth straight line lever escapement with oversprung two arm monometallic balance and blued steel balance spring. Regulator index on balance cock engraved FS AR.
Diameter 51.1mm

Believed to be a bomb aimer's stop-watch. George Jacob Catalogue 1949, Model 19.14, Nr 60023 and 60024.

PLATE 22
Swiss Made

¹⁄₁₀ second timer with 5 minute recorder. Nickel open face case, snap-on bezel, hinged back and cuvette. White enamel dial, lower centre inscribed THE WHIPPET SMITHS SUNDERLAND. Seconds dial 0-10 seconds with black arabic numerals 0-9 and 5 minute recorder dial inset upper centre. Keyless wound. Slide acting on edge of balance. Top plate stamped SWISS MADE PATENT + 86583. 15 jewel. Half plate club tooth and straight line lever movement with oversprung monometallic balance with blued steel balance spring. Regulator index on balance cock engraved FS AR with arrow between.
Diameter 51.3mm

+ 86583 not found (Swiss patent 86583 dated 15 January 1920 refers to ED Fierz Eirr Kontroll Apparatus). George Smith, 23 West Street and 27a High Street, West Sunderland, is listed as watch and clock retailer, jeweller in Kelly's Directory for 1913.

Chronographs

PLATE 21A

PLATE 21B

PLATE 22A

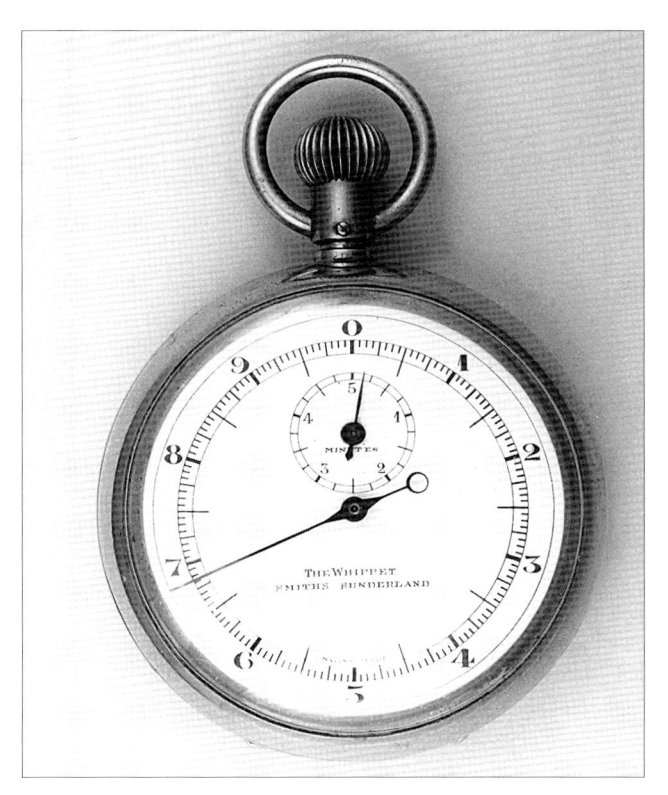

PLATE 22B

297

Chronographs

PLATE 23A

PLATE 23
Excelsior Park (founded 1866), **St Imier, Switzerland**

⅕ second timer with 30 minute recorder stop-watch. Nickel chrome open face case, snap-on bezel, hinged back and cuvette. Slide in band stamped at side GO ON STOP and operating on edge of balance. White enamel dial, black arabic numerals 5-55, 60 red, blued steel hands. Sweep seconds dial calibrated ⅕ seconds up to 60 seconds, 30 minute recorder inset upper centre, black arabic numerals 3-27, 30 red. Top plate stamped No 3364. Pillar plate spotted. Keyless wound. Stop/start at slide in band, push piece at pendant to reset. 9 jewels. Three-quarter plate club tooth and tangential lever escapement with oversprung three arm brass balance and white balance spring. Regulator index on balance cock stamped SF RA.
Diameter 55.5mm

PLATE 23B

PLATE 24A

PLATE 24
Findlay & Co (retailer), **London W6, Excelsior Park, St Imier, Switzerland**

¹⁄₁₀ second timer with 30 minute recorder. Chrome on brass open face case, snap-on bezel, hinged back and cuvette. White enamel dial inscribed FINDLAY & CO LONDON W6 SWISS MADE, black arabic numerals. Seconds dial 10 divisions in 60 seconds. Inset 30 minute recorder upper centre. Keyless wound. Slide in band. Barrel bridge stamped No 3364 1130785. Jewelled club tooth straight line lever escapement with oversprung two arm flat rimmed balance and blued steel balance spring. Regulator index on balance cock engraved RA SF.
Diameter 49mm

Findlay & Co (A Findlay, AMI, Mech E), Broadway Chambers, Hammersmith, London W6. Catalogue No 1A lists Chronographs, Timers, Time recorders, Electric Clocks, Instrument parts, Instrument gears and Industrial jewels. See also Pringle & Sons, List No 975/761, August 1966, Calibre JBI. Excelsior Park Movement with spare parts compartment. Swiss Patent No 100230, 24 May 1922, Publié 16 Juillet 1923. German Patent DRP 388992. Patented USA.

PLATE 24B

Chronographs

PLATE 25A

PLATE 25B

PLATE 25C

PLATE 25D. *Kelly's Directory, 1913*

PLATE 25
S Smith & Son Ltd (retailer), Trafalgar Square, London WC

⅒ second timer and 15 minute recorder stop-watch **c.1918.** Oxydised steel open face case, hinged back and cuvette. Engraved on back cover arrow S SMITH & SON/T P No 114/1918. White enamel dial inscribed S SMITH & SON LTD, TRAFALGAR SQUARE, LONDON WC No 249 182 SWISS MADE, black arabic numerals 0-30 with every fifth in red. Inset 15 minutes recorder, black arabic numerals 0-15 with 15 in red. Barrel bridge stamped No 3364 and top plate 385455. Keyless wound. 7 jewels. Half plate club tooth straight line lever movement with oversprung two arm flat rimmed brass balance and white balance spring. Regulator index on balance cock engraved RA SF.
Diameter 59mm

Excelsior Ebauche Calibre 114. S Smith & Son Ltd who were at 9 Strand, Trafalgar Square were later to become the great Smith's Industries. Advertisement from 1929 Horological Journal *carries warning to imitators of Excelsior Park facility for spare parts being placed inside the movement. They also state that all of their watches bear No 3364 on the movement.*

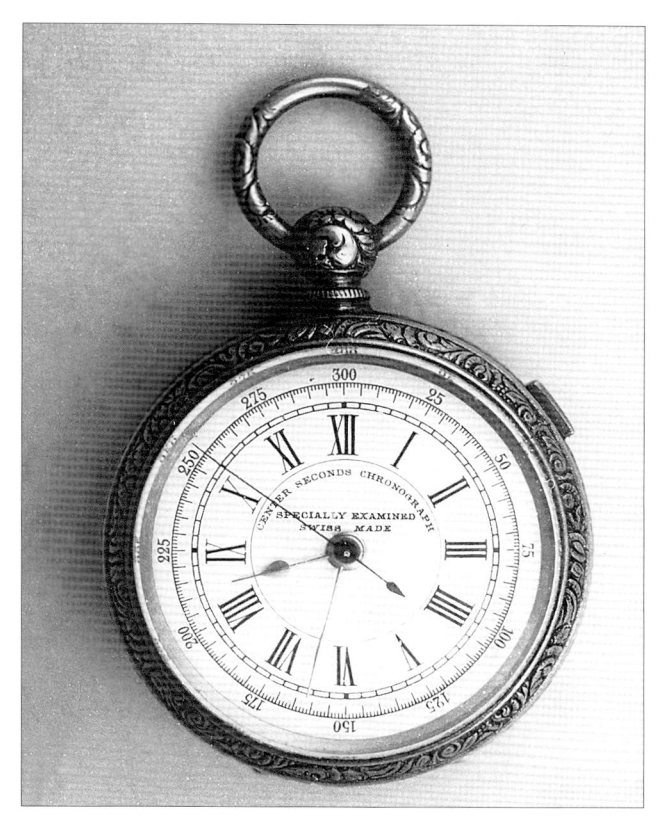

PLATE 26A

PLATE 26
Jequier Frères, Fleurier, Switzerland

⅕ centre second chronograph. Brass open face case, snap-on bezel, hinged back and cuvette, decorated cast pattern to bezel. Slide in band at two o'clock position. White enamel dial inscribed upper centre CENTER SECONDS CHRONOGRAPH SPECIALLY EXAMINED SWISS MADE, black roman numerals, gold spade hour and minute hands with blued steel spade centre sweep seconds hand. Key wound and set. 4 jewels. Three-quarter cylinder movement and regulator index on balance cock engraved FAST SLOW.
Key size 7. Diameter 49mm

Kelly's Directory 1913, advertisement p26.

PLATE 26B

Chronographs

PLATE 27A

PLATE 27B

PLATE 27
Swiss

⅕ seconds chronograph. 0.935 silver open face case, Swiss hallmark bear, hinged bezel, back and cuvette with engine turned barleycorn pattern to back. White enamel dial with sunk centre inscribed 95877, black roman numerals I to XII and centre sweep seconds with ⅕ second markings and small arabic numerals 25-300 outside. Key wound and set from the back. Slide at side of case at 2 o'clock position for stop/start. Pillar plate stamped . . . . 10 jewels. Three-quarter plate cylinder movement with oversprung three arm brass balance and blued steel balance spring with regulator index on balance cock engraved FAST SLOW.
Key size 7. Diameter 50.3mm

PLATE 28
Jequier Frères & Co, Fleurier, Switzerland

Nickel open face case, snap-on bezel, hinged back and cuvette. White enamel dial inscribed upper centre BEST CENTRE SECONDS CHRONOGRAPH REFMETER (TRADE MARK PAT APPD), below centre SWISS MADE, black roman numerals, sweep centre seconds dial calibrated 0-300 in ⅕ seconds. A three-quarter annular black celluloid dial covering 45 minutes of anticipated play is located by a pin on the underside of the bezel which, when rotated, can set dial to appropriate time of kick-off of football match. Keyless wound, push piece to set at XI position. Top plate stamped SWISS MADE. Pillar plate stamped P alongside balance. Slide at 1 o'clock position in band. Unjewelled. Three-quarter plate Roskopf pin pallet tangential lever movement with oversprung monometallic balance and index on balance cock stamped RA SF.
Diameter 54.7mm

See Kelly's Directory 1897, p795 illus., 1901, p952 illus. Jequier Frères made a speciality of the TELL Best Centre Seconds Chronographs, of which this appears to be a specialised example.

PLATE 28A

PLATE 28B

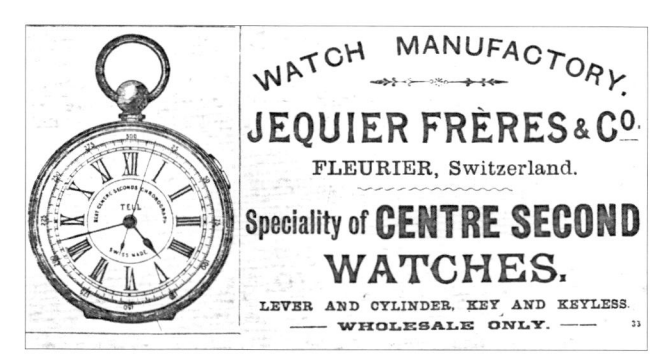

PLATE 28C. *Kelly's Directory, 1897*

PLATE 29

PLATE 29
Jequier Frères & Cie Fabricant, Fleurier Watch Co, Switzerland

⅕ second TELL Best Centre Seconds Chronograph. Pure white metal open face case, snap-on bezel, hinged back and cuvette. Case stamped F W Co within separate shields, cuvette and top plate with TELL registered trade mark (Registered 26 Aout 1895 No 7718) 754 - 80. White enamel dial, black roman numerals, sweep centre seconds calibrated ⅕ seconds 0-300, gilt spade hour and minute hands, blued steel centre seconds hand. Keyless wound, pull to set hands. Slide in the band at II o'clock position. Top plate numbered 754.30. 4 jewels three-quarter cylinder movement with oversprung mono-metallic balance and blued steel balance spring. Polished index on balance cock stamped SF RA.

Diameter 55.5mm

Chronographs

PLATE 30A

PLATE 30B

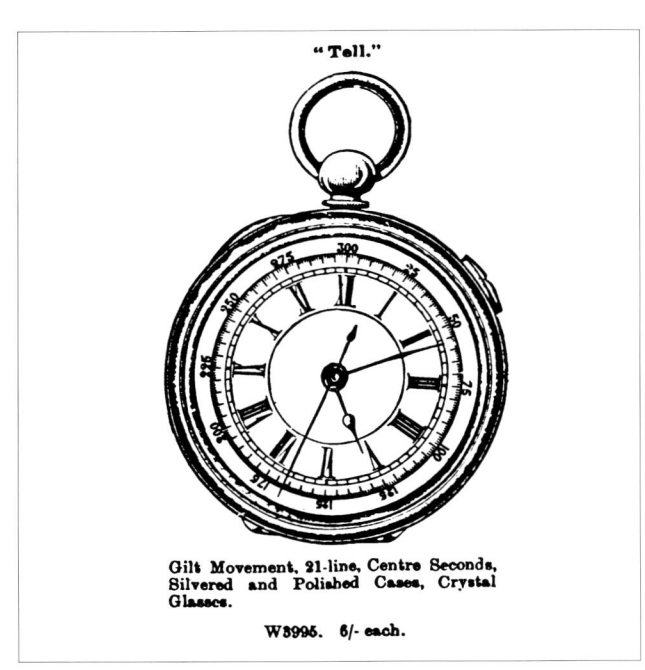

PLATE 30C. *Hirst Bros. Catalogue,* 1910

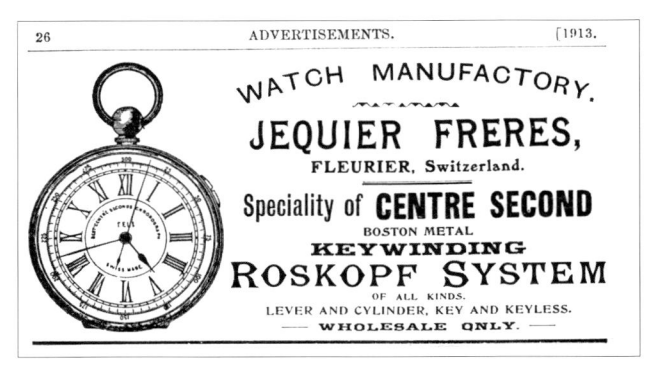

PLATE 30D. *Kelly's Directory,* 1913

PLATE 30
Jequier Frères, Fleurier, Switzerland

⅕ second TELL Best Centre Seconds Chronograph. Silvered brass open face case, snap-on bezel, hinged back and cuvette. Back of case stamped plain shield with a looped belt, inside case stamped JF and five-pointed star within separate shields. White enamel dial inscribed TELL BEST CENTRE SECONDS CHRONOGRAPH SWISS MADE, black roman hours chronograph, seconds dial calibrated ⅕ seconds, 0-300 marked in periods of 25 in small arabic numerals, gilt spade hour and minute hands, blued steel centre seconds. Key wound and set. Top plate stamped SWISS MADE and REG. William Tell signature on the barrel bridge. 4 jewels. Three-quarter plate cylinder movement with oversprung three arm (brass) monometallic balance and blued steel balance spring. Regulator index on balance cock stamped SLOW FAST.

Key size 6. Diameter 54.8mm

Kelly's Directories 1897, p793, illus. advertisement; 1901, p952, illus. advertisement; 1913, p26, illus. advertisement. Hirst Bros 1910 catalogue.

Chronographs

PLATE 31A

PLATE 31B

PLATE 31
Richard Fennell (watch manufacturers),
Portland Terrace, Chapel Fields, Coventry

⅕ seconds centre seconds chronograph. Silver open face case hallmarked Chester **1889**, hinged bezel, back and cuvette. Casemaker AG. Light cream enamel dial with sunk centre inscribed CENTRE SECONDS CHRONOGRAPH upper centre, PATENT No 8727 lower centre, black roman numerals, gold spade hands, chronograph sweep centre seconds calibrated ⅕ second division 0-300. Top plate engraved E Wise MANCHESTER (retailer) No 24667. Key wound and key set, push at pendant to stop and start chronograph. 13 jewel. Three-quarter plate pointed tooth tangential lever fusee movement with oversprung cut bimetallic balance and blued steel overcoil balance spring. Regulator index on balance cock engraved with scrolls and SF with Coventry star between.
Key size 6. Diameter 59.8mm

This interesting patent should be studied in relation to the more familiar Coventry English lever centre second chronographs which used a simple slide on the band. Fennell sought to give his type of watch the look of the more expensive chronograph operating from the pendant. Reference UK Patent No 8727 27 May 1889, Richard Fennell, 2 Mount Street, Chapel Fields, Coventry. Listed in Post Office Directory 1882.

8727. Fennell, R. May 27. **1889**

Chronographs. — A key-winding centre-seconds watch is stopped and started from the pendant. The pusher A acts on a spring driver B which advances a wheel C of twelve teeth bearing six studs. A curved spring D has a tooth which presses against the studs, and a pin which stops the escapement roller when the tooth falls between the studs.

PLATE 31C

305

Chronographs

PLATE 32A

PLATE 32B

PLATE 33A

PLATE 33B

PLATE 34A

PLATE 34B

PLATE 32
American Watch Co, Waltham, Mass, USA

⅕ centre seconds chronograph. Silver open face case, gold hinges to bezel, back and cuvette. Hallmarked Birmingham **1888.** Casemaker AB (Alfred Bedford, 40 Terrace Road, Handsworth, Birmingham, mark registered Birmingham Assay Office 12 March 1879). White enamel dial with sunk centre inscribed CENTRE SECONDS CHRONO-GRAPH, black roman numerals, gold spade hour and minute hands with blued steel spade sweep centre seconds hand. Keyless wound at pendant with lever under hinged bezel at 5 minute past position to set hands. Top plate engraved PAT Sep 28.80 Eng Pat No 3224 AWCo WALTHAM MASS. 17 jewels. Going barrel club tooth straight line lever movement with oversprung overcoil cut bimetallic balance. Regulator under engraved cock with SF.
Diameter 56.3mm

UK Patent No 3224 September 28 1880 Groth LA (Lugrin HA and Nordman PO August 6 1880). Waltham records: Waltham production 3161001-3162800 14 size, Model 84, 3/4 plate, 13 jewels, open face, stem wind.

PLATE 33
Le Phare, Le Locle, Switzerland

⅕ second time of day chronograph with 30 minute recorder. Nickel chrome open face case, hinged bezel back and cuvette. White enamel dial inscribed SWISS MADE, blued steel spade hands, black roman hours and small arabic numerals 0-60 minutes. Chronograph sweep seconds calibrated in ⅕ seconds with minute reading dial 0-30 minutes inset upper centre. Running subsidiary seconds inset lower centre. Keyless wound pull to set hands, push piece in pendant button to operate chronograph stop/start and reset. Top plate stamped SWISS MADE. Pillar plate numbered 324839. Three-quarter plate jewelled club tooth straight line lever movement with oversprung monometallic balance with compensation screws and pale yellow overcoil balance spring. Regulator index on balance cock engraved AR SF.
Diameter 51.7mm

George Jacob Catalogue 1949, p299, lists 19/114 VCC.

PLATE 34
Minerva, Switzerland

⅕ second chronograph with 30 minute recorder. Gold filled open face case (Geneva Watch Case Co – Warranted 10 years), snap-on bezel, hinged back and cuvette. White enamel dial, black roman numerals, gold spade hour minute and seconds hands, blued steel spade centre seconds hand, blued steel arrow hand to 30 minute recorder dial upper centre and inset running 0-30 seconds dial lower centre. Keyless wound and pull to set hands, push piece in pendant button stop/start and reset. Top plate stamped SWISS MADE and pillar plate numbered 1014543 below balance rim. 17 jewel club tooth straight line lever movement with oversprung cut bimetallic balance. Blued steel overcoil balance spring with regulator index on balance cock engraved FS AR.
Diameter 51mm

Ebauche SA Vol 2 (1949), p94 19 chronograph 9CH.

Chronographs

PLATE 35A

PLATE 35B

PLATE 35
Minerva, Switzerland

⅕ seconds time of day chronograph with 30 minute recorder. Oxydised steel open face case, snap-on bezel, hinged back and cuvette, engraved on back GPO (General Post Office) CG31 10, 194 No 1 and inside SWISS MADE. White enamel dial, black arabic numerals, blued steel spade hands, sweep seconds 0-60, inset running seconds 0-60 lower centre and 30 minute recording dial upper centre. Keyless wound and pull to set hands, chronograph push piece to stop/start and reset within pendant button. Top plate stamped SWISS MADE and pillar plate 1407128. 17 jewel club tooth and straight line lever movement. Oversprung cut bimetallic balance and blued steel overcoil balance spring. Regulator index on balance cock engraved FS AR.
Diameter 50.9mm

Watchmaker, Jeweller, Silversmith & Optician Jan 1931, p14.

PLATE 36
Excelsior Park, St Imier, Switzerland

⅕ seconds and 30 minute recording chronograph. Nickel open face case, snap-on bezel, hinged back and cuvette. White enamel dial, sunk centre, black roman hour numerals, small red arabic numerals at 5 minute intervals, 30 minute recording dial upper centre and running seconds 0-60 lower centre dial. Keyless wound and pull to set. 19 ligne 1st quality 17 jewel movement with club tooth and straight line lever escapement. Oversprung monometallic balance with white balance spring. Regulator index on balance cock engraved AR FS.
Diameter 50.7mm

Spare parts compartment stamped SPARE PARTS IN HERE SWISS MADE. This is the subject of Swiss Patent No 100231 24 May 1922, Les fils de Jeanneret-Brehm, St Imier, Switzerland. George Jacob Catalogue No 50067 chronograph, Excelsior Park Catalogue, section La Classification Horlogère, 2nd edition.

PLATE 37
Omega, Louis Brandt Frères, Bienne, Switzerland

⅕ seconds time of day chronograph with 30 minute recorder. Silver 0.935 open face case, hinged bezel, back and cuvette, hallmarked Swiss bear. Inside back stamped OMEGA trade mark. White enamel dial, black roman numerals, gold spade hands, sweep seconds dial calibrated ⅕ seconds, running seconds dial 0-60 lower centre and inset 30 minute recorder upper centre. Keyless winding at pendant, pull to set hands, chronograph push piece within pendant button. Club tooth and straight line lever movement with oversprung cut bimetallic to balance with compensation screws and pale yellow overcoil balance spring. Polished whiplash regulator and index on balance cock engraved FS AR.
Diameter 53.6mm

George Jacob Catalogue, Leipzig 1949 (movement corresponds to Omega 19 Nr 50089, p302).

Chronographs

PLATE 36A

PLATE 36B

PLATE 37A

PLATE 37B

309

PLATE 38A

PLATE 38B

PLATE 38
Anon, Switzerland

⅕ seconds chronograph 30 minute recorder. Silver 0.800 open face case, hinged bezel back and cuvette, Swiss woodcock silver mark and also (German) Deutsch silber Stempl crescent and crown. Casemaker's mark JP within oval. White enamel dial, black roman numerals, blued steel spade hands. Minute recorder 0-30 inset upper centre and running seconds dial 0-60 inset lower centre. Keyless wound with push piece to set hands at 3 minute past position. Top plate stamped with D within square within 8 pointed star. Three-quarter plate jewelled club tooth and straight line lever movement. Oversprung cut bimetallic balance. Blued steel overcoil balance spring. Regulator index on balance cock engraved RA.
Diameter 52.4mm

Movement not identified but bears many features similar to a Longines ébauche.

PLATE 39
Anon, Switzerland

⅕ seconds patent chronograph. Silver 0.800 open face case, engraved and hinged bezel back and cuvette, cuvette engraved PATENT CHRONOGRAPH. Hallmarked Swiss woodcock. White enamel dial inscribed upper centre Patent Chronograph, black roman numerals I-XII, small black arabic 25-300, gold spade hands, inset sunk running seconds. Keyless winding and push piece at 5 minutes position to set hands and push piece within the winding button at pendant to operate chronograph. 15 jewel. Half plate club tooth straight line lever movement. Oversprung cut bimetallic balance. Blued steel overcoil balance spring. Polished pointed regulator index on balance cock engraved Advance Retard. Cock to chronograph wheel engraved SWISS MADE.
Diameter 51.5mm

Trade mark stamped ℬ within a garland inside cover.

PLATE 40
Anon, Swiss

⅕ second 30 minute recorder ADEON chronograph. Gold 14ct open face case, hinged bezel, back and cuvette, cuvette engraved T/M a bem type nocmabuyukr dhopa evo/erp beaurecmba – Russian translation indicates Supplier to his Majesty's Court (Tsar Nicholas II 1894-1917). White enamel dial, black roman numerals, gold spade hour and minute hands, centre sweep seconds, blued steel chronograph hand and 30 minute recording dial upper centre with running seconds 0-60 lower centre. Keyless wound and set. 17 jewels. Half plate club tooth straight line lever escapement. Oversprung cut bimetallic balance and blued steel balance spring. Regulator index on balance cock engraved AR.
Diameter 53mm

The full details on the manufacturer of this watch are not known but Mr TE White, FBHI states that it is identical to ADEON chronographs sold by Goldsmiths and Silversmiths and supplied for police work.

PLATE 39A

PLATE 39B

PLATE 40A

PLATE 40B

Chronographs

PLATE 41A

PLATE 41B

PLATE 41
Marcks & Co Ltd (retailer), **Bombay and Poona, India**

⅕ second chronograph 30 minute recorder. Nickel open face case, snap-on bezel, hinged back and cuvette. White enamel dial, SWISS MADE across centre, MARCKS & CO LTD BOMBAY & POONA within inset minute recorder upper centre, black roman numerals, broad blued steel spade hands, centre sweep seconds dial calibrated in ⅕ seconds, small arabic numerals 25-300 outside chapter, running seconds 0-60 inset lower dial. Keyless wound and push piece to set. 19 jewels. Half plate club tooth and straight line lever movement. Oversprung cut bimetallic balance and blued steel balance spring. Regulator index engraved SF.
Diameter 52.6mm

Compare movement with Montbrillant 19 No 50114, George Jacob Catalogue, 1949, p302, which is very similar.

PLATE 42
Sir John Bennett Limited (retailer), **65 Cheapside, London**

Silver 0.935 open face case hallmarked Swiss bear, cuvette engraved MAKERS TO HER MAJESTY THE QUEEN HONI SOIT QUI MAL Y PENSE DIEU ET MON DROIT SIR JOHN BENNETT LTD 65 CHEAPSIDE LONDON SWISS MADE. White enamel dial, black roman numerals, gold spade hands, inset running seconds dial at 9 o'clock and recording minute dial at 3 o'clock, sweep centre seconds chronograph. Keyless winding, push piece to set hands at 2 o'clock, push button to stop/start and reset at XII. Top plate engraved SWISS MADE Sir John Bennett Limited 65 Cheapside London No 252225. 13 jewels. Three-quarter plate pointed tooth escape wheel with right-angle lever. Oversprung with overcoil cut bimetallic balance. Yellow/white balance spring. Regulator index on balance cock engraved FAST SLOW.
Diameter 52.3mm

A Swiss-made watch in the English style for the English market. Known to have been at this address (65 Cheapside) 1836 to 1889, Sir John Bennett was knighted in 1872 and Queen Victoria died in 1902 – therefore this watch must have been manufactured between 1872 and 1889.

PLATE 43
Lemania (?), Switzerland

⅕ seconds 30 minute recorder chronograph. Nickel open face case, hinged back and cuvette. White enamel dial inscribed WHITEFIELD CO IMPROVED CHRONOGRAPH, black roman numerals, blued steel spade hands, centre sweep chronograph seconds with subsidiary sunk minute recording dial 0-30 minute upper centre and running seconds dial 0-60 seconds lower centre. Keyless winding and push piece to set at 5 minute position. Three-quarter plate. 19 jewels. Club tooth and straight line lever movement. Oversprung uncut bimetallic balance. Blued steel balance spring. Regulator index on balance cock engraved RA SF.
Diameter 51.3mm

PLATE 42A

PLATE 42B

PLATE 43A

PLATE 43B

Chronographs

PLATE 44A

PLATE 44
Stauffer & Co, La Chaux de Fonds, Switzerland, 13 Charterhouse Street, London EC

Nickel open face case, snap-on bezel, hinged back and cuvette. White enamel dial, black roman hour numerals, gold spade hour and minute hands, centre sweep seconds dial with ⅕ second markings and calibrated 5-60 seconds in small black arabic numerals outside chapter ring, 30 minute anticlockwise minute recorder upper centre of dial with running seconds 0-60 inset lower dial. Keyless wound, push piece to act at 4 minute past position. 18 jewel. Half plate club tooth and straight line lever escapement stamped on top plate S & Co below crown. 316995 stamped on minute recording cock. Cut bimetallic balance. Blued steel oversprung balance spring. Regulator index on balance cock engraved RA SF.

Diameter 54mm

Chronograph watches of this type were manufactured for Stauffer by Atlas Watch Co, La Chaux de Fonds. Annuaire de l'horlogerie suisse, *1945*; Horological Journal *Vol XLVIII, Sept 1905*; Watchmaker, Jeweller, Silversmith & Optician, *1886*; Directory of Gold and Silversmiths *by J Culme 1987*.

PLATE 44B

314

Chronographs

PLATE 45A

PLATE 45B

PLATE 45
AH Arnold, FH Huguenin (Patentees)

⅕ seconds 30 minute recording railway chronograph. Silver open face case hallmarked London **1885**, snap-on bezel, hinged back and cuvette. Casemaker AN (? Alexis Nicole, 7 Carlisle Street, Soho). White enamel dial inscribed For speed of Railway Trains Miles to ¼ Mile distances, black roman numerals, blued steel spade hands, chronograph minute recording dial 0-30 minutes inset upper centre and inset sunk running seconds dial lower centre. Keyless winding, push piece at 5 minutes past position. Top plate engraved PATENTED MAY 3d 1883. 15 jewels. Three-quarter plate straight line lever movement. Oversprung uncut bimetallic balance, blued steel overcoil balance spring. Regulator index on balance cock hand engraved SF.

Diameter 51.7mm

1 mile per minute = 60 mph; ¼ mile post in 15 seconds = train travelling at 60 mph. UK Patent No 2258 3 May 1883, Arnold AH and Huguenin FH, for chronograph mechanism.

PLATE 45C

2258. Arnold, A. H., and Huguenin, F. H. May 3. *1883*

Chronograph mechanism.— The wheel E of the minute indicator is caused to advance by a pawl *e* on the heart cam D, one tooth at every turn of the centre-wheel C, during the period of observation, being raised into gear with the said pawl by means of a bevelled upper surface on the hammer B, acting upon the bevelled underside of the wheel E, when the said hammer is withdrawn from its cam. The heart cam *d* of the minute indicator and its hammer *b* are also bevelled in such a manner that, when it is desired to set the hands at zero, the hammer, coming in contact with the cam, depresses it, together with the wheel E, which is thereby thrown out of gear again.

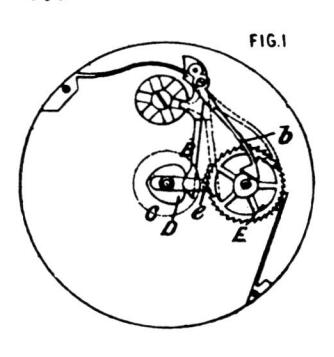

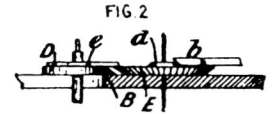

FIG.1

FIG.2

315

Chronographs

PLATE 46A

PLATE 46
Edward White (retailer), 20 Cockspur Street, Pall Mall, London SW

⅕ seconds chronograph with 30 minutes recorder. Silver 0.935 open face case, Swiss hallmark, snap-on bezel, gold hinges to back and cuvette, gold olivette, cuvette engraved Examined by E. White Cockspur St London. White enamel dial, black roman numerals, blued steel spade hands, inset seconds dial lower centre with 30 minute recorder upper centre. Keyless wound, push piece to set at 57 minute position. 15 jewels. Three-quarter plate club tooth lever movement with tangential lever steel balance spring. Regulator index in balance cock engraved FS.

Diameter 52mm

Edward White, 20 Cockspur Street, Pall Mall, SW listed in directories from 1863-1900. Watchpaper No 8393 in back. Thomas Lumsden Brown, St Catherine Street, Cupar, Fife, Scotland, listed 1897, corresponds with similar number scratch mark on inside of cuvette.

PLATE 46B

Chronographs

PLATE 47A

PLATE 47B

PLATE 47
Henchoz Frères, Le Locle, Switzerland

⅕ seconds 30 recording chronograph (The Ascot). Oxydised steel open face case, snap-on bezel, hinged back and cuvette. White enamel dial, black roman numerals, blued steel spade hands, centre seconds chronograph with ⅕ seconds markings, inset running seconds lower dial and recording 0–30 minute dial upper centre. Keyless wound, push piece at 5 minute past position to set. Top plate engraved SWISS MADE THE ASCOT PATENT 20th JUNE 1887. Half plate straight line lever with club tooth scape wheel. Oversprung monometallic balance with compensation screws and white balance spring. Calibrated snail micro-regulator.
Diameter 53.2mm

UK Patent No 8843 20 June 1887, Ernest De Pass, 62 Fleet Street, London for Henchoz Frères, Le Locle, Switzerland.

PLATE 47C. 1887

8843. Pass, E. de, [*Henchoz frères*].
June 20

Chronographs.—The wheel H on the axis D of the chronograph hand is driven by a chronograph wheel B on the axis of the second wheel, and is thrown into and out of gear by a cam G actuating a lever F. The arbor D is carried at one end by the lever F, and is capable of an oscillatory movement within the centre pinion through which it passes, a friction spring K stopping rotation when the wheel is out of gear. The lever is forced against the cam by a spring M, and carries an eccentric I which is acted on by the lever J when the hand is to be set at rest. The cam G is acted on by the push-piece in the ordinary manner.

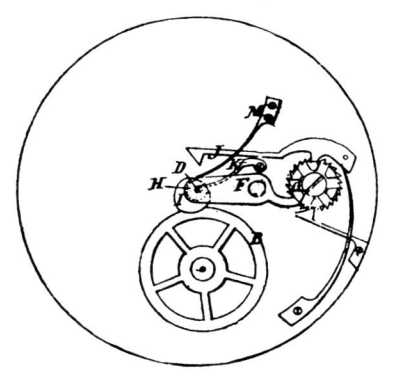

317

PLATE 48A

PLATE 48B

PLATE 48
Ed Heuer & Co, Bienne, Switzerland

⅕ seconds 30 minute recorder. Oxydised steel open face case, snap-on bezel, hinged back and cuvette, gilt olivette. White enamel dial, black roman numerals, blued steel spade hands, inset sunk running seconds 0-60 dial lower centre and inset recording minutes 0-30 upper centre. Keyless wound, push piece at 1 o'clock to set hands. 19 jewels. Half plate club tooth and straight line lever movement. Oversprung and overcoil white balance spring to monometallic balance. Regulator index on balance cock engraved Fast Slow.
Diameter 52.4mm

1882 Heuer started production of pocket chronographs and invented telemeter dial for measuring the trajectory of artillery shells. 1899 silver medal awarded at Paris World Fair for pocket chronographs with minute recorder and split second. 1908 Invented and patented pulsometer dial. 1916 First split second stop-watches indicating 1/100 second. 1920, 1924, 1928 Heuer timed Olympic Games.

PLATE 49
Swiss

⅕ seconds time of day chronograph and 30 minute recorder. Silver open face case, snap-on bezel, hinged back and cuvette, engine turned barleycorn design on back. Casemaker JW. White enamel dial inscribed CHRONOGRAPH COMPTEUR PATENT, black roman numerals, sweep centre seconds calibrated ⅕ second, subsidiary inset anticlockwise 30 minute recording dial. Keyless wound, push piece at 1 o'clock. Top plate stamped Ste J & H DEPOSE B. 11 jewel. Three-quarter plate. Club tooth straight line lever movement with oversprung cut bimetallic balance and blued steel. Overall balance spring. Regulator index on balance cock engraved SF.
Diameter 53.6mm

Early hand finished movement. The only other chronographs known with anticlockwise minute recorder are Stauffer chronographs manufactured by Atlas Watch Co, La Chaux de Fonds, but in that instance the recording dial is placed upper centre.

PLATE 50
Lemania (?), Switzerland

⅕ second 60 minute and 10 hours recorder timer stop-watch. Nickel open face case, snap-on bezel, hinged back and cuvette. White enamel dial inscribed SWISS MADE, black arabic numerals, sweep centre 0-60 seconds. Bottom centre recording minute dial 0-60 (10-50 black arabic, 60 red arabic numerals) inscribed Minutes. Top centre recording hour dial 0-10 (0-9 black arabic, 10 red arabic numerals) inscribed Hours. Keyless wound, push pendant button at side to stop/start and reset. 7 jewel. Half plate club tooth and straight line lever movement. Oversprung monometallic balance. Regulator index on balance cock engraved RA SF.
Diameter 49.8mm

Chronographs

PLATE 49A

PLATE 49B

PLATE 50A

PLATE 50B

319

Chronographs

PLATE 51A

PLATE 51
Lemania Watch Co, Lugrin SA, Orient, Switzerland

1/100 minute timer 30 minute recorder OMEGA stopwatch. Open face die-cast body with black matt crackle paint, screw back with two rubber seals (circular in cross section inside back and flat in cross section outside thread). Black sunk centre dial inscribed OMEGA with trade mark SWISS MADE and white sweep seconds annular dial calibrated 1/100 minutes, white arabic numerals and white pointer hands, 30 minute recorder dial inset upper centre. Keyless wound, side button twist to stop movement, pendant button zeros on pressing and movement starts on releasing. Top plate stamped 2279046 9300 and Omega trade mark. Pillar plate stamped LWC within cross (Lemania Watch Co). 7 jewel club tooth straight line lever movement with oversprung monometallic balance with shockproof suspension. Blued steel balance spring. The regulator index designed with three notches at side acting as a form of index pointing to + - scale on balance cock.
Diameter 54mm

PLATE 51B

PLATE 52A

PLATE 52
Lemania Lugrin SA, Switzerland

12 hour stop-watch chronograph. Nickel open face case, snap-on bezel, hinged dome and cuvette. White enamel dial inscribed NERO SWISS MADE, black arabic numerals, inner dial indicating elapsed hours with minute dial outside and outermost seconds dial. Keyless wound. 11 jewels. Straight line lever club tooth escapement with flat monometallic balance with white oversprung and overcoil balance spring. Regulating index on balance cock. Stop/start through winding crown. Return to zero possible whilst hands are running through the right-hand push button.

Diameter 65mm

Imported by Guinard and Goley, 42 Brent Street, Hendon, London NW4.

PLATE 52B

Chronographs

PLATE 53A

PLATE 53
Lemania, Lugrin SA, Switzerland

$1/100$ second 3 second recorder timer stop-watch. Chromium plated nickel open face case, snap-on bezel and back. White enamel dial inscribed NERO LEMANIA upper centre and SWISS MADE. Outer dial calibrated 0-3 seconds in $10/100$ and $1/100$ divisions. Inner dial recording 0-3 minutes in three periods of 60 seconds. Blued steel hands. Keyless wound. 15 jewel. Half plate club tooth straight line lever escapement. Oversprung monometallic balance with toothed edge to balance to assist in action of stop/start detent. White balance spring. Pillar plate engraved LWC within cross (Lemania Watch Co) 1370, LEMANIA SWISS on top plate and numbered 1235848. Regulator index on balance cock engraved SF RA, back engraved.

$$N \overset{\uparrow}{O} \overset{O}{O} \; 2 - 64 \uparrow$$

MAXIMUM RUN 30 MINUTES

A similar watch is illustrated Journal Suisse d'Horlogerie *August 1952 in the report of the timing at the Olympic Games, Helsinki, snap shooting events timed with Omega $1/100$ second type MG 1136 timers.*

PLATE 53B

PLATE 54A

PLATE 54
S Smith & Son (Retailer), London

1/100 second chronoscope two minute recorder. Nickel open face case, back cover stamped government property mark /1) No 252-146. White enamel dial inscribed CHRONOSCOPE S SMITH & SON LTD LONDON, black roman numerals, outer scale calibrated 1/100 seconds with inset minute recording dial in three seconds steps up to two minutes, blued steel spade hands. Keyless wound, press pendant button stop/start and reset. 17 jewels. Half plate club tooth and straight line lever movement. Oversprung monometallic balance and white balance spring. The compensation screws on the balance are in the same axis as the balance staff to make possible the action of arresting and starting the detent on the side of the balance. Regulator index on cock engraved RA SF. Top plate stamped SWISS MADE. Pillar plate stamped 53206.
Diameter 58.8mm

The manufacturers of this watch were Heuer. Watchmaker, Jeweller, Silversmith and Optician, March 1931, p292. George Jacob Catalogue Leipzig, 1949, p297, No 50007. Horology – Eric Haswell 1928/1947 ed, p213.

PLATE 54B

Chronographs

PLATE 55A

PLATE 55
The Richmond Time Recording Co (retailer), Liverpool

Time of day stop-watch. Chrome open face case, snap-on bezel and back, back engraved K686. White enamel dial inscribed THE RICHMOND TIME RECORDING CO NON MAGNETIC LIVERPOOL QUICK ACTION SWISS MADE, black arabic numerals, gilt spade hands. Keyless wound and pull to set, stop/start slide in band of case acting on edge of balance rim. Top plate stamped NON MAGNETIC 7 JEWELS K686 SWISS MADE. Pillar plate stamped 8. Club tooth and straight line lever movement with oversprung monometallic balance and white balance spring.

Diameter 51mm

An interesting feature is the spring loading of the movement in the case as a shock-absorbing device.

PLATE 55B

PLATE 56A

PLATE 56
The Richmond Time Recording Co (retailer), Birmingham

Time of day stop-watch timer. Chrome on brass open face case, screw bezel and back, back engraved THE RICHMOND TIME RECORDING CO BIRMINGHAM NO 2342 15 JEWEL NON MAGNETIC SCREW BACK, inside back stamped M & J within lozenge and numbered 117. White enamel dial inscribed THE RICHMOND TIME RECORDING CO BIRMINGHAM NON-MAGNETIC SWISS MADE, black arabic numerals, blued steel skeletonised cathedral hands, inset seconds dial 0-60. Keyless wound, pull to set, slide in band of case acting on balance rim edge stop/start. 15 jewel club tooth and straight line lever with oversprung two arm monometallic balance and white non-magnetic balance spring. Regulator index on balance cock engraved AR FS.
Diameter 50.7mm

Marchant & Jobin Ltd were advertised as importers of Majex, Majorex, Waltham and Minerva Recorders.

PLATE 56B

325

PLATE 57A

PLATE 57
OF Domon (Patentee), Belfort, France

Centre seconds chronograph. Silver open face case, hinged bezel, back and cuvette, hallmarked Birmingham **1883**. Case mark and pendant mark JN. White enamel dial, black roman numerals, chapter ring with ⅕ second markings calibrated in arabic numerals in periods of 25 ⅕ seconds, blued steel hands, inset subsidiary seconds dial and centre sweep seconds chronograph. Keyless winding with push piece to set at 57 minutes position and chronograph pusher at 5 o'clock position. Bar movement engraved on centre bar CHRONOGRAPH PATENT with PP on the wings of a butterfly. 15 jewel. Club tooth straight line lever movement with oversprung uncut bimetallic balance. Blued steel balance spring. Regulator index on balance cock engraved AVANCE RETARD.
Diameter 52mm

An extremely unusual chronograph subject to UK patent No 4536 Sept 23 1882 with the patentee Ovide Ferdinand Domon, Belfort, France. JN is possibly Jules Nordman, 11 Hatton Garden, London whose case mark was registered 31 January 1833 as a manufacturer and importer of watches and music boxes. Trade mark PP on butterfly so far not identified, but perhaps a pun on papillon?

PLATE 57B

Plate 58
William Williams, 32 Rock Street, Bury, Lancashire

Chronograph stop-watch – beats ⅟₁₆ seconds. Silver open face case hallmarked Chester **1887,** hinged bezel back and cuvette. White enamel dial inscribed upper centre BEATS ⅟₁₆ BY WILLm WILLIAMS BURY, black roman numerals, gold spade hands. Inset seconds dial marked in ⅟₁₆ seconds and indicating two seconds in one revolution, centre sweep seconds marked in ⅕ seconds (1 revolution in 60 seconds). Key wound and set, a slide at 2 o'clock position operating a lever acting on edge of balance roller to provide stop/start action. 17 jewels. Three-quarter plate fusee movement engraved on the top plate Wm Williams Bury 1809. Pointed tooth scape wheel with tangential lever. Oversprung three arm gold balance. Blued steel balance spring and regulator index on balance cock engraved with Coventry star. *This particular watch was developed for whippet (dog) racing.*
Key size 8. Diameter 57.3mm

UK Patent   323 (Prov Prot)   25 Jan 1881
              4496   20 September 1883
              4762   26 March 1890

The first patent related to a method of enabling watches to register distinct ⅛, ⅟₁₀ or ⅟₁₂ of a second by mean of the wheel train while the 1890 patent related to an escapement that could beat ten, twelve or sixteen times a second.

Plate 58a

Plate 58b

Chronographs

PLATE 59A

PLATE 59
Anon, Switzerland

Chronograph. Silver open face case stamped inside back JST. White enamel dial indicating ½ seconds with 5 seconds black arabic numerals, inset recording minute dial 0-30 arabic (anticlockwise) inscribed TIME upper centre, black roman hours. Key and keyless wound and push piece at 5 minutes past to set hands. 15 jewels. Half plate club tooth and straight line lever movement with oversprung cut bimetallic balance. Blued steel balance spring. Index regulator on balance cock engraved FS. Top plate stamped HUGUENIN'S PATENT Aug 29th 1882 and trade mark of knight's visor and feathers surmounting shield within which a fleur-de-lis is depicted. This minute recorder for chronograph is actuated by the vertical action of the wheel of the indicator.

Key size 5. Diameter 51mm

UK Patent 4132 27 August 1883 (Prov Prot for dial chronographs) Arnold Huguenin, Chaux de Fonds. UK Patent 2258 3 May 1883 chronograph mechanism, Auguste Henchoz Arnold and Fritz Henchoz Huguenin, both of Geneva.

PLATE 59B

PLATE 60A

PLATE 60
Alfred Stradling (Retailer), 78 Northbrook Road, Newbury, Berkshire

Double sided chronograph. Silver double sided silver case, snap-on bezel. White enamel dial inscribed STRADLING NEWBURY upper centre, black roman numerals, blued steel spade hands, inset sunk running seconds. Subsidiary dial at front with silvered annular dial, arabic numerals 0-60, chronograph centre seconds numerals and ⅕ second markings. Keyless wound with push piece to set hands at 4 minutes past hour position. Jewelled club tooth lever movement.

Diameter 53mm

Stradling is listed in the Post Office Directories for 1892 and 1897. It is interesting to note that an Alfred Stradling MBHI was writing in the Horological Journal of 1942 describing the making of a James Harmonograph, the inference being that he had but recently retired to live in Lechlade.

PLATE 60B

Chronographs

PLATE 61A

PLATE 61B

PLATE 62A

PLATE 62B

COLOUR PLATES 14 and 15 (page 9) and PLATE 61
Anon, Switzerland

Double sided centre seconds chronograph stop-watch. Oxydised steel case, front bezel snap-on, rear bezel hinged to give access to regulator index at side of chronograph dial. Front white enamel dial inscribed PATENT + 44244, short black roman numerals, inset seconds and inset 30 minute dial with each of the first six two-minute periods being indicated by a series of coloured segments, blue, red, black, green, brown, yellow, MENSOR DEPOSE inscribed above centre of reverse dial, seconds being indicated by anticlockwise motion of blued steel spade centre seconds hand over a dial calibrated spirally. Keyless lever movement.
Diameter 55.5mm

'*The MENSOR (registered) speed indicator chronograph for the use of travellers, railway officials, sportsmen etc To show without any calculation the rate of speed of Motor Car, Train, Carriage, Bicycle or other Conveyance Given a measured distance of any kind (miles, kilometres, verstes etc) this watch shows the exact rate of speed per hour, attained between the starting and stopping points.*' Advertisement placed by Weill & Co in the Watchmaker, Jeweller, Silversmith and Optician December 1906, p29. A slightly different model was reviewed in the Horological Journal in January 1903.

PLATE 62
Bovet, Fleurier, Switzerland

Chronographie compteur. Silver .800 open face case hallmarked Swiss woodcock, hinged bezel back and cuvette. White enamel dial inscribed MONTRE UNIVERSELLE PATENTEE CHRONOGRAPHIE COMPTEUR MINUTES HEURES, narrow black roman numerals, blued steel hands. Three subsidiary inset sunk dials for elapsed minutes hours and seconds. The small chronograph hand is concentric with the subsidiary seconds hand. Keyless wound and push piece to set hands at four minutes past position. The push button at 9 o'clock resets the three hands to zero, slide at 3 o'clock operates chronograph. 15 jewels. Gilded club tooth lever movement engraved on the top plate BREVETE S.G.D.G. D.R.P. ANG PATENT APPL. FOR Bovet trade mark. Straight line lever oversprung cut bimetallic balance with blued steel balance spring. Regulator index on engraved balance cock. *A rare and unusual chronograph.*
Diameter 54mm

Edouard Bovet, born 1797 and died 1849. Founded Edouard

11,625. Pass, E. de, [*Bovet, H.*]. Aug. 11.

Chronographs. — Relates to mechanism for indicating the duration of an experiment, a race, a cab journey, &c. from one second up to twenty-four hours, which mechanism may be combined with watches. The ordinary seconds - arbor P has an additional hand S on a friction-tight cam R, and the movement is provided with an extra minute - arbor and hour-arbor, not shown, also having their hands borne by friction-tight cams. These three cams may be acted on by a lever T with three beaks t^1, t^2, t^3, so that their hands are set to zero. The lever T is operated by a push-piece and an intermediate lever W. A spring U restores it when released. To set the minute-hand with precision to 60°, a brake Y is applied to the pinion on the arbor by an intermediate lever X. The second cam, instead of being friction-tight, may be loose, and with springs S^1, S^{11} pressing it down ; a pin on its underside may engage a fluted wheel p. By two levers X^1, Y^1, and a push-piece V^1, the cam can be raised and the hand stopped at any place. In returning to zero by the lever T, a branch of the lever X raises the cam off the wheel p.

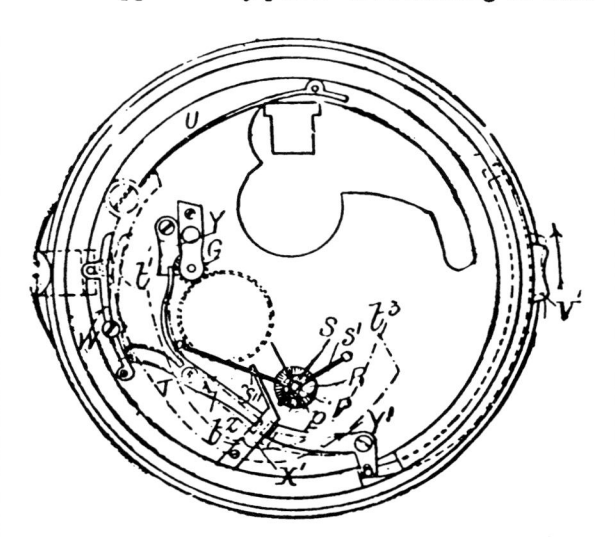

PLATE 62C

Bovet 1820 for manufacturing watches especially for the Chinese market. The Bovet trade mark was purchased for an extremely high figure by César and Charles Leuba Frères who subsequently sold it on in 1918 to Ullmann & Co. UK Patent No 11625 11 August 1888 taken out in the name of E de Pass for H Bovet. Relates to mechanism for indicating duration of an experiment, a race, a cab journey etc from one second up to twenty-fours hours — were cabs timed for reasons of personal curiosity or in order to berate the cab driver?

PLATE 63A

PLATE 63B

PLATE 63
Charles Churchill & Co Ltd (retailer), Manchester

⅕ second production timer 30 minute recorder. Oxydised steel open face case, snap-on bezel, hinged back and cuvette. White enamel dial inscribed CHARLES CHURCHILL & CO LTD MANCHESTER SWISS MADE, arabic numerals 1800-60 black, 30-59 red, 5-60 black, inner scale 0-60 seconds and inset 30 minutes recording dial upper centre. Keyless wound. 6 jewels. Top plate stamped SWISS MADE. Pillar plate stamped 1095653. Three-quarter plate cylinder movement with oversprung three arm monometallic balance with white balance spring. Regulator index on balance cock engraved RA SR.
Diameter 52.9mm

Example of use – One produced in 15 seconds = 4 per minute = 240 per hour (outer scale). Red scale reversed seconds scale, ie time left in seconds from minute period.

PLATE 64
Le Phare, Fabrique Le Phare SA, Le Locle, Switzerland

⅕ second chronograph production timer 30 minute recorder. Nickel open face case, snap-on bezel, hinged back and cuvette. White enamel dial, arabic red and black numerals, blued steel spade hands, centre sweep seconds and small inset running seconds dial. Keyless wound. Pillar plate stamped 343 671. Jewelled club tooth straight line lever movement oversprung monometallic balance with compensation screws. Overcoil balance spring. Regulator index on balance cock engraved AR FS. Time to produce one item on inner scale shows number on outer scale which would be produced hourly.
Diameter 51.5mm

Le Phare registered trade mark of C Barbezat-Baillot fabricant, Le Locle, Switzerland, registered 4 May 1897, No 9230.

PLATE 65
Sewill (retailer), Liverpool, Ed Heuer & Co SA (Manufacturer est 1860), Bienne, Switzerland

⅕ second chronograph miles per hour. Gilt open face case, snap-on bezel, hinged back and cuvette. White enamel dial inscribed SEWILL LIVERPOOL, black arabic hours, sweep centre seconds chronograph dial outside chapter ring 0-60 seconds x 4, outer ring red arabic numerals 15-120 mph. Small sunk seconds dial 0-60 seconds inset lower centre dial, blued steel hour and minute hands, gilt centre seconds hand. Keyless wound, push piece to set. Top plate stamped SWISS MADE. Pillar plate 48.363. 17 jewels. Half plate club tooth straight line lever movement with compensation screws to oversprung monometallic balance and pale yellow overcoil balance spring. Regulator index on balance cock hand engraved SF RA.
Diameter 53.3mm

Example, ¼ revolution of centre hand = 60 seconds = 1 mile = 60 mph. Joseph Sewill is listed at 61 South Castle Street, Liverpool 1848-51, chronometer maker, and in the Post Office Directory of 1897 at 15 and 16 Canning Place, Liverpool.

Chronographs

PLATE 64A

PLATE 64B

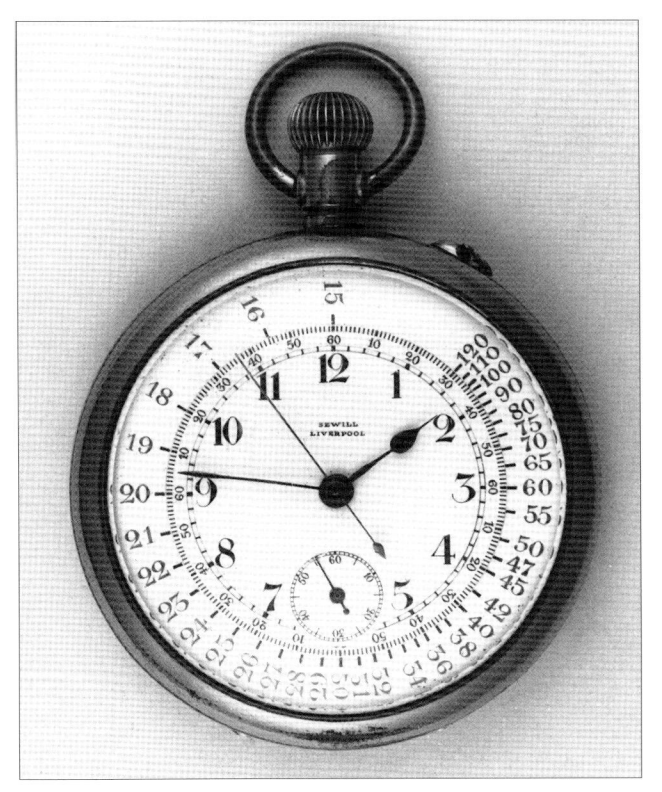

PLATE 65A

PLATE 65B

333

Chronographs

PLATE 66A

PLATE 66B

PLATE 66

John Taylor, 36 Drake Street, Rochdale, Lancashire
Chronograph stop-watch. Silver open face case hallmarked London **1913**, snap-on bezel, hinged back and cuvette. Casemaker WB (William Thomas Bullock, 4 Waveley Road, Coventry), pendant hallmarked Birmingham **1902**, CH (Charles Harrold, 2 & 3 St Pauls Square, Birmingham). White enamel dial inscribed WILLIAMS PATENT NO 4762-90 SOLE MAKER JOHN TAYLOR ROCHDALE, sunk second dial 0-60, black roman numerals, blued steel spade hands, centre sweep seconds with dial marked 1-16 twice indicating $\frac{1}{16}$ seconds (1 revolution in 2 seconds). Keyless wound at pendant button but setting by key at back, silver slide in band at 2 o'clock position activating stop/start lever on edge of balance roller. 15 jewels. Three-quarter plate going barrel pointed tooth scape with straight line lever. Oversprung three arm round rimmed steel balance with blued steel balance spring. Polished steel index on balance cock engraved with Coventry fleur-de-lis. Inscription on backplate John Taylor 36, Drake Street ROCHDALE. 38953. *Used for timing whippets (dogs).*
Diameter 59mm

Taylor is listed in Post Office Directory for 1890. UK Patent No 4762 March 1890, William Williams, 32 Rock Street, Bury, Lancashire, watch manufacturer.

PLATE 67
Smith's Industries

Football timer. Chrome open face case, snap-on bezel and dome. Cream painted dial divided into five minute red segments for 45 minute period and inscribed REFEREE, black arabic numerals for minutes with inset seconds dial with miniature red arabic numerals. Keyless wound, slide to start stop-watch at 5/10 minute position of dial. Single black painted hand in centre to indicate minutes. Top plate stamped MADE IN GT BRITAIN. Unjewelled. Pin pallet lever. Index on to scale stamped on top plate.
Diameter 51.2mm

PLATE 68
Anon, Switzerland

Football $\frac{1}{5}$ seconds timer. Nickel open face case, snap-on bezel, hinged dome and cuvette. White enamel dial inscribed FOOT-BALL, black arabic numerals with inset 60 minute dial divided into 45 minute period, blued steel centre seconds, full moon hands. Centre second and inset dial. Keyless wound and set, slide to stop-watch side of case, winding button to stop/start and reset. Half plate movement. Recoil click. 7 jewels. Straight line club tooth Swiss lever escapement with oversprung monometallic balance. Index on cock. Movement stamped 30590. SWISS MADE. Identified as Valjoux movement. Calibre 68.
Diameter 49.2mm

Chronographs

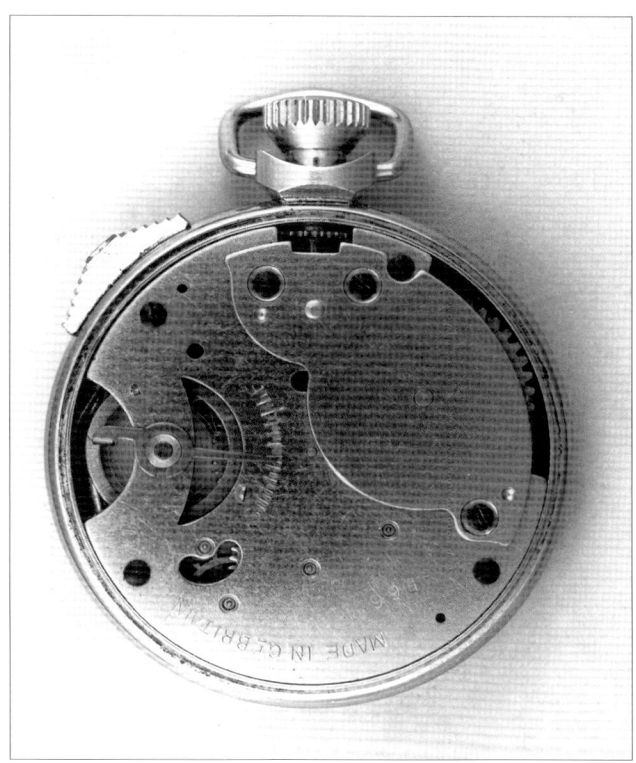

PLATE 67A

PLATE 67B

PLATE 68A

PLATE 68B

335

Chronographs

PLATE 69A

PLATE 69B

PLATE 69
Dimier Frères & Co Ltd, Switzerland

Football ⅕ seconds timer. Nickel open face case, snap-on bezel, hinged dome and cuvette. White enamel dial inscribed FOOT-BALL, black arabic numerals, 60 minute inset dial divided into 45 minute period, blued steel full moon hands. Keyless wound, stop slide side of case, push pendant button to stop/start and reset. 4 jewels. Half plate movement stamped on top plate DF & Co in oval (trade mark of Dimier Frères & Co Ltd). Engine turned pattern movement and stamped 102088. Recoil click. Club tooth straight line lever movement with oversprung monometallic flat rimmed balance. Regulator index on cock stamped RS FR.
Diameter 51mm

★DIMRA★ *engraved on barrel wheel.*

PLATE 70
Smith's Industries Ltd

⅕ seconds yachting timer. Open face case, Cuprel Regd Dennison Watch Case Co Ltd in England 1622 stamped inside snap-on back, snap-on bezel. Cream painted dial inscribed SMITHS YACHTING TIMER MADE IN ENGLAND, black arabic numerals in yachting sector, black pointer hands. Keyless wound and push to stop/start and reset. Pillar plate stamped B and BG either side of balance. 7 jewels. Half plate club tooth and straight line lever movement. Oversprung monometallic balance and white balance spring. Regulator index on balance cock stamped FS.
Diameter 50-8mm

PLATE 71
Smith's Industries Ltd

1/10 second yachting timer. Chromed open face case in rubber outer cover stamped inside back snap-on cover COVEX BASE METAL MADE IN GREAT BRITAIN SMITHS INDUSTRIES LIMITED. Painted dial inscribed YACHTING TIMER 1/10 SECONDS JEWELLED GREAT BRITAIN, blue centre and white chapter ring, small arabic numerals 0-10 seconds and 15, 20, 25 (anticlockwise) with larger arabic numerals 35, 40, 45, 50, 55 within chapter marking. Yachting sector 0-10 minutes (anticlockwise) upper centre. Keyless wound, push button at side to stop/start and reset. Top plate stamped SMITHS INDUSTRIES LIMITED ONE (1) JEWEL UNADJUSTED GREAT BRITAIN. Pin pallet escapement with straight line lever. Oversprung monometallic balance with white balance spring. Brass regulator index on to scale on top plate.
Diameter 57.4mm

PLATE 70A

PLATE 70B

PLATE 71A

PLATE 71B

Chronographs

PLATE 72A

PLATE 72B

PLATE 72
Weill & Co (Importers), Switzerland

Yachting timer. Nickel open face case, snap-on bezel, hinged dome and cuvette. Dome stamped WEILL & CO SWITZERLAND. White enamel dial inscribed THE PAGET REGd No 4530, black arabic numerals 55-5 clockwise with red 60. Upper centre of dial an inset dial calibrated in 5, 4, 3, 2, 1 periods clockwise. Keyless winding at pendant with stop/start button through winding button. Movement stamped WEILL & CO SWITZERLAND 13 THIRTEEN JEWELS 3 ADJS 330 584. Straight line lever with club tooth scape wheel. Monometallic balance with compensation screws. Oversprung. Overcoil white balance spring. Index regulator on balance cock engraved AF RS.
Diameter 67.2mm

This is a 19 Le Phare movement. The trade mark THE PAGET is recorded in Marque de Fabrique Suisse registered mar No 9277 22 May 1987 Weill & Co Fabricants, Chaux de Fonds, Switzerland. Weill & Co were established in 1863. Louis Weill died in 1919 but the firm continued trading from 111 Hatton Gardens, London as Arthur Mayer. An advertisement selling off this old line appeared in the Watchmaker, Jeweller, Silversmith *and* Optician, *September 1930.*

COLOUR PLATE 16 (page 9) and PLATE 73
Waltham Watch Co, Waltham, USA

6 seconds stop-watch timer. Chrome on base metal Keystone open face case, screw back and bezel snap-on cuvette, back cover engraved PATT 6 /!)U3536. Silvered dial inscribed WALTHAM USA ADMIRALTY PATTERN NO 6, arabic numerals 0-5000 yards, black 0-6 seconds red with 1 second markers, blued steel centre pointer hand. Keyless winding, press to stop/start and reset. Top plate stamped 31455749 WALTHAM USA 16 43 TA NINE JEWELS. 9 jewels. Three-quarter plate club tooth straight line lever escapement oversprung monometallic balance with serrated edge to balance to make more positive action stop and start. Regulator index on balance cock engraved FS.
Diameter 57.2mm

Waltham records list 31451001-31461000, 16 size, Time Grade, 9 jewels, Positive setting, Style OF, Serial number corresponds to manufacturing period 1942/1945. Believed to be ASDIC timer (Allied Submarine Detection Investigation Committee). Patt 6 ↑ U3536.

Chronographs

PLATE 73

PLATE 74A

COLOUR PLATE 17 (page 9) and PLATE 74
Charles Frodsham, 27 South Moulton Street, London W

Micrometer chronograph for timing flight of projectiles. Nickel open face case, snap-on back, bevelled edge glass. Matt silvered recording seconds dial, arabic numerals 0-60 in five seconds steps with arabic 0-100 subsidiary inset dials in $\frac{1}{100}$ second, blued steel pointer hands. Dial inscribed Chas Frodsham 27 South Molton St London No 09571. Keyless winding with turn button projecting through back cover, turned anticlockwise when viewed from the back (in direction of arrow), reset by push pendant button. 6 jewels. Three-quarter plate cylinder movement with oversprung monometallic balance, movement number (under dial) 13574. Regulator index on balance cock engraved FS.

Diameter 65mm

Made for Frodsham by Nicole Nielson, 14 Soho Square, London. These quick acting micrometer chronographs were extremely accurate for short trials up to 60 seconds. The first models (1895) showed seconds on outer circle and 20ths of a second on small circle but by 1899 these had improved so that

PLATE 74B

339

Chronographs

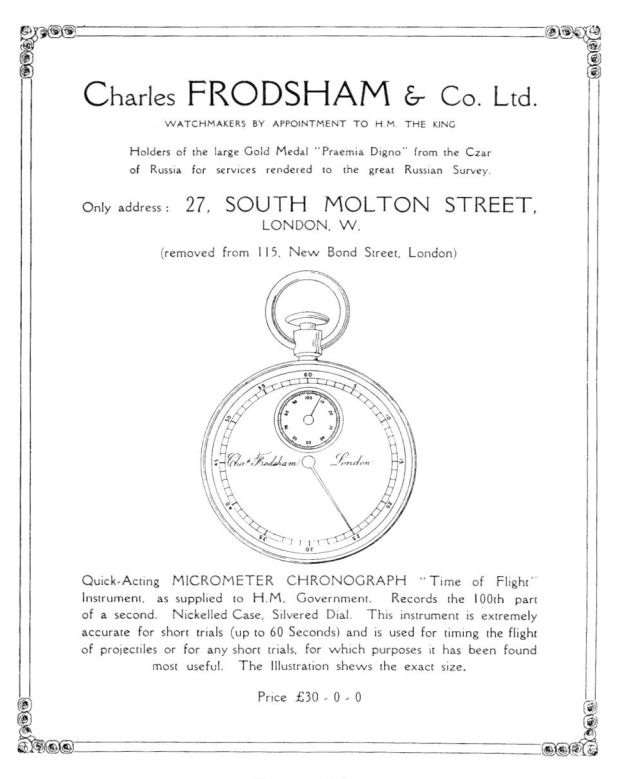

PLATE 74C

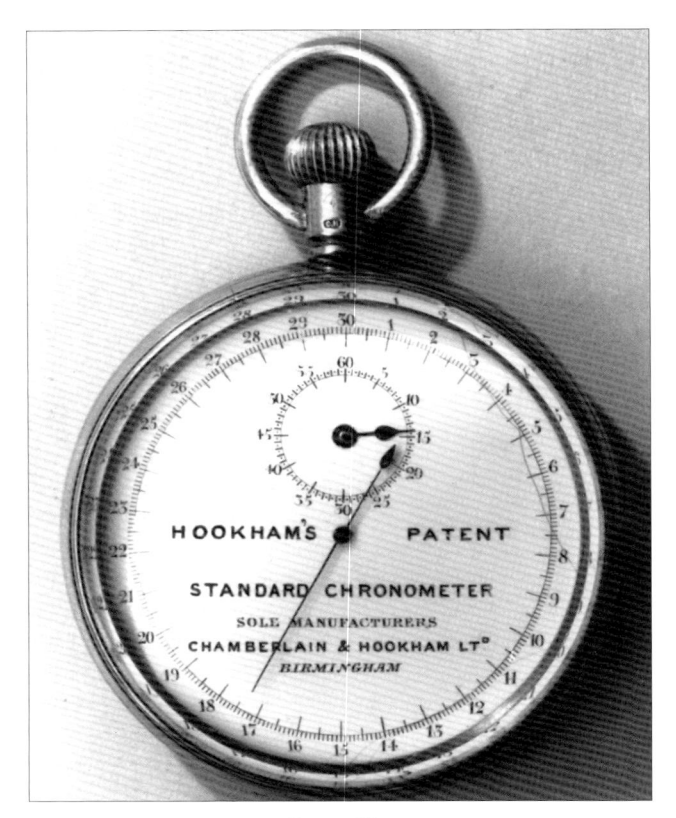

PLATE 75A

small circle read 100th part of a second. Manufactured for £20, they sold for £30. The Frodshams The Story of a Family of Chronometer makers, by Dr Vaudrey Mercer.

PLATE 75
Chamberlain & Hookham Ltd, Birmingham

Nickel open face case engraved on back of case BROMPTON & KENSINGTON ELECTRIC Co. Casemaker WHC (William Henry Christie, 102 Spon End, Coventry – Kelly Directory 1896/1904). Pendant CH (Charles Harrold). White enamel dial inscribed HOOKHAM'S PATENT STANDARD CHRONOMETER SOLE MANUFACTURERS CHAMBERLAIN & HOOKHAM LTD BIRMINGHAM, black arabic numerals, sweep second centre dial 0-30 seconds with minute recorder dial inset 0-60 minutes. Key wound and set. 7 jewels. Three-quarter plate English lever movement with pointed tooth scape wheel and tangential lever. Oversprung monometallic balance. Top plate engraved No 4001. Regulator index on balance cock hand engraved FAST SLOW.
Key size 8. Diameter 72mm

It is believed that this chronograph was for checking speed of AC/DC motors. It is of interest that G Hookham did have an electric clock patent, UK No 353 February 22 1900.

PLATE 75B

CHAPTER NINE

Novelty Watches

It could be argued that the very origin of the watch lies in its being nothing more or less than a novelty. The early watches were certainly not precision timekeepers and their owners acquired them for effect rather than use, a trend that was to continue long after technical advances had ensured a watch capable of keeping accurate time at an acceptable price. For example, around the turn of the eighteenth century the ability to produce smaller movements resulted in the Geneva watchmakers in particular displaying their ingenuity in the making of form watches, many of these embodied in the shapes of fruit, musical instruments, butterflies, etc.

What is the appeal of a novelty watch for the present-day collector? Much the same as it was for its original purchaser! The fact that it is unusual and different from the ordinary run of everyday watch design, whether this be in a technical feature, a dial or case advertising either an event or a particular product. In some instances watches were made to commemorate a particular personality or grand occasion of national importance.

Watches where the escapement was visible were extremely popular and appeared in great variation by a number of manufacturers.

More often than not the intrinsic value of the materials used and quality of finish were of secondary importance. They were often produced as toys and certainly not

PLATE 1. Advertising card of F. Riondey (see Plate 2 for patent detail)

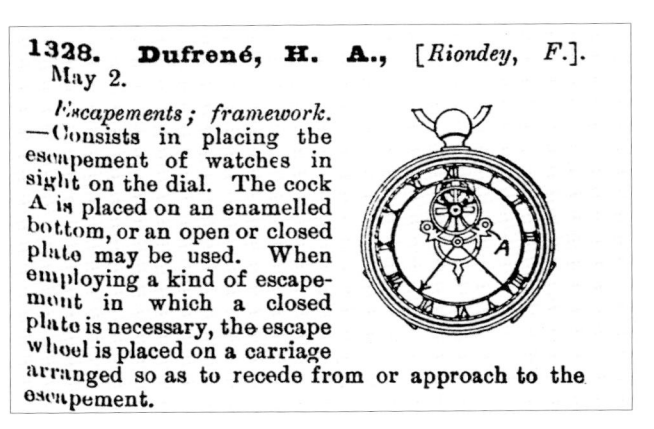

PLATE 2

intended to have a long and useful life as serious timekeepers. This has often resulted in signs of wear and tear or even neglect. They are more highly valued in today's market, but good condition is of vital importance.

The range of novelty features appears to be endless. Think of an idea and it has probably appeared at some time or another. The highly competitive nature of this end of the market resulted in many of these concepts being covered by patent. There does, however, seem to be an ill defined no man's land between the sophisticated, technically interesting watch and those that rely on pure gimmickry for their market. This is the main area for present-day collecting. What is the true novelty of tomorrow? If the trends observed in the window of the High Street jeweller of today is to be any guide it is highly possible that it will be the watch with the plain white dial, free of embellishment apart from clear black arabic numerals and black spade hands – in short, a watch capable of telling the time with unmistakable clarity!

'Dudley and His Masonic Watch', William Henry, *NAWCC* Whole No 136, October 1968

"English Made Masonic Watches', *NAWCC* Whole No 204, February 1980

PLATE 3. UK Patent Specification Abridgement 1885

Watches of Fantasy 1790-1850, Oswaldo Patrizzi, Fabienne X Sturm

PLATE 4. *L'Horloger,* 1912/13

Novelty Watches

PLATE 5

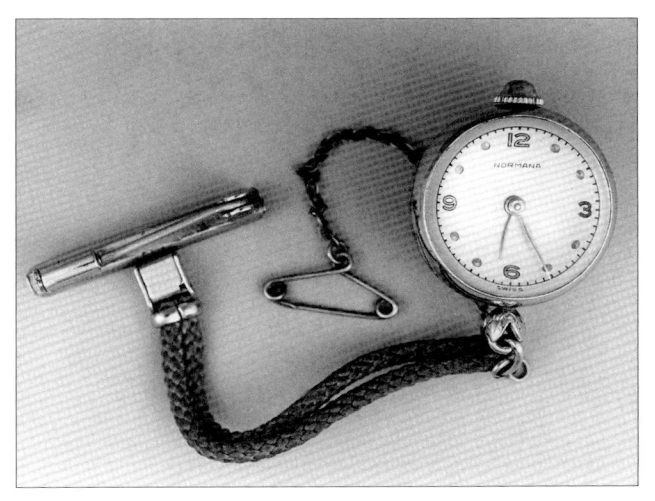

PLATE 6A

PLATE 6B

COLOUR PLATE 18 (page 10)
Skeletonised Reverse Movement

Oxydised steel open face, Borgel case, removable brass bezel. Annular white dial, decorated skeletonised plates, black arabic numerals, narrow blued Breguet hands. Keyless winding and set. 15 jewels. Bar movement with straight line club tooth lever escapement. Blued steel balance to monometallic balance with compensation screws.
Diameter 50.6mm

Swiss Patent No 96828.

COLOUR PLATE 19 (page 10)
Twenty-four Hour Dial

Nickel open face case, snap-on bezel, hinged back and cuvette, engraved inside back. White enamel dial inscribed BREVET 56177, black arabic numerals 1-12, inset dial with red arabic numerals 13-24, narrow blued steel hands. Keyless winding, pull to set. 15 jewels. Half plate with straight line lever escapement. Compensated monometallic oversprung balance.
Diameter 50.6mm

PLATE 5
Buttonhole Watch

Oxydised steel case, gilt bezel, button and olivette, numbered 3602 inside, snap-on back. White enamel dial. Swiss 10 jewel cylinder. Keyless bar movement, push piece to set hands at 5 minute past position.
Diameter 29.3mm

PLATE 6
Ball Fob Watch, Normania & Co,
La Chaux de Fonds, Switzerland

Chrome case and bezels. Silvered dial NORMANIA upper centre, gilt arabic numerals 3, 6, 9 and 12. Swiss 15 jewels club tooth lever escapement with straight line lever. NORMANIA on barrel wheel. Complete with matching fob brooch and safety chain.
Diameter 22.5mm

343

Novelty Watches

PLATE 7A

PLATE 7B

PLATE 7
Golfer's Watch

Cast silver full hunter case, inside front cover stamped REGd No 705616 Swiss Made BREV + 107244, inside back cover stamped USA No 65351. White enamel dial, SWISS MADE lower edge, black arabic numerals, blued steel hands, inset sunk seconds. 15 jewel straight line lever club tooth escapement. Monometallic balance with compensation screws. White oversprung and overcoil balance spring. Index on balance cock.

Diameter 37mm

Swiss Patent 107244, Reichenberg & Cie, Tallis Watch Co, La Chaux de Fonds, Suisse Montre Savonette (Golf Ball), Demande déposé 23 Oct 1923, Publié 16 Oct 1924. Advertised Practical Watch and Clockmaker March 1928. Reichenberg Co (The Tallis Watch Co), established 1900, importers, 7 Hatton Garden, London EC1. Factory 66 Rue Leopard Robert, Chaux de Fonds, Switzerland. Advertised with best 'Hexameter' movement Model T3633 and T3872 (wrist).

Registered No		
	Great Britain	73872
	Switzerland	41406
	France	21885
	USA	Applied for

COLOUR PLATE 20 (page 10) and PLATE 8
Instrument Panel Watch

Nickel open face case, hinged back and cuvette. White enamel dial inscribed MARCONI WIRELESS TELEGRAPH CO Ltd upper centre and Z – MIDNIGHT M–MIDDAY lower centre. The dial, in addition to black roman numerals, is inscribed with telegraphic alphabet for sending time signals by Morse code. Thin blued steel spade hands. Half plate keyless Swiss club tooth 7 jewel lever movement with straight line lever. Pillar plate stamped H.

Diameter 49.3mm

JF Hills, Post Office, Sudbury, Suffolk exhibited an English keyless lever watch with specially designed dial showing the code letters for the use and convenience of postmasters and telephonists at the 1885 International Inventions Exhibition (Group XXVII, South Central Gallery, Cat No 2113).

Novelty Watches

PLATE 8A

PLATE 8B

PLATE 8C

PLATE 8D

Instruction card for sending Morse time signals

345

Novelty Watches

PLATE 9A

PLATE 9B

PLATE 10

PLATE 9
Commemorative Pocket Watch, Columbia Time Products, La Salle, Illinois, USA

Nickel open face case. White enamel dial utilising 25th ANNIVERSARY in place of numerals and inscribed N.A.W.C.C. INC. PHILA. CHAP. 1 in upper centre with the Association's logo below. Pin pallet unjewelled keyless wound movement stamped 667 on pillar plate, Model No 20001 Made in USA on barrel bridge. Note open barrel and method of fixing movement. Complete with original box and guarantee card.

Diameter 50.2mm

As indicated by details on watchpaper and dial, this watch was specially commissioned to commemorate the 25th Anniversary of the Philadelphia Chapter 1 of the National Association of Watch and Clock Collectors June 19-23 1968.

PLATE 10
Commemorative Pocket Watch, Westclox

Chrome open face case. White painted dial utilising Worlds Fair 82 in place of numerals, sunk seconds with Knoxville, Tenn and the centre dial with decoration and the words ENERGY TURNS THE WORLD USA. Black matt hands. Keyless wound. Unjewelled pin pallet movement stamped 4 on pillar plate. Complete with original box with printed guarantee by Wagner Time Inc, Barrington, Illinois, USA.

Diameter 50mm

This was the official commemorative pocket watch for the USA World Fair held May to October 1982.

Novelty Watches

PLATE 11A

PLATE 11B

PLATE 12A

PLATE 12B

PLATE 11
Braille Watch, Tavannes Watch Co, Switzerland

Nickel full hunter case with hinged bottom and cuvette. White enamel dial with braille markers for hour position and TAVANNES WATCH Co Swiss Made, strong gilt brass spade hands with minute hand screwed to centre arbor. Keyless wound with recoil click. 15 jewel Swiss bar movement with club tooth lever escapement straight line pallets. Monometallic balance with compensated screws and white overcoil and oversprung balance spring.

Diameter 51mm

PLATE 12
Blind Man's Watch

Nickel open face case. Metal engine-turned dial bare to quarter position and dots at five minute intermediate position, DEPOSE upper centre. strong blued steel hands. Keyless wound and set. 15 jewel movement. Monometallic oversprung balance with compensation screws fitted. Index on balance cock.

Diameter 44m

Novelty Watches

PLATE 13A

PLATE 13B

PLATE 14A. *Journal Suisse d'Horlogerie et de Bijouterie,* 1951

PLATE 14B. *Journal Suisse d'Horlogerie et de Bijouterie* No.11-12, November 1954, p.25

Plate 13
Blind Man's Watch, Swiss

Full hunter nickel case. Brass dial, three dots at 12 o'clock position with two dots at quarters with raised dots at intermediate positions, heavy screwed-on brass hands. Keyless wound with push piece at 55 minute position. Half plate Roskopf pin pallet movement stamped 62 on pillar plate under edge of balance. Jewelling to balance staff lever and scapewheel, tangential pin pallet lever, brass three arm oversprung balance. Blued steel balance spring. Regulator index on small balance cock engraved Avance Retard.
Diameter 53.4mm

Plate 14
Calendar Pocket Watch, Movado, La Chaux de Fonds, Switzerland

Chrome case covered with crocodile leather with safety chain loop attached. Opening and shutting case winding. Small folding strut at back of case for alternative use as desk/table clock. Silver dial with gilt applied arabic numerals for hours and small black arabic numerals for date with aperture for day, month and moon phase, gilt hands. High quality Movado movement.
Width 52mm, depth 35.6mm, thickness 16.8mm

Design subject to Swiss patent 146907 (convention date), Huguenin Frères & Co, Fabrique Niel Le Locle, Suisse Montre avec étui. See 'Hermetic Watches for Collectors', Rita Shenton Antique Dealer & Collectors Guide May 1986.

Plate 15
Travel Watch/Clock, PONTIFA, 2316 Les Ponts de Martel, Switzerland

Dark green leather case with gilt bands. Cream dial with textured surface and gilt 12, hour bars and hands. 17 jewel Swiss keyless lever movement.
Width 54mm, depth 36mm, thickness 15.5mm

Subject of USA Patent 2640668 filed 5 April 1951, 2719402 filed 9 November 1953 by Karl Otto Schmidt, Idar-Oberstein, Germany. Henry Stauffer, watchmaker, founded the company in 1850 in a small mountain village, Les Ponts de Martel, in the canton Neuchatel. He was succeeded by his son Timothée who in turn was succeeded by his eldest son Marc Timothée. The latter visited England, China and Australia in order to gain commercial experience. In 1904 he registered with the local chamber of commerce under the name of MT Stauffer-Jeune. The trade name of Pontif was registered in 1911 followed by Pontifa in 1925. Marc Leuthold, Stauffer's nephew, joined the firm in 1927 and took over after MT Stauffer's death in 1950. His children continued the business, now a limited company.

See Colour Plate 21 (page 10) for three travelling watches

PLATE 15

Novelty Watches

PLATE 16A

PLATE 16B

PLATE 17

PLATE 18

PLATE 16
Hermetique Folding Watch, Swiss

Square chrome open face movement case in outer folding case with snake patterned leather covered case. Case opens in camera style, operated by compression of the ends of the watch. Silvered dial and silver arabic numerals at quarters with black arabic numerals at other positions. Inscribed KIRCHHOFER upper dial, SWISS MADE lower edge. Inset seconds dial, blued steel swell hands. Keyless wound and set Swiss lever movement.

Width 5.3mm, depth 31mm, thickness 10.8mm

Case design Swiss patent No 146907, Huguenin Frères & Co. Fabrique Niel Le Locle, Suisse, Demande déposée 6 mars 1930, Brevet enregistre 15 mai 1931. Publié le 16 juillet 1931. Watches to this design were being made in 1930 by Charles Tissot SA, Le Locle.

PLATE 17
Folding Sports Watch

Square nickel open face movement case in outer snakeskin leather covered case. Case stamped with heart pierced by arrow. Square spring loaded movement swinging out of case when case opened by pressing. Silvered dial, luminous hands and arabic numerals. Keyless wound and set Swiss lever movement.

Width 37.2mm, depth 37.2mm, thickness 11mm

Both the September 1930 and April 1931 issues of Watchmaker, Jeweller, Silversmith and Optician show similar watches made by Tallis Watch Co factory at 66 Rue Léopold Robert, Chaux de Fonds, Switzerland (Agents Reichenberg & Co, 7 Hatton Gardens, London EC1).

Novelty Watches

PLATE 19

PLATE 18
Folding Travelling Watch, Vertex, Swiss

Square Swiss silver (0.925) open face movement case, London import hallmark 1936, marked Engl Pat Reg No 365646. Brown snake leather outer case sliding downwards to open the spring loaded square movement rising out of case for winding (by button) or display. Silvered dial with VERTEX upper centre and SHOCKSORBER lower half, black arabic numerals with quarters, black pointer hands. 15 jewel Swiss lever movement.

Width 43.7mm, depth 30.8mm, thickness 11.3mm

Patent Registered No 365 646, Frederick Baumgartner, 13 Coulouvrenière, Geneva, 24 Oct 1930.

PLATE 19
Folding Travelling Watch

Square chrome open face movement case in red leather camera style folding outer case. Silvered dial inscribed Cortébert upper centre, luminous hands, arabic numerals at quarters. Cortébert (Juillard et Cie Switzerland) movement calibre 575.

Width 53.5mm, depth 35mm, thickness 13.8mm

PLATE 20
Keyless Pocket Watch

18ct gold open face case, hinged dome and cuvette engraved Spiral Breguet Remontoire No 12227 XV JOYAUX ANCRE LIGNE DROITE. White enamel dial, roman hour numerals and arabic minute markings at five minute intervals, inset seconds dial, blued steel narrow spade hands. 15 jewels. Bar straight line club tooth movement. Oversprung uncut bimetallic balance. Balance cock engraved ADVANCE RETARD.

PLATE 20A

PLATE 20B

Diameter 47.5mm

UK Patent No 3100 December 1864, Charles Lehmann, Watch manufacturer, Paris. Awarded Prize Medal at International Exhibition 1885. See also UK Patent No 544 PD Azemar February 1862 for a worm gear.

351

Novelty Watches

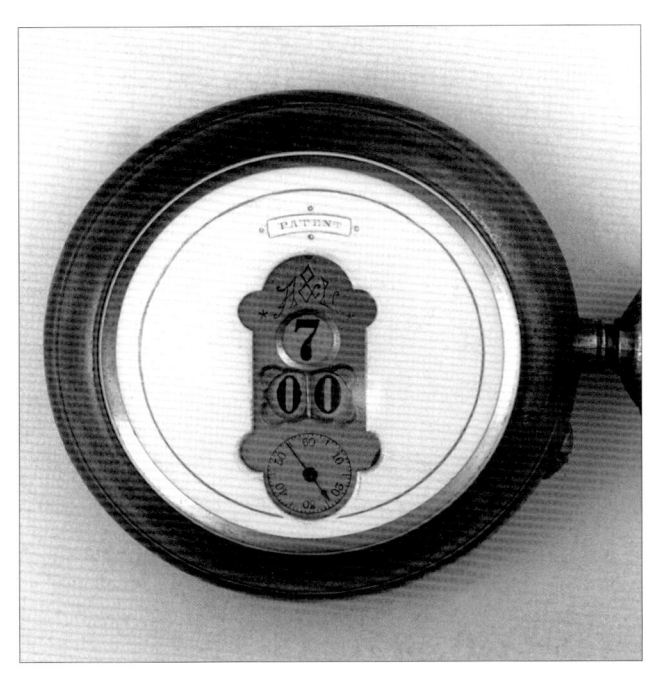

PLATE 21A

PLATE 21B

PLATE 22A

PLATE 22B

352

Novelty Watches

PLATE 23A

PLATE 23B

**COLOUR PLATES 22 and 23 (page 11) and PLATE 21
Digital Pocket Watch, Swiss**

Oxydised open face steel case stamped ACIER GARANTI with arrows pointing towards initials ER within a shield and D without. Glazed cover to movement. White painted dial inscribed PATENT with inner digital dial with inset seconds inscribed ★A & L★ (five pointed star). Keyless wound, push piece to set hands. 15 jewel Swiss lever movement. Club tooth scape wheel. Index on balance cock. Bimetallic uncut 'compensated' balance fitted with compensation screws.
Diameter 54.5mm

A & L could possibly be Alphone Thommen and partner Louis Schopp. Alphone was brother of Gideon Thommen who died in 1890. Clocks 1986, p8. Pocket Watches by Reinhard Meis, Figs 713-716.

**PLATE 22
Digital Pocket Watch, Swiss**

Swiss silver (0.800) open face case stamped 58. Silver digital dial decorated with garland, stamped DOMINA trade mark and SWISS BREVET 31613. Keyless wound, hand set by push piece. 8 jewels. Half plate Roskopf pin pallet movement stamped M and trade mark of hanging electric light bulb surrounded by five-pointed stars. Barrel bridge engraved REPASSE REPASADO DE and EN SECOND SEGUNDA MANO. Balance cock engraved AR and SF.
Diameter 55mm

Swiss Patent No 31613, A Sandoz Bourcherin, fabricants d'horlogerie, 1 Rue David Parre Bourguin, Chaux de Fonds, Switzerland.

**PLATE 23
Digital Pocket Watch, A Schild Ebauche AG, Neuenberg, Switzerland**

Nickel plated open face case, snap-on bezel and back. Keyless winding, push to set hands. Plate stamped SWISS MADE. Unjewelled Roskopf pin pallet movement. Brass three arm oversprung balance. Regulator index on balance cock with curb pins. Cock stamped RA SF.
Diameter 50.3mm

353

Novelty Watches

PLATE 24A

PLATE 24B

PLATE 24
Lady's Silver Fob Watch, Swiss

Swiss silver (0.935) engraved open face case stamped 4103. The inner dome is perforated to show photograph inserted through a slit in the side at 3 o' clock position. Inside dome stamped 4103 PATENT + 11401 (0.935). White enamel dial with gilt decoration to centre and outer edge, roman numerals, blued steel hands. Keyless wound, push piece to set hands. 8 jewels. Cylinder movement.
Diameter 37mm

Swiss Patent No 11401, Albert Salchli Brugg, Biel and L Kempf & Cie, Biel, Switzerland which corresponds to UK Patent No 22614, 12 October 1896.

PLATE 24C

354

Novelty Watches

PLATE 25A

PLATE 25B

PLATE 26A

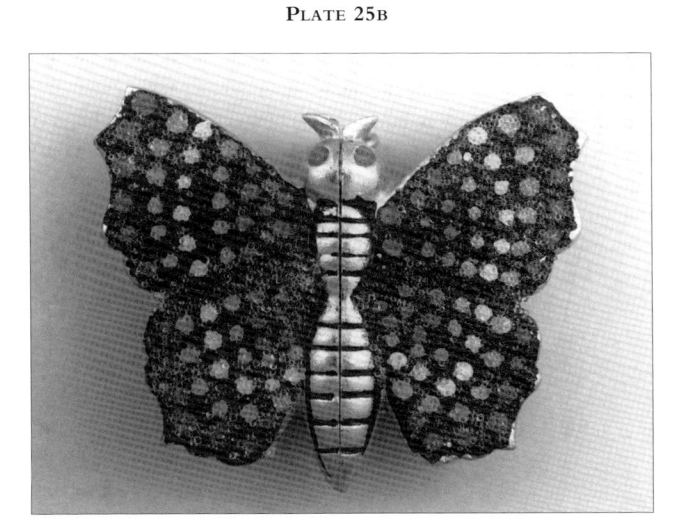

PLATE 26B

PLATE 25
Nurse's Fob Watch, Swiss

Oxydised steel open face case. White enamel dial inscribed SWISS lower edge, gilt spade hour and minutes hands with blued steel centre seconds hand. Keyless wound with push piece to set hands. 10 jewels. Three-quarter plate cylinder movement stamped 22940 within outline of cross.
Diameter 34.3mm

Swiss Patent No 22940, Alphonse Boichat & Co à Fleurier, Switzerland 9 November 1900.

TITLE-PAGE and PLATE 26
Novelty Butterfly Form Watch, Swiss

Spring-loaded hinged wings open to reveal watch. Dial inscribed ADREM 17 jewels ANTIMAGNETIC SWISS MADE. 17 jewels movement stamped WZ Swiss Made EB 836 1-67 Fabrique d'Ebauche de Bettlach Switzerland.
Width 40.5mm, depth 35mm

355

Novelty Watches

PLATE 27A

PLATE 27
Hydrographic Survey Watch, Swiss

Nickel open face case, engraved screw back H S4 and Government property arrow and numbered 449. White enamel dial inscribed 449 upper centre, black roman numerals, blued steel spade hands. Outer black ring calibrated 0–360 and adjustable in relation to time chapter ring with locking screw at 6 minute position. Keyless wound and set. Fully jewelled centre second Swiss straight line lever. Bimetallic cut compensated balance with Breguet overcoil. Plate stamped SWISS MADE.
Diameter 62.5mm

PLATE 27B

PLATE 28A

PLATE 28B

PLATE 28
Anticlockwise Seconds Dial, Kummer, Bettlach, Switzerland

Nickel open face case, snap-on bezel, hinged back with inner glazed back to movement. Cream dial inscribed OKO LEVER SPECIALLY EXAMINED upper centre and SWISS MADE lower edge, black roman numerals, inset sunk ANTICLOCKWISE seconds dial INSCRIBED BREVET + 47001. Keyless wound, push piece to set hands. Unjewelled pin pallet lever with imitation screw compensation balance.

Diameter 51mm

Swiss Patent No 47001, Ed Kummer, Bettlach, Switzerland, Mouvement de montre du genre Roskopf, 22 February 1909. Reference should also be made to UK Patent No 24164 18 October 1910.

24,164. Kummer, E. Oct. 18.

Going trains; seconds mechanism.—In a watch of the kind in which there is a train of only two wheels c, d, with their pinions, intervening between the barrel and the escape-wheel, the arbor g of the second of these wheels carries the seconds-hand, which rotates in a direction opposite to that of the hour and minute hands. The numbers of teeth in the wheels of the train may be such that the arbor g and the seconds-hand make more or less than one revolution a minute, the seconds-dial being correspondingly graduated.

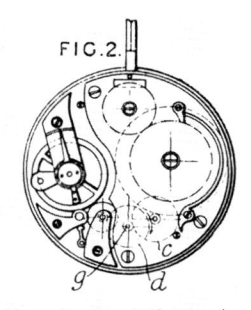

PLATE 28C. UK Patent Abridgements, 1910

Novelty Watches

PLATE 29A

PLATE 29B

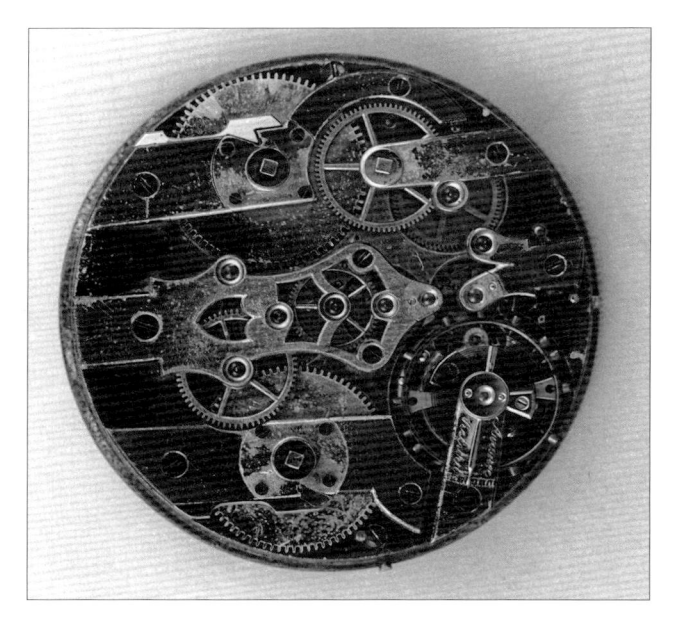

PLATE 30

PLATE 29
Anticlockwise Seconds Dial, Kummer, Bettlach, Switzerland

Pinchbeck open face case, snap-on bezel, hinged back and cuvette stamped MJ L Remontoire PERFECTIONNE lA QUALITE with nine 'medals' as design feature. White enamel dial inscribed SUPERIOR RAILWAY TIMEKEEPER LEVER Specially Examined upper centre, SWISS MADE lower edge and BREVET + 47001 within inset ANTICLOCKWISE dial. Keyless wound, push piece to set hands. Unjewelled pin pallet movement of Roskopf type.

Diameter 53mm

Swiss Patent No 47001, Ed Kummer, Bettlach, Switzerland, Mouvement de montre du genre Roskopf, 22 February 1909. Reference should also be made to UK Patent No 24164, 18 October 1910.

PLATE 30
Pouzait Type Jump Centre Seconds

White enamel dial, narrow black roman numerals, blued steel Breguet style hands. Jump centre seconds with sunk seconds dial 0-60. Key wound and set. Two train going barrel skeletonised bar movement. 31 jewels in movement plus one decorative jewel to maintain symmetry. Club tooth Swiss lever escapement. Bimetallic cut compensated balance with Breguet overcoil with regulator index on balance cock.

Diameter 42.6mm

PLATE 31A

PLATE 31
Independent Seconds Lever, Swiss

Silver open face case, cuvette engraved John Walker 68 CORNHILL (LONDON) retailer. White enamel dial, black roman numerals, blued steel spade hands, inset running seconds dial and sweep centre seconds independent beating. Double train – separate second train. Key wound and set from back. Three-quarter plate. 25 jewel tangential lever movement with pointed tooth scape wheel. Oversprung overcoil cut bimetallic balance. Blued steel balance spring. Regulator index on balance cock engraved FAST SLOW.

Key size 5. Diameter 52.5mm

A Guide to Complicated Watches *by François Le Coultre pp15-19.*

PLATE 31B

Novelty Watches

PLATE 32A

PLATE 32
Jump Centre Seconds, Louis Eugène Favre, Cormoret, Switzerland

Nickel plated open face case, hinged dome. White enamel dial, black roman numerals, blued steel spade hands, centre seconds. Keyless wind, hand set by push piece at 4 minute position. Half plate movement engraved +28265 DRP 154250. 15 jewels. Straight line lever with club tooth scape wheel. Cut bimetallic balance. Blued steel oversprung and overcoil balance spring.
Diameter 50.5mm

PLATE 32B

PLATE 33A

COLOUR PLATE 24 (page 11) and PLATE 33
Calendar Pocket Watch, Swiss

Oxydised steel case (ACIER GARANTI). Enamel dial inscribed SWISS MADE, inset subsidiary dials for day of week, day of month, month and inset seconds dial also showing moon phase, blue roman numerals with gilt decoration, blued steel spade hands. Keyless wound bar movement. 10 jewels. Cylinder escapement.
Diameter 51.5mm

PLATE 33B

Novelty Watches

PLATE 34B

PLATE 34
Double Dial Calendar Watch, Swiss

Oxydised steel case, hinged and glazed back with inner glazed bezel. White enamel dial, SWISS MADE lower edge, black roman hour numerals, blued steel spade hands. inset seconds. Dial on reverse, gilt date centre with arabic numerals, date indicated by blued steel hand. Set by lever under edge of bezel. Outer date ring is an annular day dial set by toothed wheel under bezel with similar wheel to set month and number of days per month. Swiss lever movement.
Diameter 52.5mm

An advertisement for these watches appeared in 1902 Horological Journal – Baume & Co 21 Hatton Garden London (see page 363).

PLATE 34A

PLATE 34C. *Horological Journal* Vol.44 (back cover), June 1902

PLATE 34D. *Horological Journal* Vol.44, August 1902

Novelty Watches

PLATE 35A

FRONTISPIECE and PLATE 35
Animated Pocket Watch – Cycling

Brass open face case. Enamel dial with centre decoration displaying three cyclists racing on penny-farthings with spectators and building with three flags hoisted in the background. The leg of the cyclist in the foreground animated. Arabic numerals, blued steel spade hands. Keyless wound, hand set by button on centre arbor inside back cover. Movement shielded by protective cover. Stamped 158452. Unjewelled. Pin pallet movement. Regulation index adjustable through aperture at edge of movement.

Diameter 51.5mm

PLATE 35B

Novelty Watches

PLATE 36A

PLATE 36
Animated Pocket Watch – Football, Ingersoll Ltd England

Chrome open face case. Painted dial depicting football players and goal. Leg of player in foreground animated. Ball in centre of inset seconds rotates. MADE IN GT BRITAIN INGERSOLL LTD LONDON around rim of inset seconds dial. Black minute hand, luminous tipped hour hand. Unjewelled. Pin pallet movement. Top plate stamped MADE IN GT BRITAIN 5410M.
Diameter 51.3mm

PLATE 36B

Novelty Watches

PLATE 37

PLATE 38

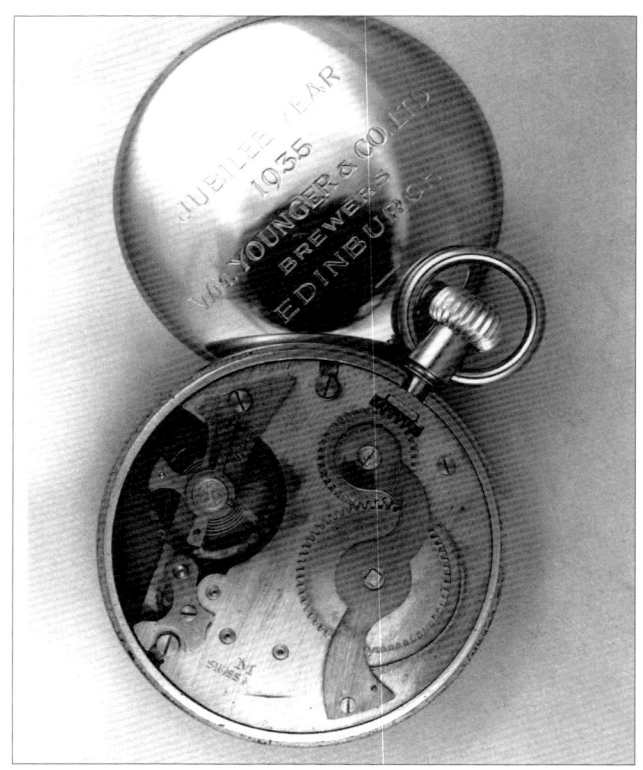

PLATE 39A

PLATE 39B

PLATE 40A

PLATE 40B

COLOUR PLATE 25 (page 11) and PLATE 37
Animated Pocket Watch – Scouts Jamboree, Smith's Industries, England

Chrome open face case, snap-on bezel and back cover. Painted dial depicting Scout Jamboree with moving left arm of pack leader. Red arabic numerals, luminous tipped spade hands. Keyless winding, push piece to set hands. Unjewelled. Pin pallet lever. Top plate stamped 532M.
Diameter 51.3mm

PLATE 38
Animated Pocket Watch – Ranger

Chrome open face case, snap-on bezel. Painted dial depicting cowboy and horse against prairie scene and herd with moving leg of rider. WATCH MADE IN GREAT BRITAIN lower edge. Keyless. Unjewelled movement. Pin pallet lever. Top plate stamped 531M.
Diameter 50.3mm

Manufactured by Anglo Celtic Watch Co, Ystradynlais, Wales (Smiths Industries were major shareholders).

PLATE 39
Commemorative Pocket Watch, Swiss

Chrome open face case, snap-on back cover engraved JUBILEE YEAR 1935 WM YOUNGER & CO LTD BREWERS EDINBURGH. Centre of cream painted dial inscribed TIME FLIES BUT YOU CAN GET "YOUNGER" EVERY DAY SWISS MADE lower edge. Blued steel moon hands, black arabic numerals. Keyless winding, push piece to set. Unjewelled. Tangential pin pallet lever movement. Top plate stamped M SWISS MADE. Brass monometallic balance.
Diameter 49.2mm

PLATE 40
Advertising Pocket Watch, Swiss

Oxydised steel case stamped ACIERY GARANTI H, hinged back, snap-on bezel. White enamel dial inscribed TIT-BITS ONE PENNY WEEKLY FIRST AND BEST in red lettering, SWISS MADE lower edge, black roman numerals. Keyless wound, push piece to set at XI position. Unjewelled. Tangential pin pallet lever.
Diameter 50.5mm

367

Novelty Watches

PLATE 41A

PLATE 41B

Colour Plate 26 (page 12) and Plate 41
Time Teacher Pocket Watch

Orange plastic open face case, inside snap-on back cover moulded SMITHS INDUSTRIES LIMITED MADE IN GREAT BRITAIN A.B.C plastic. White enamel dial depicting school teacher indicating time with his hands, black arabic numerals and words PAST and TO CLOCK on dial at appropriate positions. Keyless wound, push piece to set hands. Top plate stamped 75 MADE IN GREAT BRITAIN. Barrel plate stamped SMITHS INDUSTRIES LTD No(0) JEWELS UNADJUSTED. Unjewelled pin pallet straight line lever.
Diameter 54mm

With original card display box and guarantee.

Plate 42
Commemorative – Festival of Britain 1951, England

Chromed open face case, snap-on bezel and back, back engraved with emblem of 1951 Festival of Britain. Off-white dial inscribed SMITHS centre, MADE IN GT BRITAIN lower edge, arabic numerals, blued steel hands, inset seconds with arabic numerals at quarter positions. Keyless wound and push piece to set. Top plate inscribed SMITHS INDUSTRIES LTD NO(0) JEWELLS UNADJUSTED MADE IN GT BRITAIN 73. Pin pallet straight line lever escapement. Oversprung two arm brass balance. White balance spring. Regulator index to scale in top plate.
Diameter 51mm

Plate 43
Moon Phase/Calendar/Chronograph Stop Watch, Henri Berney Blondeau, Switzerland

Base metal open face case, snap-on bezel and back. White enamel dial inscribed Henri Berney Blondeau Horloger upper centre, SWISS MADE lower edge, centre seconds chronograph with ⅕ markings and arabic numerals, roman hour markings, inset subsidiary dials. Top plate inscribed 17 JEWELS SEVENTEEN UNADJUSTED UU. Pillar plate stamped 7760. Keyless wound, pull winder to position 1 clockwise to advance calendar and anti-clockwise to advance moon phase 2 set hands. 17 jewels. Straight line lever movement. Oversprung monometallic balance. Regulator without index. Incabloc suspension.
Diameter 47mm

Valjoux ébauche, Valjoux SA Les Bioux, Switzerland.

368

Novelty Watches

PLATE 42A

PLATE 42B

PLATE 43A

PLATE 43B

369

Novelty Watches

PLATE 44A

PLATE 44B

PLATE 44
Calendar Pocket Watch – Medana, Swiss

Nickel chromium plate dress watch with engine-turned design to back. Dial inscribed MEDANA upper centre, SWISS MADE lower edge, hour markers, pointer hand, revolving disc seconds dial lower centre, calendar dial upper centre. Top plate stamped SWISS MADE. Keyless wound and set. Three-quarter plate Swiss cylinder movement. Regulator index on balance cock engraved SF RA.
Dia 48.3mm

COLOUR PLATE 27 (page 12) and PLATE 45
Cigarrillos Excelsior, Courvoisier Frères, Chaux de Fonds, Switzerland

Gilt open face case, hinged dome and cuvette, snap-on bezel. Engine-turned minute sector and digital hour dial, inscribed CIGARRILLOS EXCELSIOR with trade mark in 'shield', subsidiary seconds dial, arabic numerals, blued steel arrow minute and seconds hand, subsidiary seconds dial with blued steel hand. Top plate stamped + 27838. Keyless winding with recoil click. Setting with push piece at 23 minute position. 7 jewels. Club foot Swiss lever bar movement with straight line lever escapement. Uncut bimetallic balance. Blued steel oversprung and overcoil balance spring. Regulator index on small balance cock engraved AR.
Diameter 50mm

Swiss Patent No 27838, Gabriel Lopez Mantaras a Saragasse (Spain), 5 May 1903. Modernesto Marque deposé No 16713, 26 Dec 1903, Courvoisier Frères, Chaux de Fonds, Switzerland.

PLATE 46
Kinoba Lever, Ed Kummer SA, Bettlach, Switzerland

Nickel plated case, snap-on bezel, hinged dome. Floral decoration silver surround to eccentric dial in the upper half of dial, inscribed KINOBA LEVER AI QUALITY centre and SWISS MADE lower edge, black arabic numerals, blued steel spade hands. Inset visible balance lower half dial. Keyless wound and set. Unjewelled. Pin pallet. Roskopf type movement stamped on top plate M SWISS MADE.
Diameter 49mm

Novelty Watches

PLATE 45A

PLATE 45B

PLATE 46A

PLATE 46B

371

Novelty Watches

PLATE 47

PLATE 48A

PLATE 48B

PLATE 47
Balancier Visible, Swiss

Chrome on brass open face case, snap-on bezel, body and back of case in one piece, engine-turned pattern to a clip-on silver dial. Inscribed BALANCIER VISIBLE upper centre and SWISS MADE lower centre. Visible imitation compensation balance. Arabic numerals, blued steel spade hands. 4 jewels. Cylinder escapement.
Diameter 46.5mm

PLATE 48
Visible Balance, Swiss

Oxydised steel open face case, snap-on bezel, body and back of case in one piece stamped ACIER GARANTI. White enamel dial inscribed SLOW FAST lower edge, black roman numerals, gold minute markings, blued steel narrow spade hands. Keyless wound, push piece to set at 4 minutes past position. Swiss club tooth straight line lever with uncut bimetallic balance. Top plate inscribed ANCRE LEVEES VISIBLES DOUBLE PLATEAU 15 RUBIS BREVETE SGDG 22280 16598 BREVET +.
Diameter 50.4mm

Swiss Patent No 16598, 4 May 1898, Ernest Degoumois à St Imier, Switzerland. Montre de Hauteur minime avec balancier visible. Swiss Patent No 22280, 26 April 1901, Ernest Degoumois à St Imier, Switzerland. Raquette pour montre avec balancier visible. A similar watch is illustrated in the watch catalogue of Joseph Brown & Co, Chicago, Illinois 1907.

PLATE 49
Visible Balance, Swiss

Oxydised steel open face case, snap-on bezel and back cover. White enamel dial inscribed A R lower edge, black arabic numerals, gold minute markings, copper Louis style hands. Keyless wound and set. Pin pallet lever movement. Single jewel to balance arbor. Top plate stamped BREVET + 27692 BREVET SGDG.
Diameter 48.2mm

Swiss Patent no 27692, Onesime Favret à Tavannes, Switzerland.

Novelty Watches

PLATE 49A

PLATE 49B

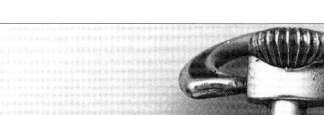

PLATE 50A

PLATE 50B

PLATE 50
Visible Balance – Amida, Swiss

Gilt brass open face case, octagonal faceted snap-on bezel, hinged engine-turned back. Inner back cover inscribed BREVET 8 DEPOSE MEDAILLIONS REMONTOIR VISIBLES ANCRE GARANTIE INTERCHANGEABLE SWISS MADE and seven reproduced 'medals'. Silvered dial offset to upper half of movement with visible monometallic balance below, dial inscribed AMIDA FAÇON 8 JOURS SWISS MADE RS AF lower edge, gilt spade hands. Keyless wound, push piece to set. 30 hour single jewel Roskopf type movement. Top plate stamped BREVET + 118260 + 118261. **Diameter 48.7mm**

This is an imitation 8 day watch in the style of Hebdomas. Swiss Patent No 118261, Baumgartner Frères SA, Grenchen, Switzerland, Demande 8 Août 1926. Publié le 16 Dec 1926.

373

Novelty Watches

PLATE 51
Commemorative – Field Marshal Sir John French, Swiss

Open face case with decorative overall pattern, snap-on bezel, hinged back with portrait of Field Marshal together with FIELD MARSHAL SIR JOHN FRENCH in raised lettering. White enamel dial, black arabic numerals, blued steel hands, subsidiary seconds dial. Keyless wound, pull to set hands. Unjewelled cylinder movement stamped M.

Field Marshal Sir John French 1st Earl of Ypres PC, KP, GCB, OM, GCVO, KCMG (1852-1925) was Commander-in-Chief of British Forces in France, First World War 1914-1915, Commander-in-Chief Home Forces 1915-1918.

PLATE 51A

PLATE 51B

PLATE 51C

Novelty Watches

PLATE 52A

**COLOUR PLATE 28 (page 12) and PLATE 52
Commemorative – Joffre/Kitchener, Swiss**

Silver open face case. Silvered dial with green hour circles, yellow arabic numerals and green dots to mark minute divisions, centre illustrated with portraits and inscribed Joffre and Kitchener SOUVENIR DE LA 1914-1915 GRANDE GUERRE, steel hands. Keyless wound. with push piece at 11 o'clock position to set hands. Roskopf pin pallet escapement.
Diameter 54mm

Joseph Jacques Césaire Joffre (1852-1931), Commander-in-Chief French Forces from 5 August 1914 and Supreme Command 3 December 1915. Horatio Herbert Kitchener, Earl (1850-1916), British Field Marshall and Statesman, 6 August 1914 took over Seals of War Office as British Commander-in-Chief. It is interesting to note that the Great War of 1914-1918 was only expected to last one year in the opinion of the watch manufacturers!

PLATE 52B

Novelty Watches

PLATE 53A

PLATE 53B

PLATE 54A

PLATE 54B

Novelty Watches

PLATE 55

PLATE 56

PLATE 53
24 Hour Dial, Swiss

Silver open face case, snap-on bezel, hinged back and cuvette stamped LF within diamond with newt between letters. White enamel dial, large black arabic 1-24 (ie am and pm), smaller arabic red 1-12 pm, narrow copper spade hands, small inset seconds dial. Keyless wound and set. 15 jewels. Swiss club tooth movement with straight line lever. Cut bimetallic balance. Blued steel balance spring. Regulator index on balance cock.
Diameter 50mm

PLATE 54
Advertising – Sunlight Soap, American

Silver engraved open face case, snap-on bezel, hinged dome and cuvette, hallmark Birmingham **1890**. Casemaker AB (Alfred Benson, 46 Terrace Road, Handsworth, Birmingham, registered Birmingham Assay office 12 March 1879). White enamel dial inscribed "SUNLIGHT SOAP" AWCo WALTHAM MASS SAVES TIME & LABOUR, black arabic numerals, narrow blued steel hands. Keyless wound, pull to set. Top plate stamped AM WATCH CO WALTHAM MASS ROYAL 4443680 SAFETY PINION. 7 jewels. Three-quarter plate club tooth escapement with straight line lever escapement. Oversprung with overcoil. Cut bimetallic balance. Blued steel balance spring. Regulator index on engraved balance cock engraved FS.
Diameter 40mm

'Sunlight' soap was introduced by WH Lever in 1884 and was the first soap to be offered as a pre-packed and branded product. This was at a time when grocers purchased their soap in 4lb bars which they cut up and wrapped as required. Sunlight's success led to the building of the world's largest soap works at Port Sunlight, Cheshire.

PLATE 55
'Watch' Key, Swiss

Chrome open face key form case, glazed bezel to back and front. Silvered dial, arabic numerals 2, 4, 6, 8, 10 and 12 with radial bars at alternate positions, blued steel pointer hands. Keyless wound and set. 13 jewels. Swiss club tooth and straight line lever wristwatch movement. Monometallic balance with blued steel oversprung balance spring. Small regulator index on balance cock.
Diameter 24mm

COLOUR PLATE 29 (page 12) and PLATE 56
Cigarette Lighter, Swiss

Brass engine-turned case. Flint petrol lighter fitted with watch movement. Gilt matt dial inscribed BEACON PARKER centre and SWISS MADE lower edge, arabic numerals, cathedral style hands. Keyless wound and set. Swiss wristwatch movement. Stamped on base LIGHTS IN A FLASH LICENCED ENGLISH PAT 143752.
Size 46.5 x 35 x 15.3mm

Patent No 143752 applies to the lighter design. Application date 17 July 1919. Complete specification left 20 Sept 1919. Complete accepted 3 June 1920. Willey Greenwood and Frederick C Wise, scientific instrument maker – patent for inclusion of a wrist watch in a cigarette lighter. Patent No 286838 was taken out for a similar concept in 1927.

Novelty Watches

PLATE 57

PLATE 57
Cigarette Lighter, Swiss

Black stove enamel open face case, gold anodised cap and bezel. Silver dial with radial rays and inscribed BENDIX 17 JEWEL, black chapter divisions, gilt pointer hands, centre seconds. Keyless wound and set. 17 jewels.
60mm x 30mm x 12.5mm

PLATE 58
Political Watch, Swiss

Nickel open face case, hinged back of case depicting map of India with Mahatma Gandhi (Mohandas Karamchand Gandhi), Hindu leader, shaking hands with Sankat Ali, Muslim leader. White enamel dial, subsidiary inset seconds dial, black roman numerals, blued steel spade hands. Keyless wound with push piece at XI o'clock position to set hands. Roskopf pin pallet movement. Inscribed on barrel bridge REPASSE EN SECOND. Tangential pallet adjustable depthing.
Diameter 45mm

A further version of this watch with identical case but without the seconds dial and with eleven portraits of political activists and

PLATE 58A

PLATE 58B

Gandhi's spinning wheel (considered by many to be the symbol of India) marking the twelfth position. See NA Bulletin Number 207, August 1980, pp376-378.

CHAPTER TEN

Repeating Watches

The production of a watch with repeating work was a natural desire following the successful use in 1675 of repeating work in clocks. This had itself been prompted by the advent of greatly increased accuracy of timekeeping in clocks, following the use of the long pendulum and anchor escapement. Naturally there was a desire to perceive this by night as well as by day. History records that Edward Barlow commissioned Thomas Tompion to make a repeating watch and indeed applied in 1675 to King James for the grant of patent. The patent was awarded instead to Daniel Quare who was able to demonstrate to the King a watch that incorporated additional features. Subsequently repeater watches were in demand but the first minute repeater did not appear till 1830. Usually the repeat mechanism was activated by pushing the pendant but after 1850, with the increasing use of the keyless mechanism, a slide in the band of the case became necessary. Daniel Delander produced a repeater watch with a slide mechanism in 1725.

Various mechanisms were used to control the rate of striking of the watch – bells at first and later a gong which enabled a thinner watch to be made. Early examples relied on the use of a variable depthing of pinion in the strike train by use of an eccentric bush and also a striking escapement akin to the ordinary alarm clock striking train escapement. However, an important development was that of Charles Barbezat-Baillot who introduced the centrifugal governor (UK Patent No 5766 1899), the outstanding feature being that the regulation was achieved almost silently in contrast to the use of a noisy recoil type escapement.

The repeating watch has much appeal and fascination for us all, but the motto *Appetetus Rationi Pareat* (Let your desires obey your reason) should be borne in mind when a tempting purchase is afoot. Treat with great caution a watch that is not working correctly in all respects. Get the advice of an expert, but the following works will be of considerable help: *Complicated Watches and Their Repair* by Donald de Carle; *A Guide to Complicated Watches* by François LeCoultre; 'A History of Repeating Watches' by Francis Wadsworth (*Antiquarian Horology* Sept/Dec 1965 and Mar/Jun 1966).

5766. Barbezat - Baillot, C. Jan. 8, [*date claimed under Sec. 103 of Patents &c. Act, A.D. 1883*].

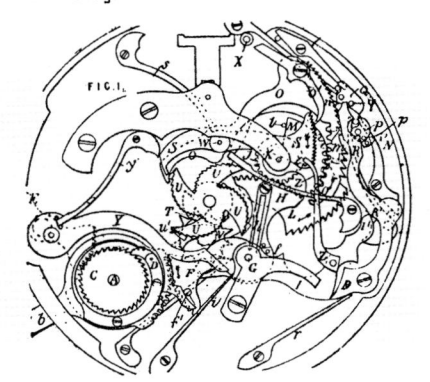

Repeating-mechanism for watches. The striking-spring is coiled round an arbor A, and is wound up by means of a push-lever b projecting from a loose disc B bearing a click D which gears with a fast ratchet-wheel C. If preferred, the disc B may gear with a toothed sector E operated by a push-pin. When the spring is allowed to unwind, the striking is occasioned, as follows:—The toothed sector B drives, in the direction of the arrow, a sector F pivoted at G. A pin f on the sector F engages a spring fork h on a concentric sector H, and takes it forward. The sector H drives a sector J fast on the arbor K of the rack L, which trips the hour pallet N. A pin l on the sector L pushes round a sector M, which brings the quarter and minute rack O pivoted at W from the quarter snail T. The quarter teeth m trip the pallets P, Q, and the minute teeth n the pallets Q. The number of minutes struck is determined by the rack S, which is drawn from the minute snail U by a pawl O¹ on the rack O. The racks are initially released, and allowed to set themselves to their snails, as follows: After turning the rack H to the hour snail, the pin f escapes from the fork h and shifts an arm I, the end I¹ of which trips a retaining-piece R which, by pressing against the pin p, q, holds the pallets P, Q. At the same time a spring Z on the arm I leaves a pin z on the rack S. The parts released are moved by springs r, o, s. The arm Y, with spring y and the pin f¹ on the sector F, serves to jerk up the surprise pieces u¹ of the minute snails. The striking is controlled by a special governor, consisting of a rotary cross, two arms of which carry loaded wings, that fly out against a friction ring, while the other two arms have springs which tend to restrain the wings.

PLATE 1. 1889

Repeating Watches

PLATE 2A

PLATE 2

Patek Philippe, Rue du Rhone 41, 1211 Genève 3, Switzerland

Gold (18 ct) open face case (case no 262456), snap-on bezel and hinged back and cuvette. Enamel dial inscribed PATEK, PHILIPPE & Cie GENEVA SWITZERLAND, small seconds dial, Breguet style arabic numerals, Louis hands. Keyless winding, pull to set. PATEK, PHILIPPE & CO NO 138345 GENEVA SWITZERLAND engraved on barrel bridge. WRIGHT KAY & CO DETROIT engraved on winding work plate. EIGHT (8) ADJUSTMENTS THIRTY ONE (31) JEWELS engraved on 3rd wheel bar. 31 jewels straight line Swiss lever movement with club tooth scape wheel. Oversprung and overcoil blued steel balance spring to cut compensated bimetallic balance. Index regulator on balance cock engraved FAST SLOW.

Diameter 46mm

Maitres Horlogers à Genève depuis 1839. Minute repeater of first quality, manufactured 1909 and sold by Patek Philippe to Wright Kay & Co, Detroit, in August 1910.

PLATE 2B

380

PLATE 3A

PLATE 3
Anon

Silver (0.935) open face case, snap-on bezel, gold hinges to back and cuvette Gold slide and olivette. White enamel dial, black roman numerals, blued steel spade hands, inset sunk seconds dial. Keyless winding with push piece to set hands. Pillar plate stamped 3524 with Swiss patent cross beside balance cock. Three-quarter plate. 15 jewels. Swiss tangential lever movement with pointed tooth scape wheel. Oversprung and overcoil blued steel balance spring to cut bimetallic compensated balance. Regulator index on balance cock.
Diameter 53.3mm

PLATE 3B

CHAPTER ELEVEN
Pocket Alarm Watches

The alarm pocket watch is not a recent innovation. It has been with us for more than four hundred years and there are many references in literature to examples dating from as early as the mid-sixteenth century.

For example, EJ Wood, although admittedly not the most reliable of sources, does record in his book *Curiosities of Clocks and Watches* that Lady Fellows had a circular gilt pedestal alarum-watch striking on a bell with the initials HSTA and date 1581 inside. He further records another inscribed 'From Alethea, Countess of Arundel for her dear son, Sir William Howard, KB 1629'. 'It is of oval form, two inches and a-half in diameter, and one inch and a-half in thickness. It strikes the hours, has an alarum, shows the day of the week… On the inside is a Roman Catholic calendar with the date 1613 and the maker's name, P Combret, A Lyons.' Tardy's *Dictionnaire des Horlogers Français* lists Pierre Combret II of Lyon, born in Engleton and known to have been working in Lyon in 1581. He died in 1622. Furthermore a very fine pierced and engraved silver alarm watch with champlevé dial by Markwick c.1685 is shown in *The Camerer Cuss Book of Antique Watches*.

Throughout the seventeenth and eighteenth centuries many travelling clock watches made in this country by such makers as Edward East, Markwick and George Graham incorporated alarm work. They were also popular with the Continental makers. Naturally over the centuries there were refinements to the original basic mechanisms. One of the most significant patents was that taken out by Joseph Anthony Berrollas of Coppice Row (Cowpers Row), Clerkenwell who in his UK patent No 3342 26 May 1810 (see Plate 1) described 'A warning watch on a new construction'.

Berrollas' specification

A warning watch on a new construction. The patentee says 'The inside of the movement is not different from a common watch, accepting a barrel, which is fixed with two screws on the underside of the top plate, as near to the mainspring as possible. The arbor of the side barrel is made in the same manner as a clock watch, has a brass wheel with sixty teeth, with a steel wheel fixed to it, which steel wheel has thirty-three teeth, cut like a ratchet, which cause the hammer to act. This hammer is placed between the main and warning barrels, and the side hammer strikes on a bell spring, which bell spring is fixed with two screws on the pillar plate. The spring in the warning barrel is wound up five times, which occasions the hammer to give one hundred and sixty-five knocks on

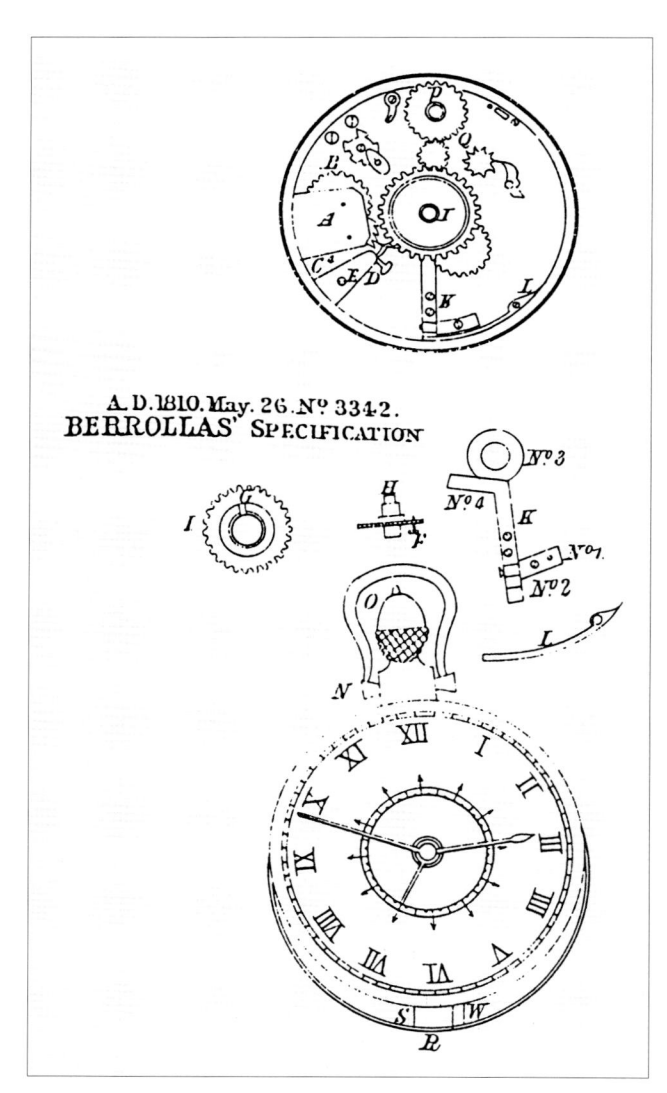

PLATE 1. Berrollas' specification

> **A.D. 1827, April 28.—Nº 5489.**
>
> BERROLLAS, JOSEPH ANTHONY.—"A detached alarum watch. The inventor lays claim:—1. To the detent, which is an oblong steel plate, spring tempered, and having a hole in its center, through which the canon pinion passes. A pin is so affixed to the detent as to slide along a plate fixed on the under side of the alarum wheel, and while doing so keep one end of the detent elevated. There is, however, a notch on the said plate into which the pin falls once in 12 hours.
>
> 2. To the elevator, which is a thin steel spring, screwed down at its thickest end, so that its other end is elevated.
>
> 3. To the propeller, which is a steel lever, having a circular head at one end, and working on a pivot at the other, and also having a projecting part which is formed into an inclined plane, which passes under the elevated end of the elevator.
>
> 4. To the locker, which is a cylindrical piece of steel, one end being smaller than the other, and projecting through a hole in the rim of the case, the larger end acting between two pins on the plate. A spring passes over the small part, serving to hold it down, and at the same time pressing against the propeller and pushing it inwards.
>
> 5. To the locking lever, which has two pins, one of which presses against an arm of steel on the arbor of the fly pinion, and prevents the alarum from running down; the other is worked on by the small end and is called the tail, and is for the purpose of discharging the alarum.
>
> The action of the above machinery is as follows:—When the detent spring is elevated, it presses down the elevated end of the elevator, which in its turn presses on the inclined plane, and thus presses the circular head of the propeller outwards, and causes the small end of the locker to protrude from the case. When the pin of the detent falls into the notch aforesaid, the whole thing is released from pressure, and the tail of the lever discharges the alarum.
>
> [Printed, 6d. See Repertory of Arts, vol. 5 (*third series*), p. 67; London Journal (*Newton's*), vol. 2 (*second series*), p. 84; Register of Arts and Sciences, vol. 1 (*new series*), p. 123; and Engineers' and Mechanics' Encyclopædia, vol. 1, p. 700.]

PLATE 2. Berrollas' further Patent

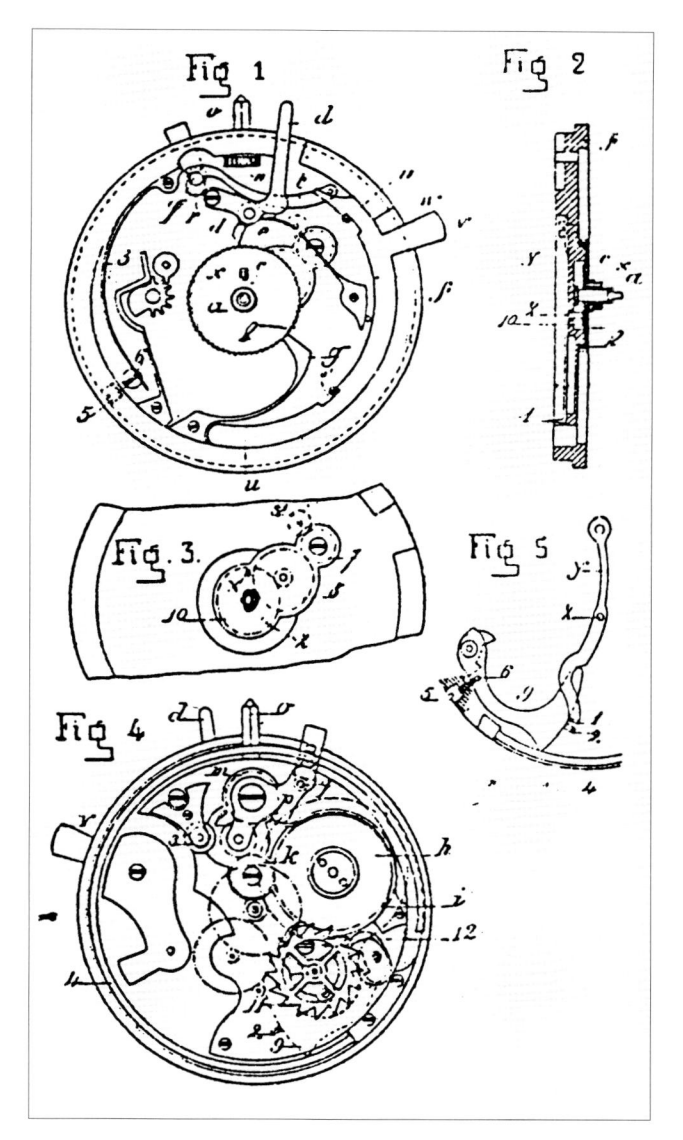

PLATE 3. Diagram from Monnier and Frey Patent

the bell spring. Opposite the hammer is a pinion with six teeth, which act in the arbor wheel; this pinion is planted on one side of the upper plate, and on the other in a bar in the back of the pillar plate. On the side pinion is a wheel with forty-five teeth, which wheel acts in a pinion with six teeth, planted in the said bar on one side, and in the pillar plate on the other. On the said pinion is a wheel with twenty teeth, like a ratchet, which acts in a pallet. These form the warning parts, the said pallet being the one acted on by the detent of the warning.'

He took out a further patent (No 5489) 28 April 1827 on a detached alarm watch, ie a lever mechanism operated the alarm (see Plate 2).

New developments in the evolution of the pocket watch, especially those related to changes in the winding mechanism and the setting of the watch, brought new challenges. At the same time, however, they opened up further opportunities for the more inventive makers to capitalise on a market that was always receptive to a new product. For many the alarm was a novelty feature and therefore desirable, but at the same time it had to be closely related to price. Many makers therefore sought to use a single power train to perform both time and alarm function and achieved this with clever use of the stopwork mechanism. One example of this was that described in the Swiss Patent (No 10461, 13 June 1895) taken out by Monnier & Frey of Bienne for a 'Montre Reveil Perfectioneé' (see Plate 3). The patent specification describes a method of using a single barrel, an alarm mechanism and means of setting the alarm dial.

The following year (18 July 1896) Wilhelm Voland of Bienne applied for a patent (No. 12660) for 'Montre de poche à reveil matin et avertisseur avec un seul barillet' – again utilising a single barrel (see Plate 4).

Watches made to either of these specifications are rarely seen – they probably priced themselves out of the market. The general quality was too high and the consequential high cost restricted sales except to an interested wealthy minority. It was left to Junghans, Schramberg, in the Black Forest area of Germany to achieve with their J5 movement (1906) what their predecessors had failed to do – manufacture an alarm watch at a price to suit the popular market. This model was subsequently upgraded to a seven jewel movement (J10) which was introduced in 1911 with a further model (J36) in 1924. The last mentioned was an attempt to keep their place in the world market in the years of depression following the First World War. Thiel (Ruhla, Germany) had also by energetic marketing brought the Champion Thiel Alarm watch to a wide market at a price the ordinary man could afford.

The Swiss, however, were not content to sit by. Nathan Weil, Chaux de Fonds, had already taken out Swiss patents and German patents (Brevet No 28106 and DRP 150253) on 7 May 1903 for a well-designed alarm pocket watch marketed under the trade name FANFAR. This was in a good solid nickel case (bearing in mind that initially nickel was an expensive 'new' material and not considered at this date to be in any manner an inferior metal). Eterna (Fabriques Eterna, Schild Frères & Co, Granges (Soleure), Switzerland) made an 18 ligne, calibre 52 under the Swiss patent (No 42203) taken out 11 February 1908 and on 29 April 1909 Schweigerische Eidgenossenschrift Patent schrift Nr 47816 Gustav Hausler of Hannover, Germany. They appear to have made the alarm a speciality using a 13 ligne calibre 68, an 18 ligne calibre 52, a 19 ligne calibre 805 and a 20 ligne calibre 51. A Schild produced the 'Konigen' calibre and at least five 19 ligne movements both with lever and cylinder escapements.

The first watch factory in the Canton of Soleure was The Langendorf Watch Co. Founded in 1873 by Johann Koffmann, they produced unfinished movements. Starting with sixty employees, the work-force had grown by 1880 to 970 and had become one of the most important rough movement factories in the world. A close collaborator of Johann Koffmann was Lucien Tieche who in 1890 took over the management of the firm till 1899 when Ernst Koffmann took control. He was succeeded by his brother Rudolf in 1944. By 1948 the firm had grown to 1,200 employees. The economic problems of 1880-1900 had wrought havoc in the Swiss horological industry generally, but Langendorfs were able to survive by hard work and strenuous efforts to market their products abroad. Having an eye to any possible new marketing feature, they took out Swiss patent 41794 27 November 1907, 'Mechanism d'arrelage pour montres-reveils' and were soon able to market an 18 ligne calibre 500 keyless alarm pocket movement with 4 jewel cylinder escapement. This movement was still being illustrated in Georg Jacob Catalogue Leipzig (1949) and Flume (1947) and is easily recognisable by the flat spiral alarm stopwork in the top plate of the movement (Plate 5).

It is interesting to note that Joseph Brown & Co Chicago catalogue 1907 illustrated an alarm watch which appears to be identical!

Longines, St Imier, were not to be left out. They competed with a quality two train alarm watch calibre

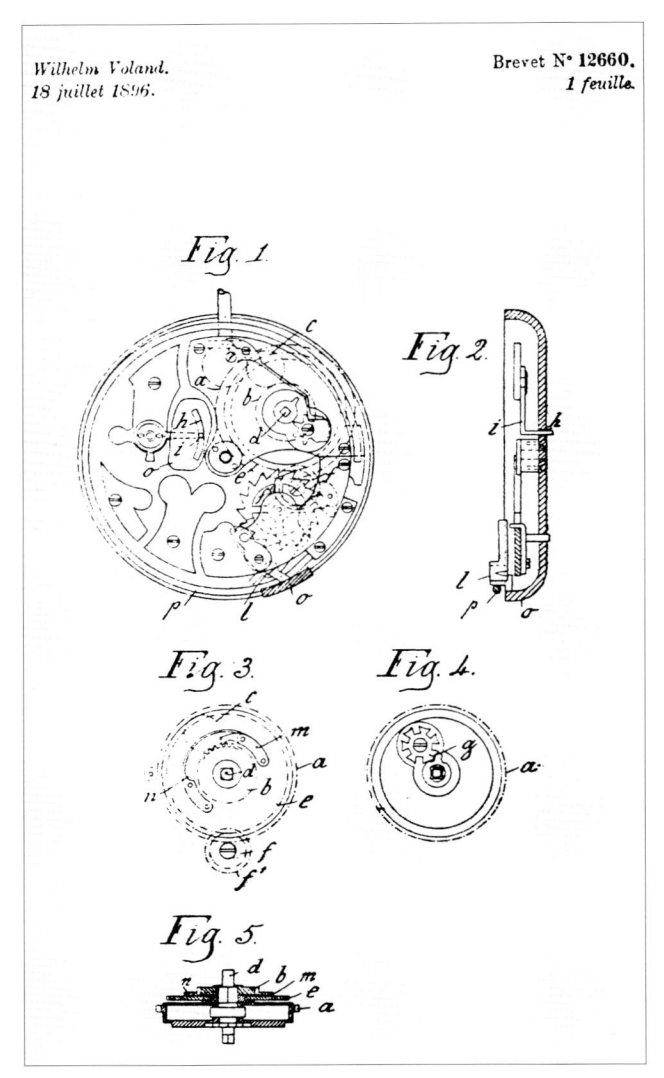

PLATE 4. Diagram from Voland Patent

19.65 produced c.1920 while Zenith (Le Locle) were at the same time advertising a two train quality alarm pocket watch. A newcomer to the field was Baumgartner Frères, Grenchen, Switzerland (founded 1899), registering the following patents:

Brevet 112427 demande 3 March 1925 for 'Mouvement de montre-reveil à un seul barillet'.

Brevet 124160 demande 10 Feb 1927 for 'Mouvement de montre Roskopf à montre de longue durée'.

Brevet 227383 demande 7 September 1942 for 'Mouvement de montre reveil à un barillet' (which appears to be detail modification of 112427).

This company was apparently also supplying a 19 ligne calibre 222 4 jewel pin pallet lever to smaller finishing companies, e.g. Du BARRY WATCH CO. is found stamped on movement of Baumgartner *ébauche*. MENTOR appears to be the registered mark of Bader, Hafner & Holderbank, Soleure, Switzerland again using Baumgartner *ébauche* and SIMPLON pocket alarm, registered trade mark of César Renfar Abrecht, Lengnau près Bienne, used 7 jewel pin pallet movement stamped by Abrecht with the 1942 Brevet 227383 of Baumgartner Frères.

The following list of manufacturers of alarm watches given in *Le Livre d'Or de l'Horlogerie* c.1930 gives some insight into the Montres-reveils/Alarum watches (Alarms)/Weckeruhren.

Audemars, Piquet & Cie, le Brassus.
H Barbezat Bole SA, le Locle.
Elida Watch Co. SA, Fleurier (Neuchâtel).
Fabriques Eterna, Schild Frères & Co., Granges.
Fabrique Excelsior Park, Saint-Imier.
Fabriques Le Phare, le Locle.
Fabriques des montres Zenith, le Locle.
Françillon & Co. SA, fabrique des Longines, Saint-Imier.
S.A. des usines fils de Achille Hirsch & Cie, la Chaux-de-Fonds.
Lemania Watch Co., Lugrin SA, l'Orient et la Chaux-de-Fonds.
Manufacture des montres Doxa, le Locle.
A Schild SA, Granges
SA Vve Chs-Léon Schmid & Cie, la Chaux-de-Fonds.
Société Suisse d'Horlogerie, fabrique de Montilier, Montilier.
Stolz Frères, fabrique Angelus, le Locle.

Journal Suisse d'Horlogerie January 1932 includes an advertisement listing ten major producers active at that date – Dimier SA (George Dimier, Chaux de Fonds), Fabrique Excelsior Park, Saint Imier, Fabrique d'horlogerie de Montilier, Montilier près Morat, Jura Watch Co., Delemont, Liga SA, Soleure, Doxa, Le Locle and Chs Léon Schmid & Cie, Chaux de Fonds, as well as manufacturers previously mentioned.

The variations in design to be found in these watches demonstrate yet again that there is a degree of interest to be found in the later mass-produced watches. In the instance of alarm watches, the manufacturers had not only to find a movement that was reliable but they also had to find one that was best suited to low cost production. As each manufacturer strove for a solution – each with his own characteristic approach – so he provided today's collector with a range of collectable watches.

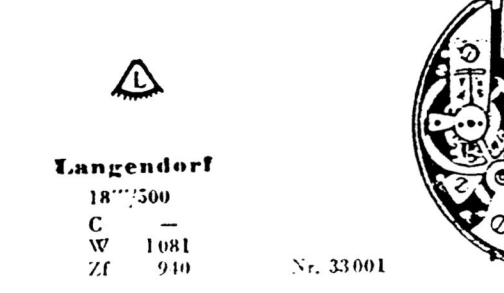

PLATE 5. Langendorf movement taken from 1949 catalogue

PLATE 6A

PLATE 6B

PLATE 6
Charles Edward Viner, New Bond Street, London

Silver open face alarm case hallmarked London **1810**. Casemaker LG. Hinged back and bezel, single case, gilt dome engraved Cha Edwd Viner New Bond Street, London. Round silver bow with gold pin to oval shaped pendant and acorn button. Bull's eye glass. White enamel dial with black roman numerals and black diamond-shaped half hour markers for alarm dial, pale gold spade hour and minute hands, steel pointer for alarm hand. The plate engraved Viner Royal Exchange & New Bond St London. Foot of cock engraved PATENT 135 (movement number). Key winding through the dome for both trains, setting time by key cannon pinion square, alarm set by turning acorn button on pendant. Fusee and chain verge movement with flat steel undersprung balance. Polished and blued steel Bosley regulator to engraved graduations on top plate. Alarm sounding on gong – the lever at the six o'clock position and free when moved clockwise.

Key size 5. Diameter 57mm

There appear to have been two generations bearing the name of Charles Edward Viner spanning the years between 1776-1842 (a firm of this name continuing to 1869). They are listed as watch, clock and chronometer makers at 151 New Bond Street and Royal Exchange as well as branches at 8 Sweeting Alley, etc. Cecil Clutton, writing on the Viners, describes a gold watch hallmarked 1814 (No 317), movement with duplex escapement, half quarter repeating train and alarm – both alarm and repeat using wire gong coiled around edge of case. He also describes a watch (No 350) from the Cuss Collection similar to that shown here but in a full hunter silver case. Mention is made of other examples with duplex escapement and pull wind (No 5011 hallmarked 1848) and Savage two-pin escapement, again with pull wind. The word PATENT on the cock refers to the Berrolas patent relating to the alarm mechanism. UK Patent No 3342 26 May 26 1810, Joseph Berrolas Coopers Row, Clerkenwell. 'Charles Edward Viner and His Time' by Cecil Clutton, Horological Journal *Vol 123 April 1981, pp8-10.* The Camerer Cuss Book of Antique Watches *by T P Camerer Cuss, 1987.*

PLATE 7A

PLATE 7B

PLATE 7
JB Yabsley, 72 Ludgate Hill, London EC

Swiss silver (0.935) open faced case, snap-on bezel, hinged back and dome stamped within F RP & Cie. The white enamel dial inscribed across centre JB YABSLEY LONDON SWISS MADE, black roman numerals and chapter ring with outer red quarter hour chapter ring outside, blued steel spade hour and minute hands, alarm hand gilt, inset and sunk seconds dial. Keyless winding by pendant button, push piece at 4 minutes past position to set hands, alarm set by rotating bezel. Top plate engraved Brevete + 12660. 15 jewel. Half plate lever movement with club tooth scape wheel and straight line pallets. Bimetallic oversprung balance. Blued steel balance spring. Regulator index on small cock engraved Advance Retard. The alarm wind-down controlled by stopwork to retain power for the going train when alarm stops striking on gong. A shut-off lever to silence alarm can be activated at edge of cuvette at 32 minute position.

Diameter 57mm

James Benjamin Yabsley was a manager for JW Benson but established himself in business in 1877 and in 1897 was recorded as chronometer maker at 72 Ludgate Hill, London EC. He died shortly afterwards with the business continuing as JB Yabsley Limited. Brevet Swiss 12660 Class 64 Juillet 1896 Wilhelm Voland à Bienne Suisse Montre du poche à reveil matin et avertisseur avec un seul barillet (single barrel).

PLATE 8A

PLATE 8B

PLATE 8
Anon

Nickel open face case, back cover opens sideways to act as a stand, glazed inner bezel to protect movement. Pin in the door to stop alarm. White enamel dial with arrow to indicate direction, black roman numerals with 5 minute markings in black, arabic numerals outside chapter ring. Keyless movement wound at pendant, alarm set by turning bezel anticlockwise, thus turning alarm pointer trapped under bezel. The blued steel hour and minute hand set by push piece at 4 minute past the hour position. Nickel movement. Stopwork to limit the amount of unwinding by alarm. 6 jewels. Cylinder escapement with three arm oversprung brass balance and blued steel balance spring.
Diameter 50mm

PLATE 9
Gala, Liengme & Co, Switzerland

Oxydised steel open face case, snap-on bezel, hinged back with straight edge at 6 o'clock position to enable back when open to act as stand. White enamel dial, black roman numerals, gilt spade hands with blued steel pointer for alarm. Keyless winding by pendant button with push piece to set time at 3 minute position, push piece at 56 minute position to set alarm. Inside back cover bronzed steel bell covering movement and pierced for regulator index access and SILENCE/REVEIL lever. Club tooth scape wheel with straight line 7 jewelled Swiss lever escapement. Uncut imitation compensated bimetallic oversprung balance. Blued steel balance spring. Regulator index on balance cock engraved F S A R. The top plate engraved + 17447 DRP 104227 GALA. The bell with additional information DRGM 131367 USP 26 XII 99 SWISS MADE Patd 113899 BTE SGDG.
Diameter 54mm

Gala was the trade name used by Liengme & Co SA, Cormoret. Brevet + 17447 30 Aout 1898 Louis Eugène Favre à Cormoret, Switzerland Mechanisme de reveil applicables à des montre de poche et pendules de tous genres (corresponds to UK Patent No 1138 Lake HH (Favre LE) 17 Jan 1899 Alarms, Barrel Ratchets) (see Plate 9D)
Brevet + 20281 12 juin 1900 Louis Eugène Favre à Cormoret Switzerland Mechanisme de remontoir pour montres et pendulettes à reveil-matin (see Plate 9C).

Pocket Alarm Watches

PLATE 9A

PLATE 9B

18,532. Favre, J. Oct. 17.

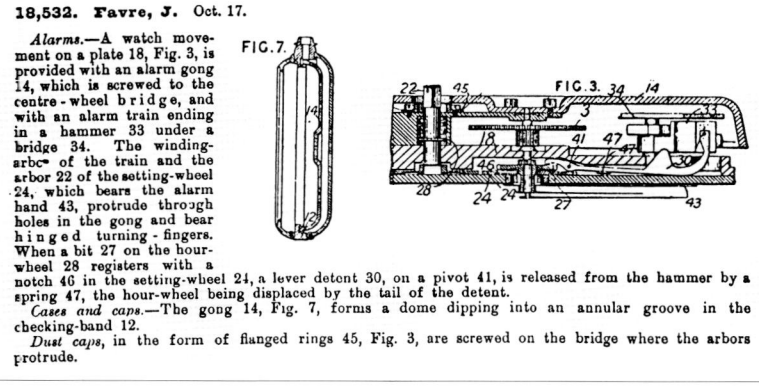

Alarms.—A watch movement on a plate 18, Fig. 3, is provided with an alarm gong 14, which is screwed to the centre-wheel bridge, and with an alarm train ending in a hammer 33 under a bridge 34. The winding-arbor of the train and the arbor 22 of the setting-wheel 24, which bears the alarm hand 43, protrude through holes in the gong and bear hinged turning-fingers. When a bit 27 on the hour-wheel 28 registers with a notch 46 in the setting-wheel 24, a lever detent 30, on a pivot 41, is released from the hammer by a spring 47, the hour-wheel being displaced by the tail of the detent.

Cases and caps.—The gong 14, Fig. 7, forms a dome dipping into an annular groove in the checking-band 12.

Dust caps, in the form of flanged rings 45, Fig. 3, are screwed on the bridge where the arbors protrude.

PLATE 9C. UK Patent Specification Abridgement 1900

1138. Lake, H. H., [*Favre, L. E.*]. Jan. 17.

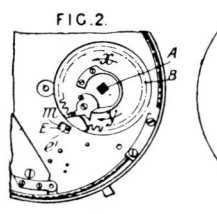

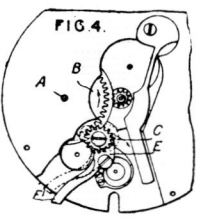

Alarms; barrel ratchets.—Alarm mechanism takes the place of the barrel ratchet, the barrel arbor A being geared through wheels B, C and others (not shown), to the hammer-propelment, which is locked in the usual way until the hour hand overtakes the alarm-hand. The intermediate wheel C is mounted upon a rocking bridge E having a pin e^1, playing in a slot m in the pillar-plate, and a tail E^1, for allowing operation by hand. The barrel wheel B bears a stop-wheel which catches on the head of the rocking bridge. The directions of rotation in running down and in winding respectively are indicated by the arrows x and y. A slide in the band of the case locks or frees the alarm hammer as desired.

PLATE 9D. UK Patent Specification Abridgement 1899

Pocket Alarm Watches

PLATE 10A

PLATE 10B

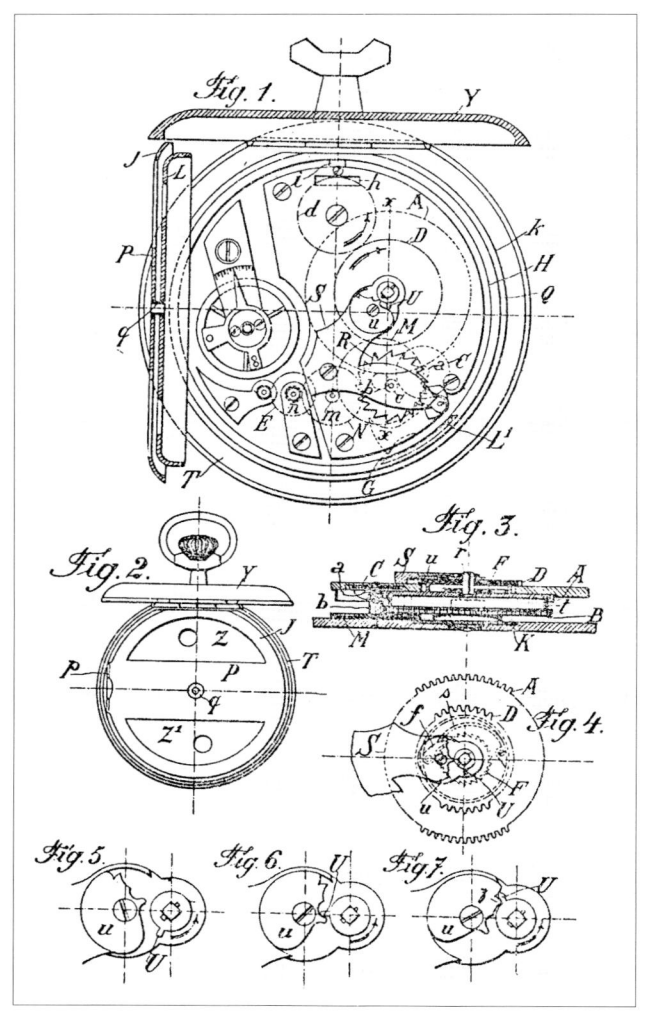

PLATE 10C

PLATE 10
Fanfar, Nathan Weil, Chaux de Fonds, Switzerland

Nickel open face case, snap-on bezel, hinged back to act as a strut clock when open, snap-on cuvette skeletonised to leave a broad central bar to which bell is fastened by a central screw and stamped DRP 150253 Brevet 28106+ Bte SGDG. White enamel dial inscribed FANFAR and arrow indicating direction, black roman numerals, blued steel hands, the minute hand screwed to the centre arbor. Alarm hand set by rotating the bezel anticlockwise. Keyless wound at pendant button with setting time by push piece at 55 minute position. Top plate stamped FANFAR and M BREVET 28106 +. 8 jewel tangential pin pallet lever movement with adjustable depthing to escapement. Imitation compensated and oversprung.
Diameter 58mm

Brevet + 28106 Nathan Weil Chaux Suisse Montre reveil. Fanfar was the marque deposé No 15635 19 mars 1903 by Nathan Weil, Chaux de Fonds.

Plate 11
Reveil Vigor, L. Sandoz-Vuille, Locle, Switzerland

Nickel open face case, snap-on bezel and hinged back with straight lower edge to provide stable support when used standing. A button is provided at 6 o'clock position to release case spring for back cover. Pale lilac enamel dial with arrow to indicate direction. Inset chapter upper centre with gilt roman numerals within ruby red glass discs, REVEIL and VIGOR at the bottom of the dial which is cut away to display the engraved bridge over visible monometallic balance with imitation compensation screws. Keyless wound and set with push piece at the 7½ minute position. Alarm set on alarm dial engraved on the bell that covers the back of the movement and pierced for regulator under at 12 o'clock position. Blued steel oversprung and overcoil balance spring. Brevet + 31781 on the movement under the balance rim.

Diameter 51.3mm

Brevet 31781, L Sandoz-Vuille, Locle, Switzerland 7 October 1904 (stamped 1st August 1905). Mouvement de montre reveil.

PLATE 11A

PLATE 11B

Pocket Alarm Watches

PLATE 12A

PLATE 12B

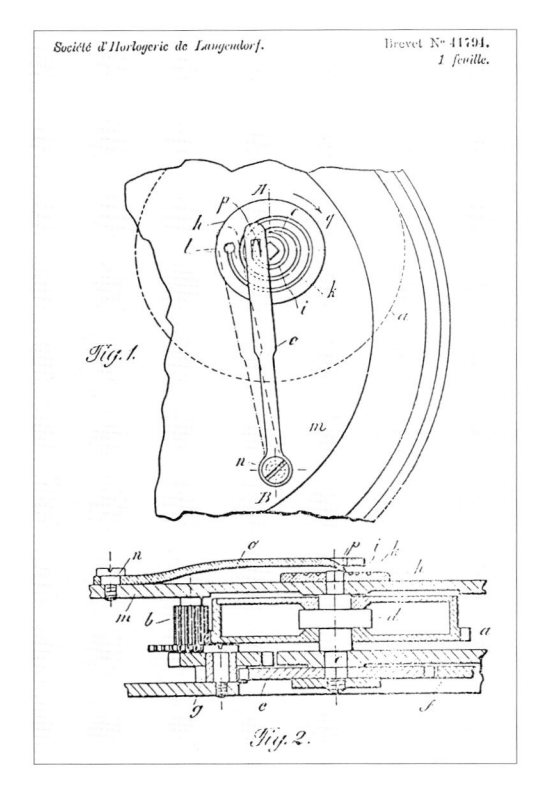

PLATE 12
Lanco, Langendorf Watch Co, Langendorf, Switzerland

Oxydised steel open face case, snap-on bezel with hinged back to act as support. White enamel dial with luminous arabic numerals and fenestrated spade hands, pointer alarm hand set by counter clockwise rotation of bezel. 6 jewel cylinder keyless wound movement covered by bell which is pierced for index regulation and REVEIL SILENCE lever and engraved M Patent 41794. Three arm oversprung brass rimmed balance. Small index on cock engraved SF RA.

Diameter 51.66mm

Lanco was the trade name for watches manufactured by the Langendorf Watch Co, Langendorf, Switzerland. Swiss Patent 41794 27h Nov 1907, Société d'Horlogerie de Langendorf, Langendorf. Mechanism d'arretage pour montres-reveils. This illustrates the snail mechanism to limit the duration of the alarm.

PLATE 12C

Pocket Alarm Watches

PLATE 13A

PLATE 13
Junghans J10, Junghans, Württemberg, Germany

Nickel chrome on brass open case, snap-on bezel with a hinged back of case that opens to act as stand. White enamel dial, luminous arabic numerals, inset seconds dial, luminous skeletonised hands. Keyless winding at pendant, push piece to set alarm at 11 o'clock position and push piece with rocking bar to set hands at 1 o'clock position. Chromed steel bell covering movement and stamped with the post 1890 Junghans 8 pointed star and pierced for index regulation Slow/Fast (Nach/Vor) and alarm on/off lever (Weckt/Still). 11 jewels. Club tooth and straight line lever movement engraved Junghans 18542. Oversprung monometallic balance and regulator index on balance cock.

Diameter 52.4mm

Manufactured by Junghans, Württemberg, Germany. The J10 calibre was introduced by them in 1911 (see Junghans Story Vol 2 *by Karl Kochmann). Movement also illustrated in* Flume Catalogue 1947, *p150 and* George Jacob Catalogue 1949, *p266.*

PLATE 13B

Pocket Alarm Watches

PLATE 14A

PLATE 14
Junghans J5 The Glow Worm, Junghans, Württemberg, Germany

Oxydised steel open face case, snap-on bezel and hinged back to form stand when open, chrome cover stamped with Junghans trade mark (post 1890 design) and inscribed Stop 7 Jewel Alarm MADE IN WURTEMBURG. White enamel dial secured two copper feet inscribed THE GLOW WORM, black roman numerals, fenestrated luminous hour and minute hands, blued steel pointer hands to alarm and seconds dial, luminous dots at each five minute position. Keyless winding by pendant button with push piece at 56 minute position to set alarm anticlockwise and to set hands at 4 minute position. Three-quarter plate Swiss 7 jewelled lever movement, 4 teeth removed from winding wheel to limit winding and double tooth to prevent unwinding fully after alarm sounds to retain power for time train. Club tooth scape wheel. Straight line pallets. Balance spring with imitation compensation screws. Balance spring stud under balance cock. Red brown oversprung balance spring. Index on balance cock. J5 Alarm movement introduced by Junghans 1906.

Diameter 58.9mm

UK Patent No 10327, Junghans O 28 April 1911 (see Plate 14B) for means of illuminating dials (by means of luminous material).

10,327. Junghans, O. April 28.

Illuminating, means for.—In a dial provided with dots or panels of luminous material, preferably containing radium, covered with pieces of opal glass or the like, the dots &c. *a* are arranged in the form of a ring at a short distance from the centre of the dial, and the hands *d* are also provided with the luminous substance *e* as far as the ring.

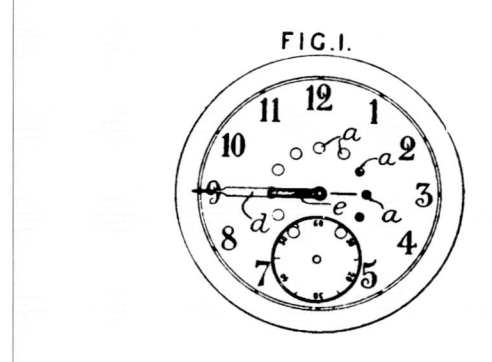

FIG. I.

PLATE 14B

394

Pocket Alarm Watches

PLATE 15A

PLATE 15
Renova, Fabrique Renova Henri Dalcher, Locle, Switzerland

Oxydised steel open face case, snap-on bezel and hinged back to form stand. White enamel dial with inset alarm dial 1-12 upper centre with RENOVA and inset seconds dial lower centre. Name of retailer Finnigans below centre. Skeletonised arabic numerals and fenestrated hands (no sign of luminous paint). Keyless wound, pull to set hands at pendant with push piece 11 o'clock position to set alarm, slide at one o'clock pushed anticlockwise for time winding and hand set and pushed clockwise for alarm winding. Half plate 15 jewel Swiss lever movement stamped RENOVA DRP (Deutsches Reiches Patent) Bte SGDG (Brevete Sans garantie du gouvernement) BREVET + (Patent Swiss). Striking on gong. Monometallic balance with imitation compensation screws. Index on balance cock engraved FS AR.
Diameter 56mm

RENOVA Marque Deposé No 28858, Fabrique Renova, Henri Dalcher, Locle, Switzerland, 9 February 1911. The Renova trade mark was also used by LIGA Watch Factory Ltd, Soleure, Switzerland.

PLATE 15B

395

Plate 16
Racine, Gallet & Co SA, Avenue Léopold Robert 66, La Chaux de Fonds, Switzerland

Oxydised steel open face alarm pocket watch with snap-on bezel and hinged back which serves as stand when watch open. Inside back stamped JG RACINE ACIER GARANT SWISS 1382334. White enamel dial inscribed ALARM and arrow indicating direction, blue arabic hour numerals, small red arabic 5 minute numerals with blued steel hour and minute hands, gilt alarm pointer, inset seconds dial. Chrome bell secured by three screws centrally placed. Stop-alarm lever at bottom of bell which is also pierced for regulator index. Keyless winding with push piece at 11 o'clock to set alarm in an anticlockwise direction, push piece at 5 o'clock position to set hands. 11 jewel club tooth lever movement with straight line lever and oversprung bimetallic balance. Index on balance cock. Movement stamped 1382334 and trade mark PROT within outline of cross.

Diameter 50.9mm

Racine was the trade name of Gallet & Co SA, Avenue Léopold Robert 66, La Chaux De Fonds (see Annuaire d'Horlogerie 1969 p263) *and Racine Fabricant d'Horlogerie, Rue de Rhone, Geneva 1828.*

PLATE 16A

PLATE 16B

Plate 17
Zenith, Le Locle, Switzerland

Silver (.925) open face case hallmarked imported Glasgow Assay Office **1917,** snap-on bezel, hinged back as stand, cuvette. White enamel dial inscribed 61 QUEEN ST KINGSWAY WC SWISS MADE, luminous arabic numerals and fenestrated hands, inset seconds and alarm dials. Push piece 11 o'clock to set hands, push piece at 1 o'clock to set alarm dial upper centre anticlockwise. 15 jewel half plate movement striking on gong with club tooth straight line lever escapement. Bimetallic oversprung balance with blued steel balance spring. Movement inscribed ZENITH SWISS MADE.

Diameter 48mm

London Illustrated News *Vol 153, No 4136, 27 July 1918, p114*. Horological Journal *December 1922, p74*.

Plate 17a

Plate 17b

Pocket Alarm Watches

PLATE 18A

PLATE 18
Gebrüder Thiel, Ruhla, Germany

Nickel open faced case with snap-on rotating bezel and hinged back cover, the bezel and back cover with beaded edges and the back cover designed with flat at the six o'clock position to provide a stable support when the watch is on standing display. White enamel dial with arrow indicating direction, luminous arabic numerals, fenestrated hour and minute hands, inset seconds dial. Keyless wound at the pendant with push piece to set hands at 4 minute position, the alarm dial hand set by rotating the bezel in an anticlockwise direction. 4 jewel pin pallet lever movement, undersprung monometallic balance with imitation compensation screws. The regulator index on balance cock engraved RA SF. Alarm striking on bell which also acts as cover for the movement and is pierced for regulator adjustment and alarm silence lever.
Diameter 49.2mm

Original watch paper with instructions for use. 'Winding. Hold the watch in your left hand, face up and turn the winding cover away from you, similar to the way you would turn a screw. To set the Hands: Depress the small projecting pin over the Figure 1 as far as it will go and simultaneously turn the winding crown until you get the right time. To set the alarm: Hold the watch in your left hand face up, and turn the milled glass ring in the direction of the arrow on the dial. This will move the gilt hand to the time at which you require the alarm to go off. On no account turn the glass ring in the opposite direction as this will injure the movement. If you require the watch to act as an alarm, open the back of the watch and move the small lever in the slot towards the engraved "Bugler". If the alarm is not required move de small lever in the opposite direction towards the engraved "Bed" which means "Silence".' Sadly the Bugler and Bed are not engraved on this watch! Practical Watch and Clockmaker *March 1928, p24.* Watch and Clockmaker *August 1934, p194 (Champion Alarm Watch – Champion was trade name of Thiel). George Jacob Catalogue 1949 No 39001 p267.*

PLATE 18B

398

Plate 19
Aeonicloc, Baumgartner Frères SA, Grenchen, Switzerland

Nickel on brass case, snap bezel rotating either clockwise or anticlockwise to set alarm hand to time. Hinged back acts as stand when open. Hands set by push piece at 11 o'clock. Paper dial inscribed AEONICLOC, inset seconds dial, arrow style hour and minute hand. The unjewelled pin pallet lever movement covered by steel bell pierced for regulator index and Stop/Alarm lever and stamped BREVET + 112427 BREVET + 124160 SWISS MADE. Movement stamped FV below rim of three arm brass balance.

Diameter 54.9mm

Brevet 112427 Baumgartner Frères SA, Grenchen, Suisse, Demande 3 Mars, Publié 2 Novembre 1925. Mouvement de montre-reveil à un seul barillet. Brevet Swiss 124160, Demande 10 Fevr 1927, Publié 2 Janv 1928.

PLATE 19A

PLATE 19B

Pocket Alarm Watches

PLATE 20A

PLATE 20B

PLATE 20
Konigen, Schild, Switzerland

Nickel open face case, snap-on bezel, hinged back acts as a stand when open. Stamped A SCHILD SWISS inside back cover and repeated on top plate. Enamel dial with arabic black numerals inscribed SWISS. Bell covering movement stamped 15 JEWELS 1 ADJ A SCHILD SWISS and bell pierced for index adjustment. Straight line Swiss lever escapement with oversprung monometallic balance with regulator index positioned over balance rim towards centre arbor with graduates engraved on top plate.
Diameter 51.5mm

Movement illustrated Flume Catalogue 1947 p151 and George Jacob Catalogue 1949 p261.

PLATE 21
Konigen, Schild, Switzerland

Oxydised steel open face, snap-on bezel and hinged back to act as stand. White enamel dial with sunk centre and sunk inset seconds dial, luminous arabic numerals and skeletonised spade hands. Push piece to set time at 56 minutes position and to set the alarm at 4 minute position. Each depression of alarm button adjusts alarm anticlockwise by ¼ hour. The push piece operating case spring to open back of the case also acts as a stand when the watch is open. 15 jewel club tooth lever movement with straight line lever and oversprung bimetallic balance with regulator index pointing towards the centre arbor with graduations on the top plate, SWISS MADE 136256.
Diameter 51.5mm

This ébauche corresponds to a movement illustrated in Georg Jacob Catalogue 1949 and Flume Catalogue 1947 but is jewelled also to the intermedial wheel arbor (15 jewels).

PLATE 22
Du Barry Grand, Du Barry Watch Co, Switzerland

Chrome open face case, snap-on bezel, hinged back acts as stand when open. Black painted dial inscribed DU BARRY GRAND ALARM SWISS MADE, luminous arabic numerals and skeletonised luminous hands with gilt arrow hand for alarm, inset seconds dial. Chromed steel bell covers movement stamped BREVET 124160 and pierced for alarm/stop lever and regulator as well as spring-loaded setting knob which pushes down when back closes. Top plate stamped DU BARRY WATCH CO ONE 1 JEWEL UNADJUSTED. Three arm brass oversprung balance. Pin pallet tangential lever movement stamped DXU. Small regulator index on balance cock.
Diameter 53.8mm

Brevet + 124160, Baumgartner Frères SA, Grenchen, Suisse. Demande deposée 10 fevrier 1927 19h, publié le 2 janvier 1928. Mouvement de montre Roskopf à marche de longue durée. This patent also used in Festiva alarm watch.

Pocket Alarm Watches

PLATE 21A

PLATE 21B

PLATE 22A

PLATE 22B

401

Pocket Alarm Watches

PLATE 23A

PLATE 23
Mentor, Bader, Hafner and Holderbank, Soleure, Germany

Chrome open face case, snap-on bezel, hinged back acts as stand, METAL CHROME stamped inside back cover. Dial cream painted with engine-turned sunray pattern radiating from centre and inscribed upper centre MENTOR 4 JEWELS SWISS MADE, luminous arabic numerals and luminous skeletonised hands with gilt alarm pointer, inset seconds dial. Striking on chromed steel bell (stamped BREVET + 227383) which covers back of movement and pierced for Stop/Alarm lever and regulator index. The alarm set button spring-loaded so that it may be depressed as case closes. Top plate stamped SWISS MADE around oil sink hole of scape wheel. Pin pallet and tangential pallet lever escapement with three arm flat rimmed brass oversprung balance.

Diameter 52.3mm

BREVET + 227383, Baumgartner Frères SA, Grenchen, Suisse. Demande depositer 7 September 1942. Brevet enregistre 15 Juin 1943. Publié 16 Aug 1943. Mouvement de montre reveil à un seul barillet (single barrel). Compare this patent with patent 112427. Meister der Uhrmackerskunst *by Abeler p52*. Georg Jacob Catalogue 1949 p267, Baumgartner 19/222.

PLATE 23B

PLATE 24A

PLATE 24
Simplon, César Renfer-Abrecht, Lengnau, Près Bienne, Switzerland

Chrome open faced case, snap-on bezel and hinged back which also acts as a stand when opened. White enamel dial, SIMPLON upper centre and 7 JEWELS below centre, luminous arabic numerals and fenestrated hands, inset seconds dial. Striking on bell which also acts as movement cover and stamped STOP ALARM R A BREVET + 227383 SWISS MADE and pierced for access stop alarm lever and regulator as well as spring-loaded alarm setting knob which is able to rise as the back cover of the watch is opened. Pin pallet 7 jewel tangential lever movement stamped BREVET + 227383 CESAR RENFER ABRECHT SWISS UNADJUSTED. Brass three arm oversprung balance with small index regulator on balance cock.

Diameter 52.3mm

SIMPLON is the trade mark of C Renfer-Abrecht, Lengnau, Près Bienne, Switzerland. BREVET + 227383 Baumgartner Frères SA, Grenchen, Switzerland. Demande deposé 7 Sept 1942, Brevet enregistre 15 Juin 1943. Mouvement de montre-reveil à un seul barillet (single barrel).

PLATE 24B

Chapter Twelve
Eight-Day Watches

Long duration watches present a challenge to the ingenuity and skill of the watchmaker both in design and execution. Long duration is of little merit if other aspects such as accurate timekeeping are consequentially unacceptable.

Surprisingly it is not just in recent times that this challenge was taken up. Nicholas Gribelin and Abbé Jean de Hautefeuille constructed an eight day watch in 1692 for which they received a Royal Patent in 1693. During the eighteenth century there are examples recorded by Massoteau (Paris), Hagen (Hamburg) and Langlois (Le Mans), to name but a few. Some watchmakers were even more ambitious – Boyer took a Swiss patent out in 1844 for a watch running for thirty-two days, Gontard of France following two years later proposed a duration of fifteen days.

Even when the problems encountered by the presence of a large space taking fusee had been overcome by the application of a going barrel, there still remained the difficulty of maintaining constant power throughout the duration of the running of the watch on one winding.

The most common methods of overcoming the problems encountered when extending the duration of a watch on one winding have been by:

1. The introduction of a second barrel
2. The introduction of an extra wheel
3. The use of an extra long mainspring taking up the width of the case

The English eight day pocket watch is exceedingly rare. One patent taken out in this country was by Robert Westwood (born 1784, died 4 June 1839), watch and clockmaker and jeweller, Princes Street, Leicester Square, London. This was Patent No 5850 awarded on 23 September 1829:

> The wheels and pinions are so arranged as to act under the bezel, that is to say between it and the dial plate, thereby admitting within the limits of a pocket watch of the usual size a maintaining power of sufficient strength with one winding up to keep a vigorous motion in the balance for 8 days or more.

In other words, room for the extra large barrel was made by arranging the train on two planes (see Plate 1).

Robert Westwood was an interesting character. He was born in Alloa in 1785 and apprenticed to Henry Redpath of Stirling. According to John Tyler's articles on Westwood (*Horological Journal,* 1969) and on Horology in Soho *(Clocks* magazine, 1986), Westwood was working at 23 Princes Street, Leicester Square, London by 1823. Later his address was given as 34 or 23 Princes Street and it was at this address that he met his death. Apparently he was a somewhat excitable character – he assaulted his wife, attacked customers and indeed died as the result of violence. His throat was cut by thieves who had broken into his shop during the night. They escaped with about £2,000 worth of stock - a princely sum for the mid-nineteenth century. Watches seen signed with his name have included a pocket chronometer and an alarm watch. The Duke of Sussex (Augustus Frederick 1713-1843, sixth son of George III) purchased one of his patented eight day watches in a gold engine-turned case thus enabling him to advertise 'By Appointment to HRH The Duke of Sussex'. Two of his watches had also been taken on the last Arctic expedition led by Sir William Edward Parry.

The difficulties encountered with the long duration watch are illustrated by the prolonged efforts in the USA by Aaron L Dennison. In 1850 he produced an eight day model (possibly with the assistance of OB Marsh from Howard's factory) but it turned out to be a mechanical failure. Two years later AL Dennison and NP Stratton remodelled this watch as a thirty hour watch and this became a Waltham Standard in 1870. Further attempts were made by the Marsh brothers but only twenty watches were ever completed.

Even though eight day watches have been manufactured by the Swiss in significant quantity, the concentration has always been on the more satisfactory thirty hour watch and therefore in relative terms their eight day watches are uncommon. The most likely found example would be one of the Hebdomas watches with the mainspring and barrel taking up the entire back of the movement and a visible balance. These originated from the design patented in 1889 by Irenée Aubry of Saignelegier (a Canton of the Jura region in Switzerland).

Arthur Graizley refined the original design in order to facilitate production, the name Hebdomas being taken when initially intending to produce a watch running for a duration of fifteen days. He went into partnership with Otto Schild early in 1900 as Graizley & Cie. This partnership was dissolved in 1906 with Schild & Co continuing to produce watches from their factory at La Chaux de Fonds.

Le Livre d'Or de l'Horlogerie for 1925 lists the following manufacturers of eight day watches:

H Barbezat-Bole SA, Le Locle
L. Brandt & Frères SA, Omega Watch Co, Bienne
Fabriques Eterna, Schild Frères & Co, Granges
Fabriques Le Phare, Le Locle
Fabriques d'Horlogerie Thommen SA, Waldenbourg
Françillon & Co SA, Fabrique des Longines, Saint-Imier
General Watch Co., Bienne
Ed Heuer & Co, Bienne
SA des Usines fils de Achille Hirsch & Cie, La Chaux de Fonds
Arthur Imhof, La Chaux de Fonds
Lemania Watch Co, Lugrin SA l'Orient et La Chaux de Fonds
Manufacture des Montres Doxa, Le Locle
Sada, Bienne
A Schild SA, Granges
Société Suisse d'Horlogerie, Fabrique de Montilier, Montilier
Stolz Frères Fabrique Angelus, Le Locle
Wyss Frères, Granges

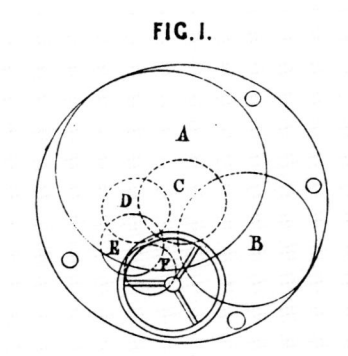

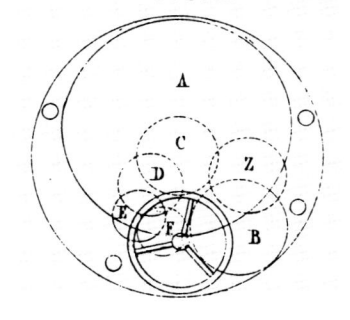

PLATE 1. Westwood's Patent

Eight-Day Watches

PLATE 2A

PLATE 2B

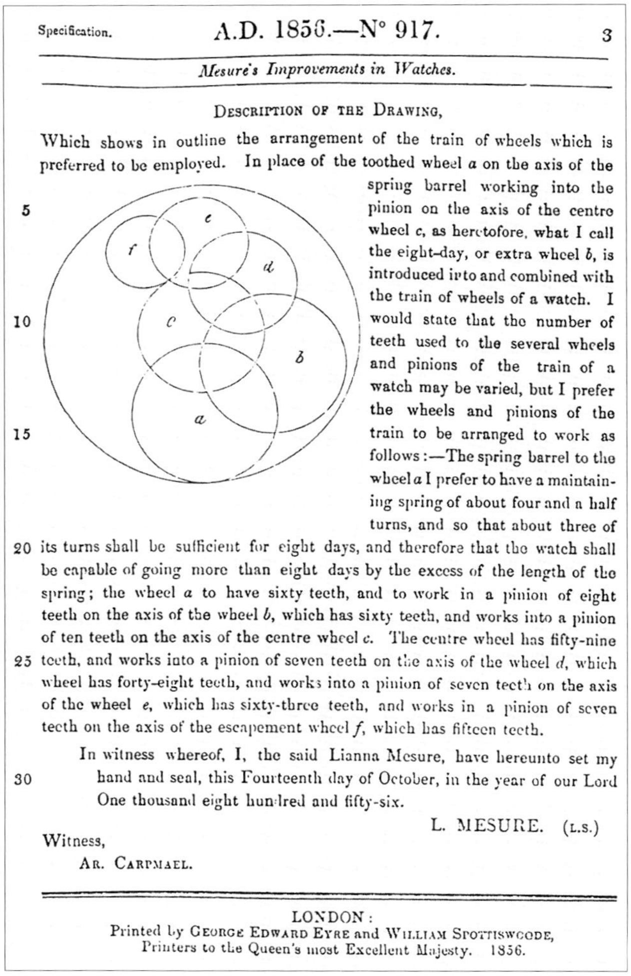

PLATE 2
Lianna Mesure
Billericay, Essex and King William Street, London

Silver double bottomed open face pocket watch with hinged bezel and back hallmarked London **1856**. Casemaker J.H (Joseph Hirst, 12 Skinner Street,

```
Specification.                A.D. 1856.—N° 917.                    3
                         Mesure's Improvements in Watches.

                        DESCRIPTION OF THE DRAWING,
    Which shows in outline the arrangement of the train of wheels which is
    preferred to be employed. In place of the toothed wheel a on the axis of the
                                    spring barrel working into the
                                    pinion on the axis of the centre
                                    wheel c, as heretofore, what I call
                                    the eight-day, or extra wheel b, is
                                    introduced into and combined with
                                    the train of wheels of a watch. I
                                    would state that the number of
                                    teeth used to the several wheels
                                    and pinions of the train of a
                                    watch may be varied, but I prefer
                                    the wheels and pinions of the
                                    train to be arranged to work as
                                    follows:—The spring barrel to the
                                    wheel a I prefer to have a maintain-
                                    ing spring of about four and a half
                                    turns, and so that about three of
    its turns shall be sufficient for eight days, and therefore that the watch shall
    be capable of going more than eight days by the excess of the length of the
    spring; the wheel a to have sixty teeth, and to work in a pinion of eight
    teeth on the axis of the wheel b, which has sixty teeth, and works into a pinion
    of ten teeth on the axis of the centre wheel c. The centre wheel has fifty-nine
    teeth, and works into a pinion of seven teeth on the axis of the wheel d, which
    wheel has forty-eight teeth, and works into a pinion of seven teeth on the axis
    of the wheel e, which has sixty-three teeth, and works in a pinion of seven
    teeth on the axis of the escapement wheel f, which has fifteen teeth.
       In witness whereof, I, the said Lianna Mesure, have hereunto set my
    hand and seal, this Fourteenth day of October, in the year of our Lord
    One thousand eight hundred and fifty-six.
                                               L. MESURE.   (L.S.)
    Witness,
       AR. CARPMAEL.

                                LONDON:
              Printed by GEORGE EDWARD EYRE and WILLIAM SPOTTISWOODE,
                    Printers to the Queen's most Excellent Majesty. 1856.
```

PLATE 2C. Mesure's Patent

Clerkenwell, PO Dir 1852, and 26 Spencer Street, Off Goswell Road, PO Dir 1859/60). White enamel dial with eccentric chapter ring signed EIGHT DAY LEVER, black roman numerals, gold spade hands, inset seconds dial. Key wound and set. Top plate engraved Mesure King William St London No 279. Three-quarter plate going barrel. 13 jewel lever movement with pointed tooth scape wheel and tangential lever. Three arm round rimmed steel oversprung balance with blued steel balance spring. Regulator index on balance cock engraved with scrolls and Fast Slow.
Key size 8. Diameter 52.5mm

Provisional specification dated 17 April 1856
Sealed 20 June 1856
Full specification signed 14 October 1856
filed 15 October 1856
Lianna Mesure, Billericay in the County of Essex, the improvement consists of fixing on the barrel containing the spring a wheel of 60 or more teeth, to work on pinion, on the axis of eight day wheel of 70 or more teeth, which works centre wheel.

Eight-Day Watches

PLATE 3A

PLATE 3B

PLATE 3
John Hickling
6 Nelson Buildings, City Road, London

Silver open face pocket watch, hinged bezel and back, hallmarked London **1868**. Casemaker PW (Philip Woodman, 10 Great Sutton Street, Clerkenwell). Dustcap with blued steel slide and screw knob to the centre to assist in removal engraved John Hickling London AD 1868. Set hands/wind up pendant hallmarked **1867**, maker WS. White enamel dial with inscription EIGHT DAY TIMEKEEPER J HICKLING LONDON, inset seconds dial, black roman numerals, narrow blued spade hands. Keywound and set at back. Inscription on top plate J Hickling London No 787. Full plate. 7 jewels. Going barrel. Lever escapement. Three arm round rimmed steel oversprung balance. Regulator index on scroll engraved balance cock engraved Faft Slow.
Key size 8. Diameter 55mm

Included in list of new members of British Horological Institute, Horological Journal Vol 8 May 1866, p108. John Hickling, 122 St John Street, West Smithfield, Clerkenwell, in 1835/1844 Trade Directory – was this his father?

PLATE 3C

407

Eight-Day Watches

PLATE 4A

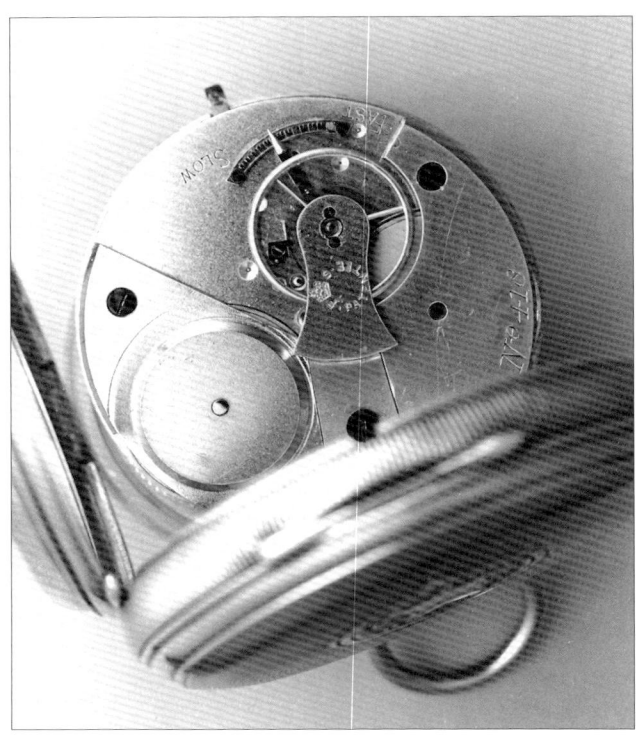

PLATE 4B

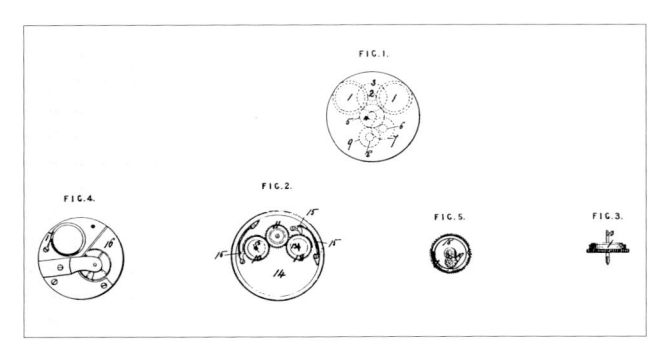

PLATE 4C. Silveston's specification

PLATE 4
Francis Silveston
2 Spon Street, Coventry, Warwickshire

Silver open face pocket watch hallmarked London **1881**. Casemaker CH, probably Charles Harris, Norfolk Street, Coventry. Pendant maker JS, hallmarked 1873. White enamel dial signed PATENT 8 DAY, inset seconds dial, black roman numerals, blued steel spade hands. Keywound and set. Full plate stamped No 418. 7 jewels. Double barrel English lever escapement. Three arm steel balance with Bosley regulator. Plain balance cock stamped F S with crown between and PATENTEE. Complete with original red leather and velvet presentation box – a valuable bonus.

Key size 10. Diameter 45mm

Taking patent date and case hallmark into account, it would appear that this was an early example. It is doubtful that the numbering of movements commenced at number one. Listed at this address in Kelly's 1880 Directory for England, Scotland and Wales. UK Patent No 440, 31 January 1880, relates to 'adapt the watch with two "barrel" which I cause to work into a common pinion' (Plate 4C).

PLATE 5
Jacot & Bovy, Switzerland

Silver open face pocket watch with snap-on bezel and hinged back, Swiss hallmark 0.935, case numbered 10100. White enamel dial, inset seconds dial, broad black roman numerals, blued steel spade hands. Keyless winding with push piece to set hands at 3 minute past hour position. The barrel, which is of contrate gear form, extends across the full width of the back of the movement and is rotated when wound by a pinion on the winding stem. The centre of the skeletonised and glazed barrel cover carries Swiss patent mark 6279 + 497.

Diameter 56.5mm

Further examples have been seen with a folding key across the back of the watch (see detail of patent specification, Plate 5C). UK patents No 4279, 26 March 1886, No 4505 31 March 1886.

PLATE 5A

PLATE 5B

4279. Jacot, C. E., and Bovy, E.
March 26.

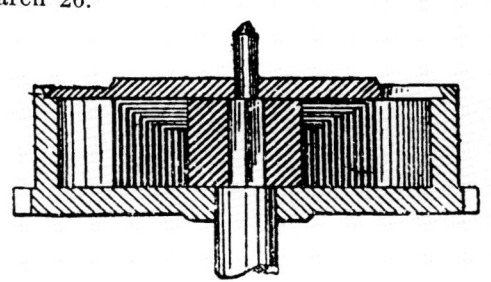

Mainsprings. — The mainspring is tapered in width gradually from its outer end to about two-thirds at its inner end, that it may give a more regular pull to the barrel. The Figure shows a section of the barrel and spring. It is applicable to clocks and watches and to musical boxes and other mechanism.

4505. Jacot, C. E., and Bovy, E
March 31.

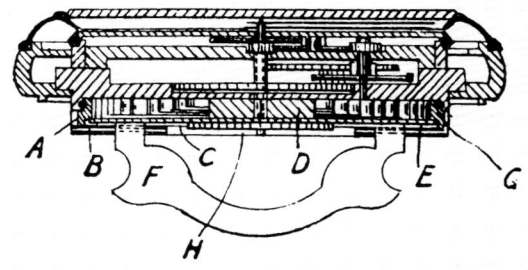

Watch mainsprings; barrels; framework; going trains.—Relates to an arrangement of movement to go ten days with one winding. The mainspring tapers in width, getting gradually narrower towards its inner end. This spring is mounted in a barrel A which occupies the whole of the back of the movement. The barrel is of peculiar shape, has no bottom, and fits into a recess in the pillar plate. The flange of the pillar plate and the side of the barrel are grooved to receive a wire B by which the barrel is retained in position. The barrel is covered by a glass C. The boss D has a recess cut in its rim in which the coiled inner end of the spring is inserted. The barrel has its back formed with ratchet teeth and carries the bridge-piece E with a hinged winding-handle F. The arbor of the boss D carries the first wheel of the train. The third wheel frictionally turns a pinion G which operates the dial-wheels. The arbor of the minute-hand passes through to the back, where it carries a wheel H by which the hands can be set.

PLATE 5C. Jacot & Bovy's patents

Eight-Day Watches

PLATE 6A

PLATE 6B

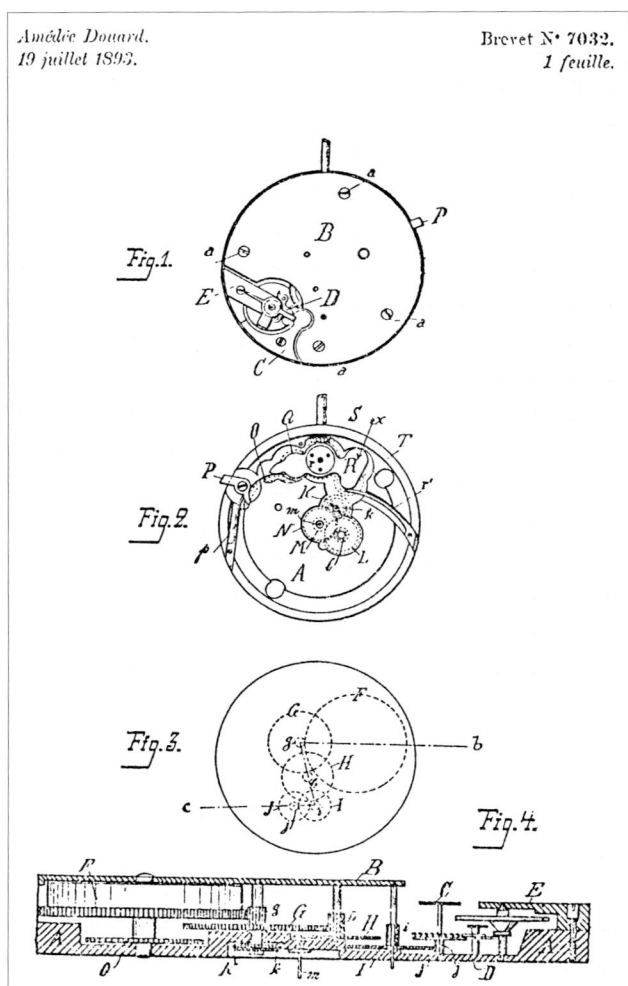

PLATE 6C. Douard's patent

PLATE 6
8 Day Swiss Lever Pocket Watch

Nickel open face pocket watch with snap-on bezel and hinged back and cuvette. White enamel dial inscribed 8 DAYS, inset seconds, broad black roman numerals, blued steel spade hands. Keyless wound and pull to set hands. Top plate stamped + 7032. Half plate. 15 jewel club tooth Swiss lever movement with straight line lever. Imitation (part cut) bimetallic balance with blued steel balance spring with overcoil. Regulator index on balance cock engraved S R F A.

Diameter 53.3mm

Swiss patent 7032, 19 July 1893, Amédée Douard à Bienne, Switzerland, Nouveau calibre de montre à huit jours de marche.

Eight-Day Watches

PLATE 7A

PLATE 7B

PLATE 7
John Ganter, 11 Queen Street, Morley, Leeds

Silver open face pocket watch, double bottomed case, engine turned back with milled body to side of case, hallmarked Chester **1900.** Casemaker CH, pendant hallmarked Birmingham **1900,** pendant maker CH (Charles Harrold, Birmingham). White enamel dial with inscription upper centre JOHN GANTERS PATENT No 19583 8 DAYS ENGLISH LEVER MORLEY YORKSHIRE, sunk seconds dial, broad black roman numerals, gold spade hands. Dust cover to full plate movement. Backplate signed John Ganter Morley 19583. Two independently wound going barrels geared to the arbor of the centre wheel. Club tooth scape wheel. Single roller. Bimetallic cut compensated oversprung balance. Engraved cock with arrow head index calibration. Index regulator on balance cock.
Diameter 59.3mm

UK Patent No 19583, 29 September 1899. Patentee claimed he could 'substitute 2 new main springs and barrels ... and add intermediate wheel (wheel C) ... converts any one day watch into a seven or eight day watch'. Listed in the Post Office Directory 1897.

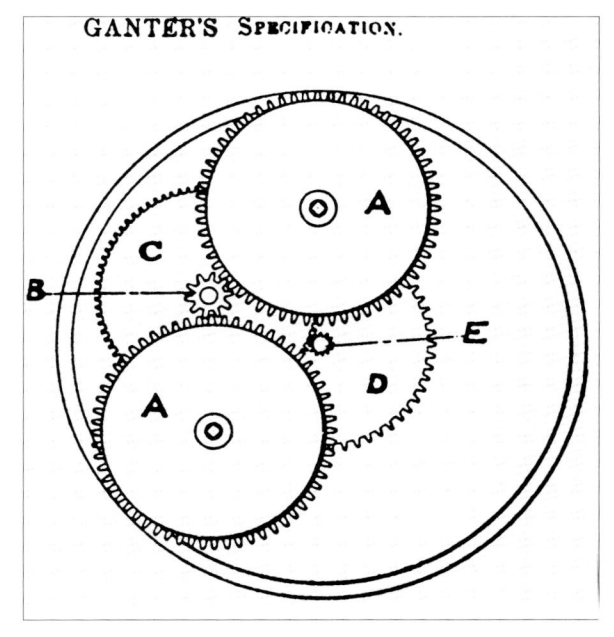

PLATE 7C

411

Eight-Day Watches

PLATE 8
Lobl's 8 Days Lever, Switzerland

Oxydised steel open face case, snap-on bezel and hinged back. White enamel dial inscribed LOBL'S 8 DAYS LEVER SWISS MADE, black roman numerals, narrow spade blued steel hands. Keyless wound and set with push piece at 56 minutes position. Visible oversprung monometallic balance with blued steel balance spring in lower cut away circular segment of dial. The regulator index across the edge of the balance at 6 o'clock position with AF SR on dial. Engraved balance bridge. Jewelled club tooth straight line lever movement. Back of barrel stencilled with floral decoration etc NOUVELLE ANCRE 8 JOURS PRATIQUE SYSTEME UNIQUE SIMPLIFIÉE + 42066 INTERCHANGE-ABILITE ABSOLUE. Centre plain. (The Swiss certainly utilised this area for self advertisement!)
Diameter 50.5mm

Swiss Patent no 42066, K Silbermann & JH Hasler, La Chaux de Fonds, 2 December 1907, Class 71a Movement de montre à marche de longue durée (see Plate 8C).

PLATE 8A

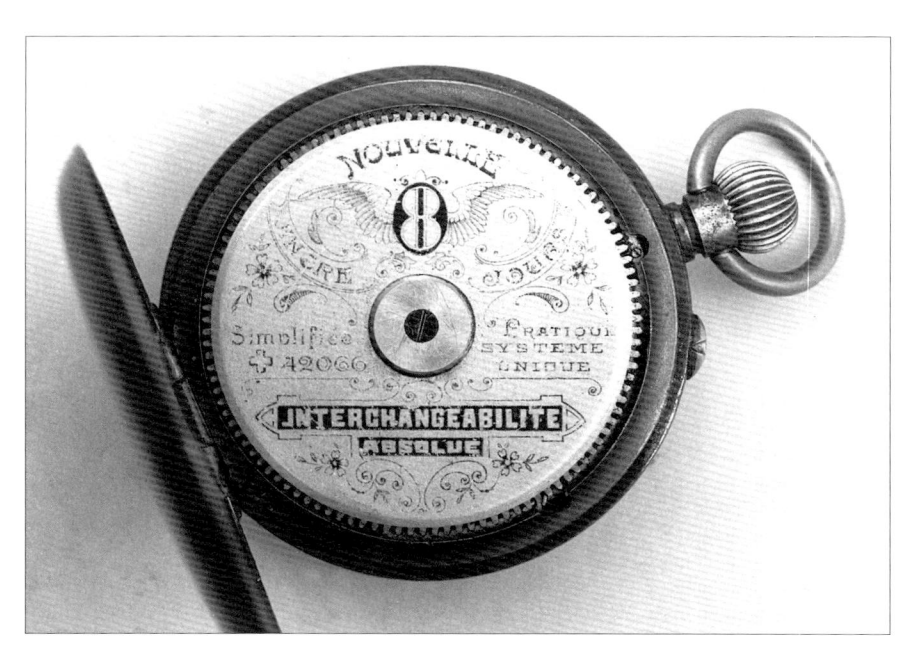

PLATE 8B

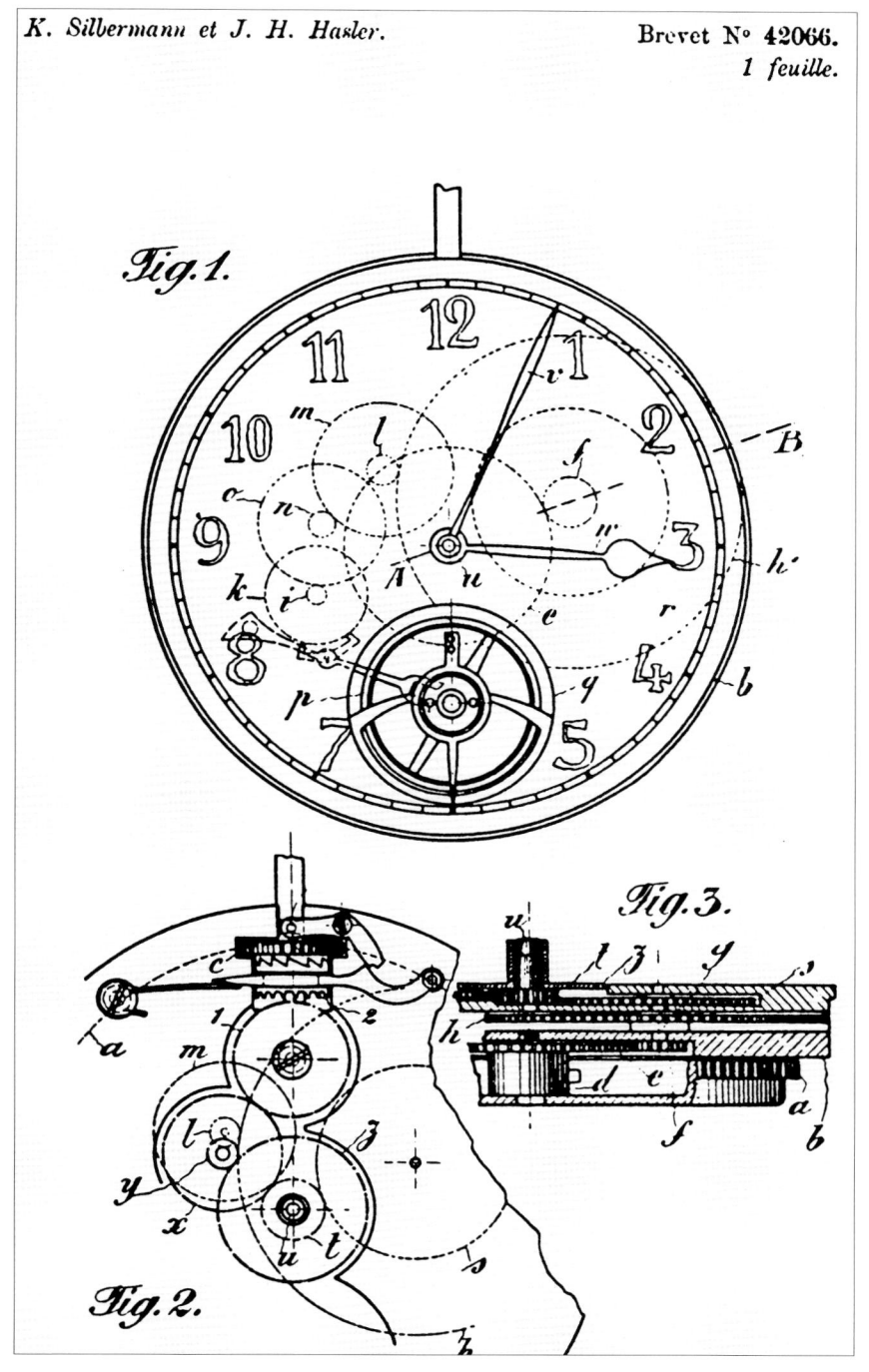

PLATE 8C

Eight-Day Watches

PLATE 9A

PLATE 9B

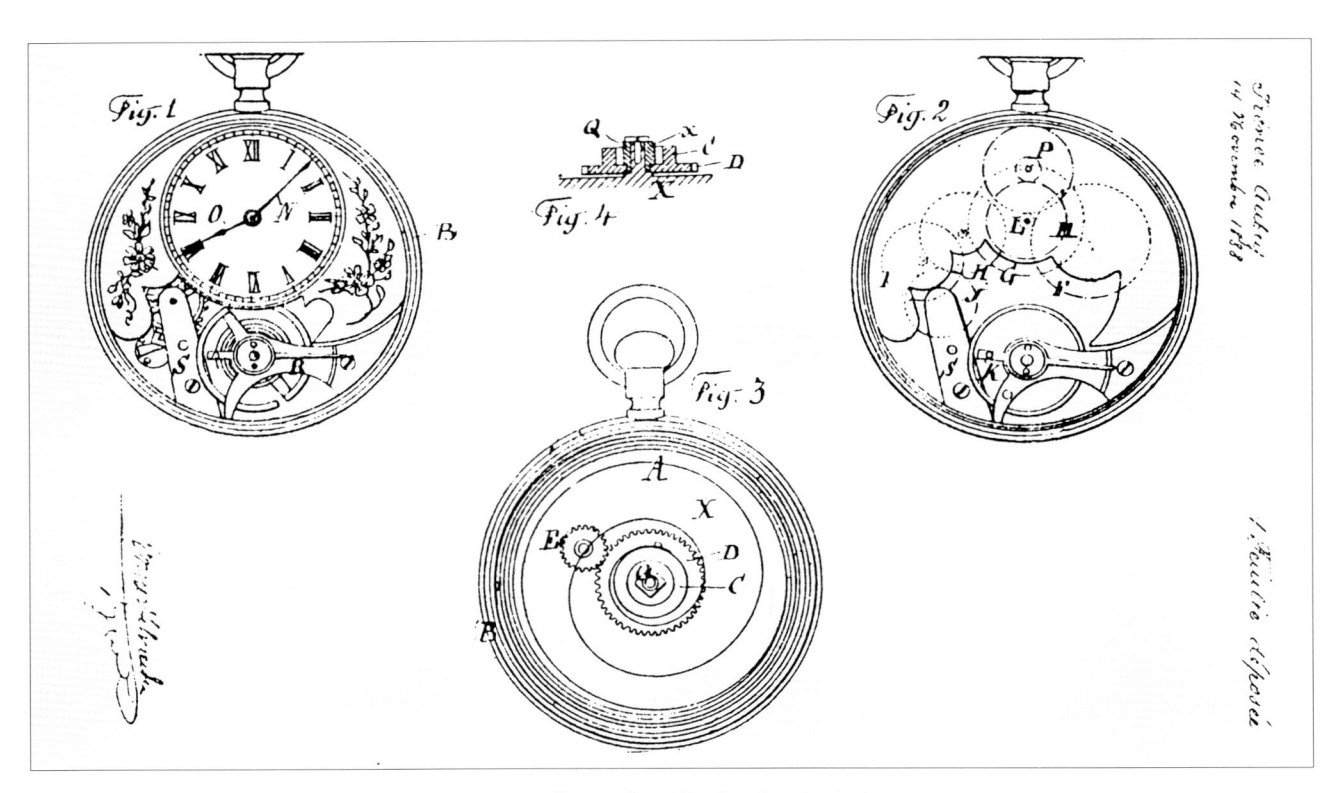

PLATE 9C. Patent Specification Irenée Aubry

414

Eight-Day Watches

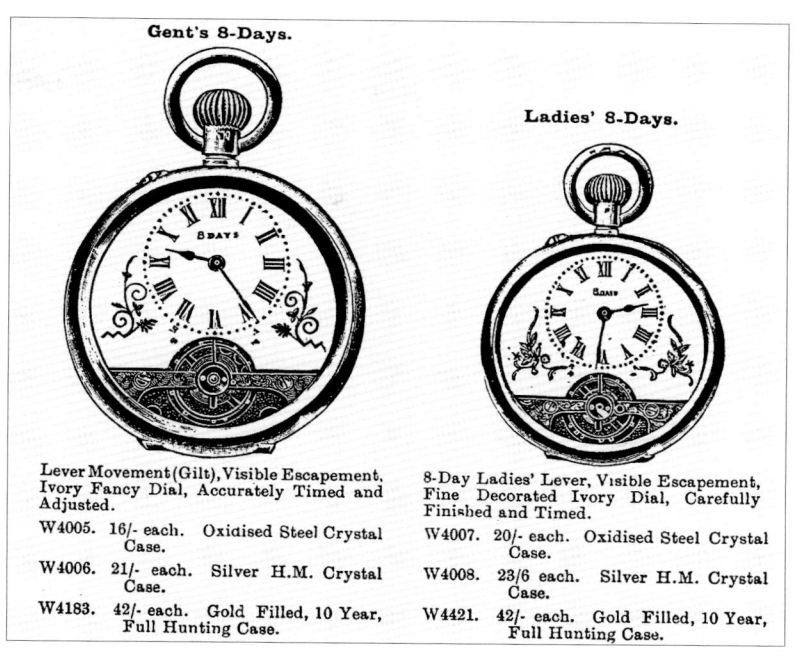

Gent's 8-Days.

Ladies' 8-Days.

Lever Movement (Gilt), Visible Escapement, Ivory Fancy Dial, Accurately Timed and Adjusted.

W4005. 16/- each. Oxidised Steel Crystal Case.
W4006. 21/- each. Silver H.M. Crystal Case.
W4183. 42/- each. Gold Filled, 10 Year, Full Hunting Case.

8-Day Ladies' Lever, Visible Escapement, Fine Decorated Ivory Dial, Carefully Finished and Timed.

W4007. 20/- each. Oxidised Steel Crystal Case.
W4008. 23/6 each. Silver H.M. Crystal Case.
W4421. 42/- each. Gold Filled, 10 Year, Full Hunting Case.

PLATE 9D. *Hirst Bros. Catalogue, 1910*

AUBRY & Cº

LA CHAUX-DE-FONDS

Manufacture de Montres
8 JOURS
Acier, Métal, Argent, Or et Fantaisie

Montres pour Automobiles

CHEVALETS □ PENDULETTES □ MONTRES PORTEFEUILLE

PLATE 9E. *Journal suisse d'Horlogerie et Inventions – Revue,* Juillet 1920

PLATE 9
Irenée Aubry, Saignelegier, Switzerland

Silver open face snap-on bezel with hinged back and cuvette, Swiss hallmark 0.800. White enamel dial with gilt floral decoration around chapter ring, black arabic numerals, fancy hands. Keyless wound with push piece to set at 55 minute position. Visible oversprung cut bimetallic balance with blued balance spring in lower cut away segment of dial. Small regulator index on balance bridge pointing to six o'clock position on dial. Dial marked S F – R A Swiss patent number (No 88 + No 2) under balance. Mainspring and barrel across full width.

Back engraved 8 JOURS BREVET S.G.D.G. No 88 + No 2 with trade mark. Jewelled Swiss lever movement. Club tooth scape and straight line lever. *This is a very rare and early example by the original patentee.*
Diameter 50.5mm

Brevet No 88, dated 14 November 1888, publié le 10 Janvier 1889, Class 123, Irenée Aubry à Saignelegier, Nouvelle disposition du mechanisme des montres de toutes dimensions particulièrement applicable aux montres-bijoux et aux montres marchant 8 jours et plus. This was for an eight day watch with visible balance. Brevet No 2 (modification to visible balance) 21 January 1889, Class 123.

Eight-Day Watches

PLATE 10A

PLATE 10
Arthur Graizley, La Ferrière, Switzerland

Silver open face pocket watch with hinged bezel, engine-turned back, Swiss hallmark 0.935. Casemaker's stamp AG (Arthur Graizley). White enamel dial, sunk seconds dial at lower centre, roman numerals, blued steel spade hands. Up and down dial inset incorporating numeral 12 position with arabic numerals 0-7. Keyless wound and set with push piece to set hands at 56 minute position. Mainspring barrel across full width of lever movement engraved 8 DAYS SWISS MADE. *A most unusual up and down dial arrangement which is set to zero manually and then the dial records the elapsed days. A rare and highly desirable piece.*
Diameter 51.5mm

Swiss Patent No 4764, 23 March 1892, Class 64, Arthur Graizley, La Ferrière, Indicateur de marche et de developpement du ressort pour montres de 1 à 8 jours de marche pouvant servir de quantième simple. Revue de l'Association Française des Amateurs Horlogerie Ancienne No 6, 1960.

PLATE 10B

416

PLATE 11A

PLATE 11
Arthur Graizley, La Chaux De Fonds, Switzerland

Oxydised steel hunter case with hinged back. Cream enamel eccentric dial with black roman numerals inscribed 8 DAYS 8 JOURS 8 TAGE, blued steel spade hands. Keyless wound with push piece at 13 minutes position to set hands. Visible oversprung monometallic balance with blued steel balance spring in lower cut-away segment of dial. Engraved balance bridge. Regulator index over balance towards scale on dial which is inscribed RS and AF. Mainspring and barrel across full width of movement silvered and stencilled in black LEVEES VISIBLES ANCRE Spiral Breguet 8 JOURS INTERCHANGEABILITE GARANTIE and various exhibition medals, 8 DAYS SWISS MADE in central position AG. Jewelled club tooth lever escapement with straight line lever.
Diameter 51.3mm

Revue de l'Association Française des Amateurs Horlogerie No 6, 1960.

PLATE 11B

PLATE 12A

PLATE 12
Swiss 8 Day Pocket Watch

Chrome on brass open face case with snap-on bezel and hinged back. White enamel dial with floral decoration inscribed 8 DAYS 8 JOURS 8 TAGE 8 GIOPNI 8 DIAS (the manufacturers obviously envisaged a wide international market for their watches). Black arabic numerals, blued steel spade hands. Keyless wound and set with push piece at 11 o'clock position. Visible oversprung monometallic balance and red brown balance spring in lower cut away segment of dial. Engraved balance bridge. Regulator index towards 6 o'clock position with RS and FA. Jewelled club tooth straight line lever movement. Mainspring and barrel across width of movement with barrel back engraved 8 JOURS LEVEES VISIBLES PLATEAU DOUBLE GARANTIE INTERCHANGEABLE ANCRE together with mock exhibition medals. Plain centre.

Diameter 50mm

This is similar to the style of the Hebdomas (Plate 15). Although the barrel engravings are superficially alike they do not incorporate the medals awarded as in the Hebdomas, Hirst Bros Catalogue 1910 and Hirst Bros 4th Wide Awake Catalogue April 1915). The barrel retaining nut is also different. It is thought that this is more in the style of an OCTO. Watches with similar dials appear in the Hirst Bros Catalogue for 1901.

PLATE 12B

Eight-Day Watches

PLATE 13A

PLATE 13
Octava, La Chaux de Fonds, Switzerland

Nickel open face case, snap-on bezel and hinged back, pendant at 6 o'clock position. White enamel dial inscribed S. SMITH & SONS (MA.) LTD. 179/85 GREAT PORTLAND ST LONDON-W-1, black arabic numerals, broad blued steel spade hands, inset seconds dial with exceptionally broad spade hand. Keyless wound and pull to set. Mainspring and barrel full width of movement. Club tooth straight line lever movement.
Diameter 50mm

Watch retailed by Smith & Sons Motor Accessories Ltd. This company was formed 15 July 1914 with Mr Allan Gordon Smith as Managing Director. Although not strictly a pocket watch, this example is included to show that the 8 day pocket watch had other applications, both as a wireless panel watch and car clock.

PLATE 13B

Eight-Day Watches

PLATE 14A

PLATE 14B

8-DAY WATCH (Interchangeable).

Interchangeable 16-Size 8-Day Nickel Movement, Open Face, White Dial, Full Jewelled, U.S.A. Patent.

W4431.	Silver Hall Marked.	59/6 each.
W4432.	10 Year, Gold Filled.	63/- each.
W4433.	20 Year, Gold Filled.	83/6 each.
W4434.	25 Year, Gold Filled.	93/6 each.
W4435.	9-ct. Hall Marked, Swing Ring.	£6/4/6 each.
W4436.	9-ct. Hall Marked.	£8/2/- each.
W4437.	18-ct. Hall Marked.	£17/10/6 each.

PLATE 14C. *Hirst Bros. Catalogue*, 1910

PLATE 14D

Eight-Day Watches

HIRST BROS. & Co. Limited, Oldham, Manchester and Birmingham. 275

OCTAVA 8-DAY WATCH MATERIALS, No. 5905.

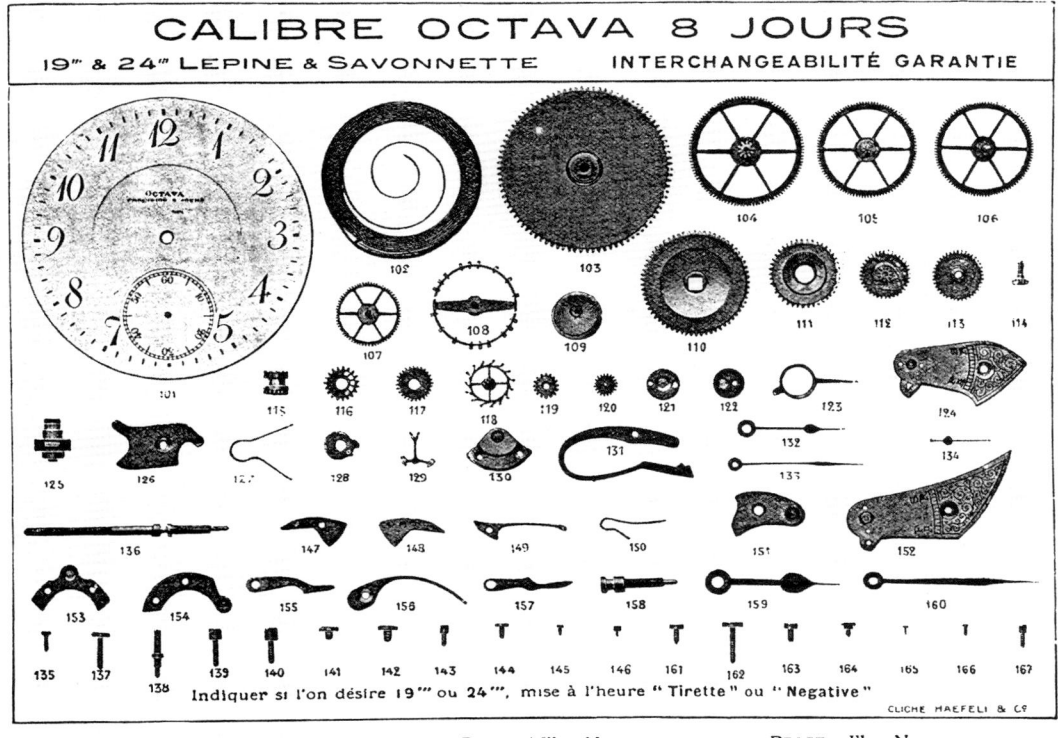

CALIBRE OCTAVA 8 JOURS
19''' & 24''' LEPINE & SAVONNETTE — INTERCHANGEABILITÉ GARANTIE

Indiquer si l'on désire 19''' ou 24''', mise à l'heure "Tirette" ou "Negative."

CLICHÉ HAEFELI & Cº

Illus. No.	PRICE.	Illus. No.	PRICE.	Illus. No.	PRICE.	Illus. No.	PRICE.
101	Each 1/3	115-116	Per doz. 3/4	129	Per doz. 10/-	149-150	Per doz. 10d.
102	Per doz. 6/8	117	„ 2/6	129 Pallet Staff only	5/-	151	„ 5/10
103	„ 5/10	118	„ 10/-	130	Per doz. 10/-	152	„ 6/8
104-107	„ 5/-	119	„ 1/3	131	„ 2/6	153-154	„ 5/-
108	„ 30/-	120	„ 2/6	132-133	„ 1/1	155-157	„ 10d.
108 Balance Staffs only	2/6	121-123	„ 1/8	134	„ 7d.	158	„ 8/4
109	Per doz. 20/-	124	„ 20/-	135	„ 10d.	159-160	„ 1/8
110-111	„ 6/8	125	„ 2/6	136	„ 3/4	161-167	„ 5d.
112-113	„ 2 6	126-127	„ 1/8	137-146	„ 5d.		
114	„ 4 2	128	„ 2/6	147-148	„ 2/6		

PLATE 14E

PLATE 14
Octava Watch Co, La Chaux de Fonds, Switzerland

Nickel open face case, snap-on bezel and back, hinged cuvette. Black dial (military issue) inscribed 8 DAY NON-LUMINOUS MARK V B.G. No 622 Swiss made, inset seconds, white arabic numerals, white spade hands. Keyless wound and pull to set hands. Top plate engraved OCTAVA WATCH CO SWITZERLAND 15 FIFTEEN JEWELS 3 ADJUSTMENTS U.S.A.P. 816321. Half plate. Club tooth and straight line lever movement Oversprung and overcoil cut bimetallic balance with blued steel balance spring. Regulator index on balance cock engraved FS AR.

Diameter 52.1mm

A further model appeared with a luminous dial. Patents taken out in name of Hartmann of Prague (assignor of one half to Josef Oliak) Prague 27 March 1906 (Plate 14D). Name Octava Registered (No 21.962) 12 April 1907, Graizley & Cie, La Chaux de Fonds, Hirst Bros Catalogue 1910 and their 4th Wide Awake Catalogue April 1915 (see Plate 14E).

Eight-Day Watches

PLATE 15A

PLATE 15B

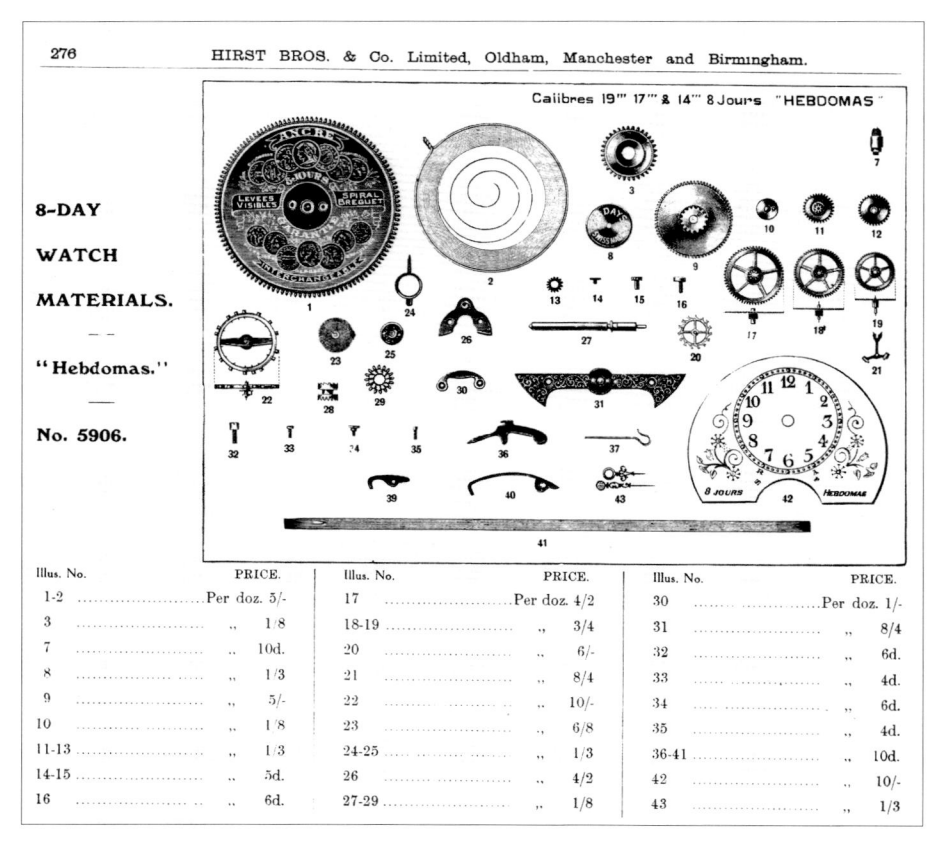

PLATE 15C. *Hirst Bros. Fourth Wide Awake Catalogue, c.1915*

FICHE TECHNIQUE HEBDOMAS 18" 1/2

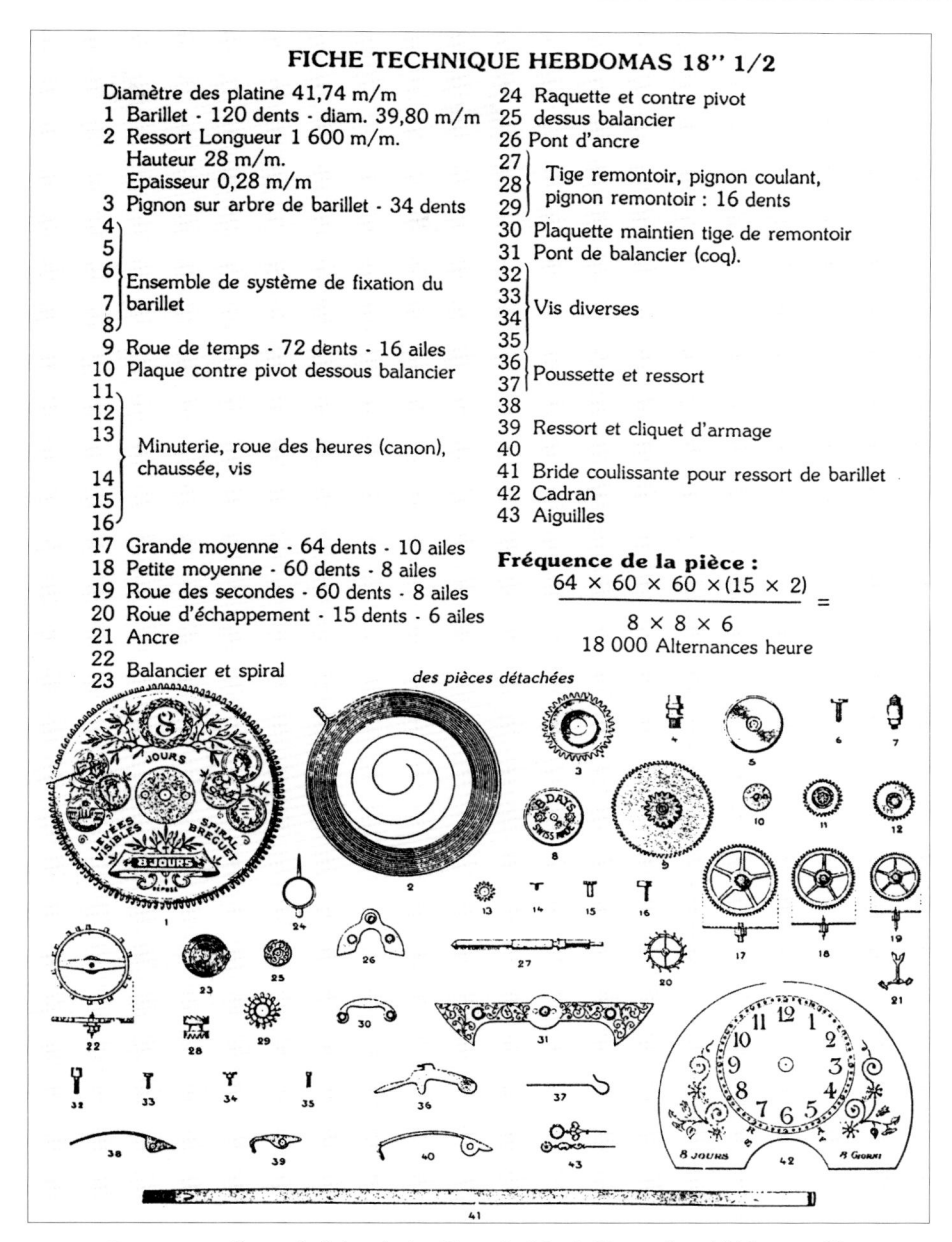

Diamètre des platine 41,74 m/m
1 Barillet - 120 dents - diam. 39,80 m/m
2 Ressort Longueur 1 600 m/m.
   Hauteur 28 m/m.
   Epaisseur 0,28 m/m
3 Pignon sur arbre de barillet - 34 dents
4 ⎫
5 ⎪
6 ⎬ Ensemble de système de fixation du
7 ⎪ barillet
8 ⎭
9 Roue de temps - 72 dents - 16 ailes
10 Plaque contre pivot dessous balancier
11 ⎫
12 ⎪
13 ⎬ Minuterie, roue des heures (canon),
   chaussée, vis
14 ⎪
15 ⎪
16 ⎭
17 Grande moyenne - 64 dents - 10 ailes
18 Petite moyenne - 60 dents - 8 ailes
19 Roue des secondes - 60 dents - 8 ailes
20 Roue d'échappement - 15 dents - 6 ailes
21 Ancre
22 ⎫ Balancier et spiral
23 ⎭
24 Raquette et contre pivot
25 dessus balancier
26 Pont d'ancre
27 ⎫
28 ⎬ Tige remontoir, pignon coulant,
29 ⎭ pignon remontoir : 16 dents
30 Plaquette maintien tige de remontoir
31 Pont de balancier (coq).
32 ⎫
33 ⎬ Vis diverses
34 ⎭
35 ⎫
36 ⎬ Poussette et ressort
37 ⎭
38
39 Ressort et cliquet d'armage
40
41 Bride coulissante pour ressort de barillet
42 Cadran
43 Aiguilles

Fréquence de la pièce :

$$\frac{64 \times 60 \times 60 \times (15 \times 2)}{8 \times 8 \times 6} = 18\,000 \text{ Alternances heure}$$

des pièces détachées

PLATE 15D. *Revue de l'Association Française* No.6. December 1960, page 29

PLATE 15
Hebdomas, Arthur Graizley & Cie, La Chaux de Fonds, Switzerland

Silver open face case, hinged bezel and engine-turned back, hallmarked London **1919** (.925). Casemaker's mark GS (George Stockwell, importer). Milk white enamel dial inscribed HEBDOMAS PATENT 8 DAYS SWISS MADE, eccentric chapter upper centre with black roman numerals and blued steel spade hands, concentric seconds, subsidiary day of week and month dials. Keyless wound and pull set. Mainspring and barrel across full width of movement and engraved SPIRAL BREGUET LEVEES VISIBLES 8 JOURS ANCRE QUALITE SUPERIEURE GARANTIE and reproduced exhibition medals. Club tooth straight line lever movement. Visible oversprung and overcoil bimetallic balance and blue steel balance spring. *This 8 day model with calendar and centre seconds is an extremely rare and highly desirable piece (see Plates 15C and D.).*
Diameter 48.8mm

Hebdomas Registered No 20669 13 June 1906, Arthur Graizley & Cie, La Chaux de Fonds.

Eight-Day Watches

PLATE 16A

PLATE 16B

PLATE 16
Hebdomas, Schild SA, Chaux de Fonds, CH 2300, Switzerland

Engraved gilt open face case, snap-on bezels, glazed back and front. Skeletonised dial with roman numerals at quarters, Louis style hands. Keyless wound and set. The dial screwed to the movement below XII. Dial side of the pillar plate engraved Hebdomas 102 (calibre) Swiss Made No 01792 and HEBDOMAS again on the left-hand side of the balance cock bridge. Skeletonised barrel across full width of movement. 15 jewels club tooth straight line lever 19 ligne movement. Oversprung monometallic balance and blued steel balance spring. Regulator without pointer on balance bridge. Limited production run in blue presentation case marketed 1982. Each model was numbered. *Value would be enhanced by documentation and original presentation case.*
Diameter 50mm

PLATE 17
Anon, Switzerland

Nickel open face case pocket watch with snap-on bezel and hinged back and cuvette. White enamel dial, blued steel spade hands. Keyless wound and push set. Top plate stamped B within square. Pillar plate stamped 413454 side of balance rim. Half plate 15 jewel club tooth straight line lever movement. Oversprung and overcoil cut bimetallic

PLATE 17

balance and blued steel balance spring. Regulator index on balance cock engraved AF RS.
Diameter 49.7mm

PLATE 18A

PLATE 18
Wittnauer Watch Co Inc, Switzerland

Chrome plated (stamped METAL) open face pocket watch with snap-on bezel and back. Case stamped WITTNAUER WATCH CO INC. SWITZERLAND, repeated on top plate. Silvered dial inscribed 8 DAYS WITTNAUER SWISS, black arabic numerals, broad blued steel spade hands, inset seconds dial 0–60 with 10 seconds arabic numerals. Keyless wound and pull to set. Top plate inscribed SWISS UNADJUSTED WITTNAUER SEVEN 7 JEWELS. Double barrel lever movement. Club tooth scape with straight line lever. Oversprung and overcoil monometallic balance and blued steel balance spring. Regulator index on balance cock engraved AF RS and also (calibre) AXA.

Diameter 50.3mm

Wittnauer subsidiary of Agassiz Watch Co SA, now owned by Westinghouse Electric Corporation. The Wittnauer trade mark No 99475 registered 28.3.41 by Wittnauer & Cie SA, Rue des Acacias, 46 Genève.

PLATE 18B

Index

Abrecht, César Renfar, 385, 403
Acme (H Williamson Ltd), 144
Adams, Stephen, 21
Adamson, Hugh, 23
'Addison' (Waterbury Watch Co), 216, 224
Adeon, 310, 311
Adrem, Title-page, 355
Aeonicloc (Baumgartner Frères SA), 399
Agassiz Watch Co SA, 425
Alert (Hoefen a Enz), 239
Ambassador (Waterbury Watch Co), 227
American Waltham Watch Co, 7, 198, 199, 200, 201, 202, 204, 205, 206, 207, 208
American Watch Case Company, 164, 182, 183
American Watch Co, 196, 197, 198, 203, 306, 307, 376, 377
Amida SA, 243, 257, 373
Angelus, 385
Anglo Celtic Watch Co, 366, 367
Ansonia Clock Company, 232, 233
Army & Navy Co-operative Society Ltd, 119
Arnold, Auguste Henchoz, 315, 328
Arnold, John, 39, 81, 82
Arnold, John Roger, 81, 82
Ascot, The (Henchoz Frères), 317
Ashley, James, 38
Astral (H Williamson Ltd), 144
Atlas Watch Co, 314, 318
Aubry, Irenée, 405, 414, 415
Audemar, Louis, 81
Audemars, Piquet & Cie, 385
Austin, John, 85
Azemar, PD, 351

Bader, Hafner & Holderbank, 385, 402
Bagdad, 240
Bain, Alexander, 217
Bank Watch (James William Benson), 112, 146, 147
Bannatyne, A, 250, 251
Barbezat-Baillot, 332, 379
Barbezat Bole, H, 385, 405
Barlow, Edward, 39, 379
Barnett, Jonah, 75
Barr, John, 101
Barraud & Lunds, 99
Barraud, Paul Philip, 99
Bartlett, PS (American Waltham Watch Co), 201, 204, 206, 207
Barwise & Sons, 217
Barwise, John, 27, 217
Barwise, Weston, 217
Baume & Co, 167, 282, 362, 363
Baume & Lezard, 47
Baume, (Joseph) Arthur, 47, 167, 282

Baume, (Louis) Celestin, 47, 282
Baumgartner Frères, 292, 293, 373, 385, 399, 400, 402, 403
Baumgartner, Frederick, 351
Baxter, Junior, 75
Beacon Parker, 12, 377
Bedford, Alfred, 306, 307
Bedford, Henry William, 198
Beesley, George & Richard, 148
Belga, Numa SA, 66, 67
Bendix, 378
Benedict and Burnham Brass Manufacturing Co, 215, 216, 221
Bennett, Sir John, 15, 99, 100, 312, 313
Benson, Alfred, 196, 197, 198, 199, 200, 377
Benson, James William, 73, 112, 113, 114, 139, 140, 146, 147, 170, 171, 283, 387
Benson, SS, 112
Berguer, Francis, 23, 27
Berrollas, Joseph Anthony, 382, 383, 386
Best Patent Lever (Jequier Frères. 276
Betima Co Ltd, The, 245
Bettlach, Fabrique d'Ebauche de, Title-page, 355
Biber, J, 259
Billodes (Zenith), 175, 180, 181
Bird, William, 120
Blondeau, Henri Berney, 368, 369
Boichat & Co, Alphonse, 355
Bonijol, Leyton, 36
Booth, Edward, 38
Borgel, François, 186, 343
Bosley, Joseph, 22
Bourne, James, 75
Bourcherin, A Sandoz, 353
Boutevile & Norton, 22
Bovet, Edouard, 330, 331, 409
Bovet, H., 331
Bovy, ML, 260
Boyer, 404
Braham, Louis, 189
Brandt & Frères, L, 405
Brandt, Louis, 173, 174, 175, 308, 309
Brandt, Robert, 53
Bravingtons Jewellers, 180, 181
Breguet, Abraham, 18, 38, 277
Breitling, Geo Léon, 292, 293, 296
Breitling Watch Co Ltd, 292, 293
British Watch Co Ltd, 146
British Watch Materials Manufacturing Co Ltd, 143
Broad, William, 24
Brompton & Kensington Electric Co, 340
Brooklyn Watch Co, 191
Brown, Isaac, 59
Brown, Thomas Lumsden, 316

Brown, William, 30
Brugg, Albert Salchli, 354
Bryan, Stephen, 76
Buck, Daniel Azro Ashley, 215, 221
Buckney, D, 223
Buffat, Eugène, 262, 276
Bullock, William Thomas, 334
Bumsel, Michael, 49
Burdess, Adam, 92-93
Buren, 144, 156, 184

Cabot Watch Co, 292, 293
Camerer Cuss (Kuss) & Co., 109
Campaign (Marcks & Co Ltd), 170
Capt, Henri and Jules, 277
Carmen, Samuel, 35
Carry, O, 342
Carter, William, 32, 35
Castelberg, Petitpierre and Co, 54
Cattin & Christian, 261, 266
Chamberlain & Hookham Ltd, 340
Champion, The (Thiel Bros Ltd), 237, 384, 398
Chasseral (Ernst Françillon), 167, 168
Châtelain, Charles Victor, 282
Châtelain, M, 260
Christie, William Henry, 340
Chronograph, centre seconds, 148-151
Churchill, Charles, 332
Claude, Francis, 288
Coastguard Watch (John Neve Masters), 103
Cocleus, Johann, 18
Cole, Thomas, 280
Collier's Friend (John Hawley & Sons), 138
Colonial (John Elkan Ltd), 186, 187
Columbia Time Products, 346
Combret, Pierre, 382
Connecticut Watch Co, 251
Copeland, Alexander, 24, 25
Cornell Watch Co, 214
Coronation Elizabeth II (Ingersoll), 256
Cortébert Watch Co, 178, 262, 292, 351
Courvoisier Frères, 12, 370, 371
Coutts-Stewart, Lord Dudley, 91
Coventry Co-operative Watch Manufacturing Society, 120, 131, 151
Cowell, Charles T, 121
Cross, John, 214, 219
Crown (Ingersoll), 250
Cuss & Co, Camerer, 109

Dalcher, Henri, 395
David, Jacques, 167
Defiance (Ingersoll), 251
Degoumois, Ernest, 372
De Hautefeuille, Abbé Jean, 70, 404

427

Delander, Daniel, 379
de Pass, Ernest, 317, 331
Dennison, Aaron L, 192, 404
Dennison, Howard & Davies, 192
Dennison Watch Case Co Ltd, 144, 145, 146, 155, 156, 158, 160, 161, 186, 187, 197, 201, 202, 204, 205, 206, 207, 336, 337
Dent, 81
Dent & Co, E, 117, 141
Dent, Edward John, 117, 141, 219, 220
Dent, John G, 117
Dimier Frères, 336
Dimier, George, 385
Dimra (Dimier Frères), 336
Ditisheim, Paul, 188, 189, 190
Domina, 353
Domon, Ovide Ferdinand, 326
Douard, Amédée, 410
Doughty, John, 25
Doughty, Thomas, 25
Dow, James, 219
Doxa, 165, 385, 405
Dreadnought (Record Watch Co), 157, 158
Drefuss Ltd, Moise, 296
Drinkwater, Alfred Henry, 116
Droz & Fils, Alcide, 176, 177
Du Barry Watch Co, 385, 400, 401
Dubois, Alex, 261
Ducommun, George, 165
Dueber Watch Case Company, 192
Dufréné, HA, 341
Durer, Emil, 237
Dutton, William, 70
Dutertre, Jean Baptiste, 215
Dyer, Elizabeth, 36
Dyer, John, 219

East, Edward, 382
Eberhard & Cie, 264
Edgar, ES, 184
Edwards, Benjamin, 28
Edwards, John Evan, 35, 36
Ehrhardt, William, 73, 103, 124, 142, 143
Ehrhardt, William junior, 143
Elgin National Watch Co, 192, 193, 208, 209, 210, 211
Elida Watch Co SA, 385
Elinvar, 188, 189, 211
Elkan Ltd, John, 186, 187
Elliott, John, 38
Emery, 23
Emery, Josiah, 70
Emmett, Thomas, 24
Erbeau, M, 260
Errington, Charles Hutton, 127, 138
Errington Watch Co, 127, 144, 145
Eterna, Schild Frères & Co, 384, 385, 405
Excelsior Park, 291, 298, 299, 300, 308, 309, 385
'Express' English Lever (John George Graves), 133, 134, 135

Fahys, Joseph, 173, 201
Fahys Watch Case Company, 192

Faivre, François, 40
Faller Bros, 114
Fanfar (Nathan Weil), 384, 390
Fatton, Frederick Louis, 277
Fattorini & Sons, 135
Fattorini, Antonio, 135
Fattorini, Thomas F, 135
Favre, Louis Eugène, 360, 388, 389
Favret, Onesime, 372
Federal (Thiel), 238, 239
Feltham, William, 34
Fennell, Richard, 305
Festiva, 400
Festonjee Franjec Bombay (Roskopf), 262
Field Watch (James William Benson), 112, 139, 140, 147
Field, Joseph, 76
Findlay & Co, 299
Findlay, A, 299
Finnigans, 395
Flinn & Co, William, 65
Fontainmelon factory, 260
Foot-ball, 334, 335
Foot-ball (Dimier Frères), 336
Forrest, John, 248, 249
Fortune, 164, 183
Fothergall, William, 28
Fowle, FB and M, 215
Fraissard, AL, 189
Françillon & Co SA, 385, 405
Françillon, Ernest, 47, 167, 168
Frary, Louis, 36
French, James Moore, 75, 118
Frodsham, Charles, 9, 82, 86, 122, 123, 339, 340
Frodsham, Harrison Mill, 122, 123
Frodsham, William James, 219
Froidevaux, S, & Block, 60
Funkley, EH, 146

Gagbenin, Numa, 66, 67
Gagnebin, E, 167
Gala (Liengme & Co), 388
Gallet & Co SA, 396
Gamel, 49
Gannay, Henry, 121
Ganter, John, 411
Gaydon, John, 111
Gee & Standley, Susannah, 115
Gee, Adam, 115
Gee, David, 115
Gee, Emily J, 115
General Time Corporation, 231
General Watch Co, 180, 181, 182, 183, 405
Geneva Watch Case Co, 307
Gibberd, Thomas, 23
Gibson, William, 115
Gilder, John, 117
Gindraux, Henri Eduouard, 259
Girard, Ulysse, 60
Girling, Barnard, 34
Girling, Samuel, 34
Godat, J, 54, 55
Goering, Vve de Louis, 164

Goldsmiths and Silversmiths, 310
Goldston, E & M, 230
Gontard, 404
Gordon, Peter, 93
Gough, Horatio, 40
Graham, 78, 79
Graham, George, 39, 70, 382
Graham, James, 20
Graizley & Cie, Arthur, 405, 416, 417, 421, 422, 423
Grant, Henry, 91
Grant, John, 40, 41, 70
Gravell & Son, William, 27
Graves, John George, 73, 133, 134, 135
Greder Frères, 261
Green, Hugh, 101
Greenwood, Willey, 377
Gribelin, Nicholas, 404
Grossman, Jules, 260
Guinard and Goley, 321
Gustave, J, 53
Guttmann Frères, 259
GWR (Rotherham & Sons Ltd), 136, 137

Hagen, 404
Hall, Charles, 216, 217
Hall Joseph, 217
Haller, Thomas, 233, 234
Hamel, Constant, 260, 262
Hamilton Watch Co, 192, 193, 211, 212
Hammon, John, 82, 85, 92
Hanrott, RC, 97
Hanson, William, 28, 29
Hargraves, James, 21
Hargreaves, Joseph & Co, 150
Harries, Charles, 121
Harris, Charles, 108, 111, 114, 129, 408
Harris & Sons, Joseph, 130
Harris, Sophia, 33
Harrold, Charles, 103, 117, 119, 120, 130, 135, 136, 138, 146, 151, 334, 340, 411
Hart, 216
Harte, Charles, 33
Hartmann, F, 420, 421
Harvey, GW, 126
Hasler, JH, 412, 413
Hatfield & Hall, 83
Hausler, Gustav, 384
Hawley, John & Sons, 136, 137, 138
Hebdomas (Arthur Graizley & Cie), 373, 405, 418, 422, 423
Hebdomas (Schild SA), 424
Helvetia (General Watch Co), 183
Henchoz Frères, 294, 317
Henlein, Peter, 18
Henriot, 43
Hermetique, 350
Heuer & Co, Ed, 291, 318, 323, 332, 333, 405
Hibbard, Sara, 21
Hickling, John, 407
Hillaby, Richard, 87
Hills, JF, 344, 345
Hirsch & Cie, les fils de Achille, 385, 405

Hirst Bros & Co Ltd, 39, 40, 50, 62, 145, 153, 170, 230, 304, 418, 420, 421, 422
Hirst, Joseph, 406
Hoefen a Enz, 239
Hogan, IB, 146
Holliday, Edward, 86
Holliday, Thomas, 86, 89
Hooke, Dr Robert, 18, 215
Hookham, G, 340
Hopkins, Jason R, 215
Horn, EH, 250, 251
Houghton, William, 38
Howard, Watch Co, E, 192, 212, 213, 404
Hubbard, Charles, 33
Huguenin, Arnold, 328
Huguenin, Fritz Henchoz, 315, 328
Huguenin Frères & Co, 349, 350
Hutchinson, George, 145
Huygens, Christian, 19

Illinois factory, 192, 193
Illinois Watch Case Co, 251
Imhof, Arthur, 405
Ingersoll Bros, Robert H, 8, 216, 250-258
Ingersoll Ltd, England, 365
Ingersoll-Trenton, 255
Ingersoll Waterbury Watch Co Ltd, 251, 256
INOK (Valjoux SA), 288, 289
International Watch Co, 184, 185, 186, 187
Invar, 188

Jackson & Son, James, 84
Jackson, James, 78, 115
Jacob, Georg, 284, 291, 296, 307, 308, 312, 323, 384, 393, 398, 400, 402
Jacot & Bovy, 408, 409
Jacot, Adolphe, 260
Jacot, Charles Edward, 215, 409
Jacot, George Favre, 175
Jaeger, Edmund, 161
Jaeger Le Coultre, 161
Jagermann, C, 162
Jant, William, 24
Japy, Frédéric, 18, 50, 51
Jeanneret-Brehm & Co, 300
Jeanneret-Brehm, Les Fils de, 291, 308
Jeanneret, Moeris et, 230
Jequier Frères, 276, 301, 302, 303, 304
Jones, John, 94-97
Jones, Robert Henry, 86
Juillard & Cie, 178, 351
Juillard, Emile, 60
Junghans, 384, 393, 394
Jura Watch Co, 385
Jurgensen, Urban, 38

Kaufleute und Fabrikanten, 191
Kempf & Cie, L, 354
Kendal, James, & Dent, John G, 116-117, 153
Kendal, James Francis, 117
Keyless Watch Case Company, 212, 213
Keystone Watch Case Company, 192, 208, 209, 210, 211, 212, 213

Kienzle Uhrenfabriken AG, 235, 236, 245
Kinoba, 370, 371
Kirchhofer, 350
Kirkman (Kirkham), Edward, 81
Kissenisky, D, 128-129
Klaftenburger, 44, 45
Kliszcewski, Wladyslaw Spiridion – *see* Spiridion
Koffmann, Ernst, Johann and Rudolf, 384
Konigen (A Schild), 384, 400, 401
Kummer, Ed, 357, 358, 370, 371
Kuss & Co, Camerer, 109

Labrador (Omega), 173, 174
Lake, HH, 388, 389
Lancashire Watch Company, 73, 133, 135, 143
Lanco (Langendorf Watch Co), 392
Langendorf Watch Co, The, 384, 385, 392
Langlois, 404
Lark, David, 91
Le Coultre, Antoine and David, 161
Le Phare, 306, 307, 332, 333, 338, 385, 405
Le Roi, Pierre, 60, 61, 215
Le Roy, Julian, 38, 70
Lee, Nathaniel or Nicholas, 74
Lehmann, Charles, 351
Lemania, Lugrin SA, 284, 285, 292, 330, 321, 322
Lemania, Nero, 284, 321, 322
Lemania Watch Co, 284, 285, 312, 313, 318, 319, 320, 322, 385, 405
Leonidas, 290, 291
Leroux, John, 70
Leuba Frères, César and Charles, 331
Leuthold, Marc, 349
Levy, Abraham, 89
Lewns, CF, 103
Lezard, Edward, 152
Lezard, Joseph, 47, 152
Liengme & Co SA, 388
Liga SA, 385, 395
Linsley, William, 20, 28
Lion's Watch, 65
Litherland Davies & Co, 74
Litherland, Peter, 70, 74
Lobl's 8 Days Lever, 412
Locke, Edwards, 215
London, The (Robinson Bros), 191
London Lever, 90
London Lever (Kendal & Dent), 153
Longines, 105, 106, 107, 167, 168, 169, 170, 171, 172, 310, 384, 385, 405
Lorimer, David (Lormier), 28
Lormier & Edwards, 28
Lorrimer, Françoise, 259
Lowe & Sons, 129
Lowe, F, GB and James F, 129
LUDGATE Watch (James William Benson), 112. 113. 114, 147
Lund Bros, 98-99
Lund, JA, 98
LWM, 248
Lyons, Claude, 160

McCabe, James, 6, 89, 215
McMillan, A, 182, 183
Mairet & Sandoz, 259
Malleray, Société d'Horlogerie de, 260, 262
Mantaras, Gabriel Lopez, 370
Marchant & Jobin Ltd, 325
Marcks & Co Ltd, 170, 312
Margetts, George, 70
Markwick, 382
Marsh, OB, 404
Martin, John, 110
Marvin Watch Co, 180
Massey, Edward, 70, 74
Massoteau, 404
Masters, John Neve, 103, 124-125
Maston, Charles, 30
Matheson, John & Co, 149
Matthews, Frederick, 81
Mathey, André, 48
Matthey, Hy, 42
Mayer, Arthur, 338
Medana, 370
Medana, Meyer & Studeli SA, 68, 69
Mensor, 9, 330, 331
Mentor (Bader, Hafner & Holderbank), 385, 402
Messagero (Roskopf), 8, 272
Mesure, Lianna, 406
Metcalfe, Mary, 31
Mickey Mouse (Ingersoll), 256
Midget (Ingersoll), 254
Miles, Arthur Fellenberg, 35
Miles, Septimus, 35
Miller, James, 30
Minerva, 307, 308
Minerva, Robert Frères, 290, 291, 295, 296, 297
Mitchell family, 31
Mitchell, James, 36
Moeris et Jeanneret, 230
Moeris, Fritz, 182, 183, 230, 231
Mohertus Trading Co, 240, 241
Mondandon, Albert, 144
Monnier & Frey, 383
Montbrillant Watch Co, 292, 296, 312
Montgolfier Frères, 260
Montilier, 385, 405
Moore, George, 89
Morgan, Samuel, 114
Morison, Thomas, 102, 114
Morpurgo, Professor Enrico, 18
Morris & Co, James W, 138
Movado, 348, 349
Mudge & Dutton, 38
Mudge, Thomas, 70
Muller, J, 46
Müller-Schlenker, 246
Murray, John, 78
Myers, John, 45
Myers, SF, 180

Nardin, Alfred, Ernest, Gaston and Léonard and sons Claude, François and Raymond, 287
Nardin, Paul D, 286, 287
Nardin, Ulysse, 286, 287

429

Nashua Company, 196
Nelson, John, 108
Nelthropp, HL, 215
Nero (Lemania Lugrin SA), 284, 321, 322
New "Castle" Lever (Reid & Sons), 156
New England Watch Company, 216, 227, 228
New York Standard Watch Co, 214
Newman, James, 108
Newman, James Thomas, 108
Niagara, 62
Nichols, B, 146
Nicole, Adolphe, 81, 148, 219, 220, 221, 277, 278
Nicole, Alexis, 315
Nicolet, Charles, 184, 288
Niel le Locle, Fabrique, 349, 350
Nielsen, Nicole, 339
Nielsen, Sophus Emile, 277
Nirvana Watch Co, 166
Nordman, 55
Nordman, Jules, 326
Normania & Co, 343
Norris, Henry John, 120
Northern Goldsmiths, 156
Norton, Samuel, 24
NOVORIS (Oris Watch Co), 245
Numa Gagbegin, Les fils de, 64, 67

Octava, 419, 420, 421
Octo, 418
Oko Lever (Kummer), 357
Oliak, Josef, 421
Olivant, Thomas & John, 78-79
Oliver, James, 88
Oliver, Richard James, 35, 36
Omega, 173, 174, 175, 288, 308, 309, 320
Oram & Son, George, 72, 73
Oris Watch Company, 236, 243, 245
Oxley Bros, 87, 99, 108, 115, 135
Oxley, WJ, 116

Paget, The (Weill & Co), 338
Parkes, CW, 214
Parkinson & Frodsham, 105, 107, 218, 219
Parkinson, William, 219
Parriott, F, 36
Patek Philippe, 45, 81, 380
Payne, William, 154, 280, 281
Pearce, Francis, 32
Pears & Boulas, 30
Peerless (Stauffer & Co), 184, 185
Pendleton, Richard, 70
Pennington, 219
Pennington, Robert, 70
People's Watch, The, 274
Perigal, Francis, 70
Perrelet, Abraham-Louis, 51
Perret, David, 66, 67
Perret, Jean, 163
Perrin, James, 76
Perron, L, 260
Perron, Louis, 229
Perseo (Cortébert Watch Co), 178

Petitpierre & Cie, 191
Philadelphia Watch Case Company, 192
Philippe, 260
Philippe, Adrien, 45, 81
Pike, RJ, 116
Piquet, M Henri, 277
Platnauer, 57
Pontif/Pontifa (Stauffer), 349
Poole, John, 102
Poole, John, junior, 102-103
Potts & Sons, William, 101, 126
Pouzait, 358
Premier (American Waltham Watch Co), 208
Presidente, La, 342
Prest, Edward, 81
Prest, Thomas, 7, 81
Pridham, Edwin, 88
Prince, John, 26
Pyott, James, 110

Quare, Daniel, 18, 379
Quillam, Samuel, 71, 84

Racine (Gallet & Co SA), 396
Racine Fabricant d'Horlogerie, 396
Raffin, E, 41
Raiguel, Juillard & Co, 260
Railway Regulator (Amida), 243
Railway Regulator (Zanco), 242
Railway Time (Thiel), 236, 237
Railway Timekeeper, 248
Railway Timekeeper, (Mohertus), 240, 241
Railway Timekeeper, Special Lever, 248
Rauschenbachs, J, 184
Raymond, BW, 210, 211
Record Watch Co, 157, 158, 159
Reconvillier Watch Co SA, 267, 268
Recordon, Louis, 23, 70
Redpath, Henry, 404
Referee (Smith's Industries), 334, 335
Reichenberg & Co Ltd, 291, 344, 350
Reid & Sons, 156
Reid, James, 128, 134, 135
Reliance (Ingersoll), 255
Renova (Henri Dalcher and Liga Watch Factory Ltd), 395
Renowned English Lever (James Reid), 135
Reveil Vigor (L. Sandoz-Vuille), 391
Revue-Thommens, 160
Richard, EA, 58, 59
Richardson, James, 128
Richmond Time Recording Co, The, 324, 325
Rieussac, H, 277
Riondey, F, 341
Ritchie & Son, James, 86
Riverside (American Watch Co), 198, 207
Roamer Watches England Ltd, 68
Robert, Françoise, 259
Robert Frères, 290, 291, 295, 296, 297
Robert, Henri, 52, 53
Robert, Louis and Edward, 52, 53
Robertson, D, 123
Robin & Cie, 280

Robinson Bros, 191
Rockford Watch Co, 214
Rolex, 191
Roper, Joseph and Margaret, 30
Roper, (George) Martin, 30
Roskell, Robert, 77, 88, 219, 220, 221
Roskopf Association, 260
Roskopf & Cie, H, 261, 266
Roskopf, Fritz Edouard, 259, 260, 261
Roskopf, Georges-Frédéric, 229, 259, 260, 261
Roskopf, Gindraux & Cie, 259
Roskopf, Johann Georg and Maria Elizabeth, 259
Roskopf SA, Louis, 267, 268
Ros(s)kopf watches, 8, 240, 259-276
Rosskopf & Co, W, 269, 270
Rosselet, M Gustave, 260
Rotherham & Sons, 21, 73, 132, 136, 137, 141
Rotherham, John, 128, 132
Rotherham, Richard Kevitt, 132
Roulet, Emile, 260
Rowlands, Christopher, 102
Russell & Co, Alfred, 105, 107
Russell & Son, Thomas, 104-106, 107
Russell, Alfred Holgate, 104
Russell, Thomas, 97, 104
Russell, Thomas Robert, 104

Sada, 405
Samuel, H, 15, 119, 231
Samuel, Harriet, 119
Samuels, 73
Sandoz, Henri, 182, 183
Sandoz-Vuille, L, 391
Savage, John, 70-71, 75
Savoye, B, 167
Schierwater & Lloyd, 203
Schild & Co, 405, 424
Schild, A, 353, 384, 385, 400, 401, 405
Schild, Otto, 405
Schindler, Paul, 63
Schmid, Charles Léon, 260, 263
Schmid, Ch Léon & Cie, Vve, 261, 263, 385
Schmidt, Karl Otto, 349
Schopp, Louis, 353
Searle & Co, 186
Sellars/Sellers, JC, 146
Serkisoff & Co, K, 180, 181
Services (Smith's Industries), 246, 247
'Services' Army (Müller-Schlenker), 246
'Services' Scout (Thiel), 238, 239
Sewill, Joseph, 332, 333
Sewill, Josh, 100
Shephard, Joseph, 78
Silbermann, K, & Hasler, JH, 412, 413
Silverine (Dueber Watch Case Company), 192
Silverode (Philadelphia Watch Case Company), 192
Silveroid (Keystone Watch Case Company), 192
Silverore (Fahys Watch Case Company), 192
Silveston, Francis, 408
Simplon (César Renfer-Abrecht), 385, 403

Simpson, Stephen, 81
Skirrow, Robert, 82
Slack, Joseph, junior, 85
Slack, Joseph, senior, 85
Slava, 284, 285
Smelt, William 85
Smith, Allan Gordon, 419
Smith & Sons, S, 73, 419
Smith & Son Ltd, S, 294, 300, 323
Smith, Ewen & Stylic Ltd, 142, 143, 157
Smith, George, 296, 297
Smith's Industries, 11, 12, 246, 247, 300, 334, 335, 336, 337, 366, 367, 368, 369
Smith, William, 31
Smith, William M, 131
Société Suisse d'Horlogerie, 385, 405
Society for Watchmaking, 160
Solatime, 246, 247
Solvil (Paul Ditisheim Co), 188, 189, 190
Souvenir European War (Ingersoll), 8, 252
Sparrow, William Henry, 153
Spence, WH, 177
Sphinx (H Williamson Ltd.), 144
'Sphinx' (P. Schindler), 63
Spiridion, Wladyslaw, 91
Squire, Reuben, 111
Stallcup, LD, 192
Standard Timekeeper (James W. Morris & Co), 138
Stauffer & Co, 184, 185, 288, 289, 314, 318
Stauffer, Henry, 349
Stauffer, Jules, 288
Stauffer, Marc Timothée and Timothée, 349
Stauffer-Jeune, MT, 349
Stockwell, George, 155, 423
Stolz Frères, 385, 405
Stradling, Alfred, 329
Stram, Alfred, 89
Stratton, NP, 404
Summersgill, Robert, 80, 81
Superior Motor Tmekeeper, 244
Superior Timekeeper, 248, 249
Suter & Co, 144
Sydenham, CS, 58, 59

Tallis Watch Co, 291, 344, 350
Talley Industries, 231
Tavannes Watch Co, 182, 183, 347
Taylor, John, 334
Tell (Jequier Frères & Co), 302, 303, 304
Thickbroom, Alfred J, 31, 95
Thickbrook, George James, 112
Thiel Bros Ltd, 237, 238, 239, 256, 257, 384, 398
Thiel, Christian, Georg and Siegmund, 236
Thiel, Ernst, 236, 237
Thiel, Reinhold, 237
Thiel, Dr Reinhold, 237
Thil, 6, 42
Thomas, John, 20
Thommen SA, 405
Thommen, Alphone, 353
Thommen, Gideon, 160, 353

Thoms, F, 119, 145
Thorneloe, R, 342
Thorpe, Thomas, 91
Tieche, Lucien, 384
Tissot, Charles, 350
Tompion, Thomas, 18, 20, 38, 379
Trenton, 158
Triumph (Ingersoll), 250
Trotti, Jacopo, 18
Trump, The (Waterbury Watch Co), 224
Tschopp, Louis, 160
Tyrer, Thomas, 215

Ullmann & Co, 331
United States Watch Company, 212
Upjohn & Brights, 143

Vale & Co (later Vale & Rotherham), 21, 29, 132
Vale & Kenyon, 21
Vale, Howden & Carr, 132
Vale, Samuel, 21, 132
Valjoux SA, 288, 289, 334, 335, 368, 369
Vanguard (American Waltham Watch Co), 7, 204, 207
Vauchay (Vve de Louis Goering), 164
Venafs, Tho, 76
Veracity Watch (John Neve Masters), 103, 124
Vertex Watch Co, 160, 351
Viner, Charles Edward, 386
Vittori & Cie, 8, 265, 270
Voiblet, JA, 260
Voland, Wilhelm, 383, 384, 387
Vulcain & Volta Ditisheim & Co, 188

Wagner Time Inc, 346
Walker, Edward, 81
Walker & Son, James, 64
Walker, John, 80, 81, 359
Waltham Watch Company, 9, 67, 135, 192, 193, 194, 195, 338, 404
Walton, Joseph, 109, 141
Ward Bros, 184
Ward, Daniel, 25
Waterbury, 215, 216
Waterbury (Ingersoll), 255
Waterbury Clock Company, 251, 255
Waterbury Watch Co, 221-227
Waterproof (Alcide Droz & Fils), 177
Watkins, A, 97
Watson, William, 217
Webb, William, 217
Wehinger, F, 251
Weil, Nathan, 384, 390
Weill & Cie, 162
Weill & Co, 331, 338
Weill, Louis, 338
Westclox, 231, 232, 233, 346
Western Clock Company, The, 231
Westinghouse Electric Corporation, 425
Westmore, Robert, 23
Westwood, Robert, 404, 405
Wheeler, PH, 214

Whibley, Ethel M, 146
Whippet, The, 296, 297
White, Ebenezer, 91
White, Edward, 316
White, George, 29
Whitefield Co (Lemania?), 312, 313
Whitehead of Sevenoaks, 103
Whittaker, Thomas, 111, 120, 121
Wichman, W, 133
Widdowson, John, 77
Wildorf, Hans, 191
Williams, William, 327, 334
Williamson, 127
Williamson Ltd, H, 63, 73, 144, 145, 156, 184
Willie Frères, 260, 263
Wilson, George, 7, 111
Windus, John, 77
Winner (Connecticut Watch Co, Ingersoll), 251
Winnerl, 277
Wise, E, 305
Wise, Frederick C, 377
Wittnauer Watch Co Inc, 425
Woerd, Charles Vander, 196
Wolfe, Elias, 108
Wolfe, Joseph, 108
Wontner, Jno, 25
Wood, William, 29
Woodhouse, James, 118
Woodman, John, 118
Woodman, Philip, 98, 99, 102, 115, 139, 407
Woog, Dr, 188
Wright Kay & Co, 380
Wyer, Joseph, 217
Wyss Frères, 405

Yabsley, James Benjamin, 387
Yankee Radiolite (Ingersoll), 253
Yates, Hannah Maria, 84
Yates, Thomas, 71, 84
Yeomans, Samuel, 126

Zanco, 242
Zenith Watch Co, 175, 176, 385, 397

431

ROTHERHA[M]

COVENTRY
WATCH MAN[UFACTURERS]

JUROR : INVENTIONS Ex[HIBITION]

INTERCHANGEABLE CASES EXTRA DURABLE AND DUST PROOF.

PARIS 1889.
○ GOLD MEDALS. ○
MELBOURNE 1888.

CHRONOGRAPHS, REPEATERS, AND ADJUSTED WATCHES IN **EVERY VARIETY**.

Price Lists of Watches, and Illustrated [...] for "Rotherha[m...]
SUPPLIED TO THE TRA[DE]

A